ELEMENTS

OF

GEOLOGY.

BY ALONZO GRAY, A.M.,

AUTHOR OF

"ELEMENTS OF CHEMISTRY" AND "ELEMENTS OF NATURAL PHILOSOPHY,"

AND

C. B. ADAMS, A.M.,

FELLOW OF THE AMERICAN ACADEMY OF ARTS AND SCIENCES, PROFESSOR IN AMHERST COLLEGE, AND STATE GEOLOGIST OF VERMONT.

NEW YORK:

HARPER & BROTHERS PUBLISHERS,

FRANKLIN SQUARE.

1863.

PREFACE.

THE work now offered to the public has been prepared with the design of presenting in a condensed form the outlines of American and European Geology.

The recent geological surveys have made us acquainted with the Geology of the American continent, especially of the United States and the British Provinces. From the great amount of material thus furnished, such selections and classifications have been made as to present American Geology in a systematic form; the subject is not treated, however, with exclusive reference to America, but the facts and deductions pertaining to each portion of the globe have been combined in one comprehensive system.

Although the book is elementary in its character, and particularly designed for students in Geology, it is not limited to the mere detail of facts or to scientific description; but, with a view of rendering the subject attractive to the general reader, the most important theories of the science are discussed, with its practical applications and relations to Natural Theology and Revelation.

The printed sheets were first used in the instruction of classes in Geology, and subjected to the criticism of scientific friends. The work has since been revised and reprinted. It is believed, therefore, that the student will find it clear in statement, and free from any important scientific errors.

Illustrations for the work have been drawn from the whole

field of geological phenomena. Many of the cuts illustrating American Geology are from original drawings; others have been taken from scientific journals and geological reports. The illustrations of European Geology have been selected from standard works. The wood-cuts, therefore, may be received as faithful representations of the actual phenomena which the crust of the earth presents.

The table of contents contains a complete analysis of each subject. Questions are added at the foot of each page, for the convenience of those who may use the work for elementary instruction.

Those portions of the work from the beginning to the tertiary, and from the triassic to the unstratified rocks, were first composed by Professor Adams, and the remainder by Professor Gray.

New York, October, 1852.

CONTENTS.

INTRODUCTION.

Section 3.—General Results of Aqueous Agencies.

PART II.—PRINCIPLES OF PALÆONTOLOGY.

CHAPTER I.—PRINCIPLES OF ZOOLOGY WHICH RELATE TO FOSSILS.

CHAPTER II.—FOSSILIZATION.

CHAPTER IV.—QUATERNARY SYSTEM

Section 2.—Antiquity of the Earth, as inferred from the Remains of Organized Beings.

Section 3.—Antiquity of the Earth, as inferred from the Mutual Relations of the Organic and Inorganic Portions of the Earth's Crust.

Section 4.—Objections to the Antiquity of the Earth considered.

CHAPTER XI.—CONNECTION OF GEOLOGY WITH NATURAL THEOLOGY AND WITH REVELATION.

Section 1.—Geology and Natural Theology.

Section 2.—Connection of Geology with Revelation.

INTRODUCTION.

1. *Geology is that Science which investigates the physical History of the Earth.*—This history is written in the layers and masses of mineral matter which constitute the crust of the earth, and becomes intelligible in the investigation of the successive changes to which the earth has been subjected.

In the study of Geology, the first step is to obtain a knowledge of the forces which are *now* active, in respect of the *manner* in which they act and of the *effects* which they produce. The second step is the application of the knowledge thus obtained to the explanation of similar effects in the earth's crust. We shall be able to understand the changes of which these effects are the memorials. If also, in the third place, by means to be hereafter explained, we can ascertain the order of time in which the events have occurred, we shall have a connected history.

It is obvious that the basis of this reasoning is the analogy between the past and the present laws of nature; for, however the circumstances may differ, the elementary forces are ever the same. With perfect confidence in their uniformity, the astronomer predicts a future eclipse, and the geologist describes extinct animals and plants.

Thus we may show that there has been a different distribution of land and water, and that mountains have arisen, in successive periods, so as to obliterate all resemblance between this truly ancient and our modern Geography. We may also see many successive races of the animal and vegetable inhabitants of the globe,

Define Geology. What is the first step in the study of Geology? the second? the third? What is the basis of geological reasoning? What are the results?

different from each other and from those which now exist, appearing, and afterward becoming extinct in an uninterrupted series, of which the beginning and end are known only to the Creator.

II. *Relations of Geology to other Sciences.*—These relations are those of mutual dependence.

1. The effects of heat on mineral substances, and the action of the substances on each other, can not be investigated without the aid of *Chemistry*. The principles of *Mechanics* are used in the investigation of the manner in which the great agents, water and heat, effect the removal of materials and elevate mountain chains, and in which the glaciers of the present and of a former period have modified the surface of countries. The *magnetic* influences which pervade the earth's crust ally Geology with another branch of science.

Without *Botany* and *Zoology*, the geologist could not interpret the wonderful records which are preserved in the form of fossil remains. These sciences enable him to perceive that most of the fossils belong to extinct species. They also enable him to understand the habits of these species, and their relations to each other, and to the varying condition of the earth's surface. *Comparative anatomy* aids in showing how, from a fragment, to reconstruct the whole animal.

On *Mineralogy* Geology depends for a knowledge of the visible composition of rocks. A knowledge of their mineral constitution often enables the geologist to understand their origin and history.

2. In return for these favors, Geology directs to the localities of useful minerals, and throws much light on their origin. It restores many species of organic bodies which have been buried in the earth, and enables the naturalist to fill up many wide gaps in the plan of the existing creation. It also carries back the

---

Of what nature are the relations of Geology to other sciences? What aid does it derive from Chemistry and Mechanics? What from Botany and Zoology? What from Mineralogy? What aid do other sciences derive from Geology?

history of the animal and vegetable kingdoms into inconceivably remote periods, and exhibits the plan of creation as extending throughout time. It throws much light on questions relating to the origin, the nature, and the destiny of species. The present distribution of animals and plants over the surface of the earth is much influenced by the nature of the geological formations.

The more intimate, therefore, the acquaintance of the geologist with these kindred sciences, the more successful and accurate will be his investigations. It is not essential, however, that his knowledge of them should embrace much more than the general results of the labors of others who are specially devoted to them. Still more ordinary attainments will suffice for the purposes of the general reader of Geology.

III. *Relations of Geology to the Arts.*—The economical relations of Geology may be chiefly comprehended under the general divisions of agriculture, mining, architecture, and engineering.

1. For the agriculturist, it detects beds of marl and peat, of limestone, of green sand, of phosphate of lime, and of gypsum. It also enables him to understand the origin of his soils.

2. For the miner, it often decides, on a moment's inspection, whether certain minerals may occur or can not occur in a given region; for many valuable substances are found only in a very limited part of the geological series of rocks. It also assists the miner to understand the mode of the occurrence of minerals. This may be in layers more or less parallel with the rocky strata, or in veins cutting across them. They may be in regular or irregular masses, or may be merely a disseminated constituent of a rock. A knowledge of these facts, and of their causes, is usually essential to success in mining.

3. To the architect Geology is of great service in the selection of building sites, and in the choice of stones, or of materials for bricks.

4. The engineer also is aided by Geology in choosing the best line in the arrangement of excavations and embankments, and in

What are the relations of Geology to agriculture? to mining? to architecture? to engineering?

the selection of the materials for the construction of roads and canals.

IV. *History of Geology.*—The remarks which have been made on the connection between Geology and other natural sciences, suggest an obvious cause of the recent origin of Geology properly so called. The sciences, without whose interpretation the facts of Geology are unintelligible, are themselves of recent origin.

Speculations on the process of creation, and idle hypotheses of the phenomena of fossils and of the stratified and the crystalline rocks, mingled with occasional glimpses of true theories, make up the history of the subject until the latter part of the last century. At that time, Werner, a German professor in a school of mines, proposed and defended the Neptunian theory. This theory accounted for all the rocks by the aqueous deposition of strata, which were supposed to be originally continuous over the whole surface of the earth. Werner was the father of Mineralogy; but the title of the father of Geology belongs to Hutton, a Scotch geologist, who soon after proposed the Plutonian theory. According to Hutton's theory, which has been established, the unstratified rocks are of igneous origin, like the lavas of the present epoch, and the stratified rocks were originally sand, clay, mud, gravel, &c., like the aqueous deposits of the present time. Hutton supposed that these deposits were derived from the abrasion of ancient continents; that some of them were rendered crystalline by the heat of protruded igneous rocks; and that in such a series of changes Geology can discover no proof of a beginning nor prospect of an end.

After much discussion, in which Werner's theory was abandoned, geologists applied themselves to laying a durable foundation for the science, in the collection and systematic study of facts. Mr. William Smith, an English geologist, although in an obscure position, was the real pioneer and head of this movement. In the latter part of the last and the commencement of the present century, he accomplished more for British Geology than all his

---

Cause of the recent origin of Geology? What was the Neptunian theory? the Plutonian theory? What was the subsequent history of European Geology?

cotemporaries. In 1807 the Geological Society of London was formed, which was followed by other European societies. The enthusiastic efforts of men of wealth and of distinguished talents, with occasional aid from governments, in almost all parts of Europe, have resulted in the collection of a vast amount of materials. These, with the legitimate deductions, constitute a science which is now in most of its parts as well established as Astronomy.

The most prominent features in the history of geological opinions, during the second quarter of this century, have been the theory of internal heat, and the question of a uniform or a diminished intensity in the action of geological agencies.

The oldest book on the geology of America was a German work written by Dr. J. D. Schöpf, and published in 1787. But the study of American Geology was effectually commenced in 1807 by William Maclure, who alone explored a large part of the United States, and published a geological map of the country. Mr. Maclure was soon followed by Dr. Bruce, Professor Silliman, and at length by a numerous body of able geologists.

In 1824 was commenced a series of geological surveys of most of the United States, made under the authority of the State Legislatures. The following is a chronological list of these surveys, with the dates of their commencement:

1824.	North Carolina;	by Prof.	Denison Olmsted.
1830.	Massachusetts;	" "	Edward Hitchcock.
1833.	Tennessee;	" "	G. Troost.
1834.	Maryland;	" "	J. T. Ducatel.
1835.	New Jersey;	" "	Henry D. Rogers.
1835.	Virginia;	" "	Wm. B. Rogers.
1835.	Connecticut;	"	Dr. J. G. Percival and Prof. C. U. Shepard.
1836.	New York;	"	Profs. W. W. Mather, Lardner Vanuxem, Ebenezer Emmons Mr. James Hall, and Mr. Timothy Conrad.

Who were the first cultivators of Geology in America? What means have been employed to develop the Geology of the United States?

1836. Maine; by Dr. C. T. Jackson.
1836. Pennsylvania; " Prof. Henry D. Rogers.
1836. Georgia; " John R. Cotting, Esq.
1837. Delaware; " Prof. James C. Booth.
1837. Indiana; " Dr. David D. Owen.
1838. Michigan; " Douglass Houghton, Esq.
1838. Kentucky; a reconnoissance only; by Prof. W. W. Mather.
1839. Ohio; by Prof. W. W. Mather.
1839. Rhode Island; " Dr. C. T. Jackson.
1840. New Hampshire; " Dr. C. T. Jackson.
1842. Louisiana; a reconnoissance; by Prof. W. M. Carpenter.
1844. Vermont; by Prof. C. B. Adams.

Geological surveys have been made, or are in progress in several of the territories, and in the British provinces.

These surveys have not only accomplished their object of developing the mineral wealth of the country, but they have also accumulated a great amount of materials for science. It is to be regretted, however, that some of the states have not published final reports, in which cases most of the results are likely to be lost.

In 1840, an association was formed for the advancement of Geology, by the gentlemen who had been engaged in the state surveys. It soon comprehended all objects of Natural History; and in 1847 it was resolved into a more general "American Association for the Promotion of Science." This association holds annual, and sometimes semi-annual, meetings in different parts of the country.

What was the origin of the American Association for the Promotion of Science?

CONSTITUTION OF THE EARTH.

I. *Chemical Constitution of the Earth.*—Of the sixty-two simple substances which are known to chemists only sixteen constitute the greater part of the earth's crust. The other forty-six exist for the most part in rare minerals, or are disseminated in very minute proportion through the more common substances.

1. *Oxygen* is the most abundant of all elementary substances. In a free state it constitutes one fifth of the atmosphere. In combination it forms eight ninths of water, and two fifths to one half of all the solid materials of the globe. It enters largely into the composition of all the earths, and of most of the earthy minerals and ores of the metals.

2. *Silicon* permanently exists in nature only in combination with oxygen, in nearly equal parts, forming silica, which is also called silicic acid. This compound constitutes forty-five to fifty per cent. of the earth's crust, one quarter of which is therefore silicon. Pure quartz is silica, and consequently silicon exists in most of the rocks and minerals.

3. *Calcium* exists chiefly in combination with oxygen, forming lime, of which it constitutes nearly two thirds. The lime exists mostly in combination with carbonic acid, forming carbonate of lime. About seven per cent. of the earth's crust is calcium.

4. *Aluminium* exists in nature only in combination with oxygen, in nearly equal parts, forming alumina. This compound constitutes one fifth of feldspar, and exists in most minerals and rocks. Probably about five per cent. of the earth's crust is aluminium.

5. *Magnesium* exists in nature mostly in combination with oxy-

Of how many substances is the crust of the earth chiefly composed? What is said of oxygen? of silicon? of calcium? of aluminium? of magnesium?

gen, forming magnesia, of which it constitutes about three fifths. Magnesia forms about forty per cent. of serpentine, and ten to twenty per cent. of dolomite. Probably about three per cent. of the earth's crust is magnesium.

6. *Iron* is rarely, if ever, found native, except in meteoric stones. It is mostly combined with oxygen, frequently with sulphur, and sometimes with carbon. It forms nearly two per cent. of the crust of the globe.

7. *Carbon* exists mostly in combination with oxygen, forming carbonic acid, in the proportion of six parts of carbon and sixteen parts of oxygen. This gas constitutes about $\frac{1}{3000}$th of the atmosphere, but is chiefly locked up in a solid state in combination with lime, magnesia, &c., forming the carbonates of those bases. Carbon also exists in a free state in the various kinds of mineral coal. Nearly two per cent. of the earth's crust is carbon.

8. *Potassium* exists in nature almost wholly in combination with oxygen, forming potassa, of which it constitutes about five sixths. Potassa exists chiefly in feldspar and clay, in the soil, &c. Potassium forms nearly five per cent. of the unstratified rocks, being about one tenth of the feldspar in them, and constitutes about one per cent. of the total of the earth's crust.

9. *Hydrogen* resides chiefly in water, of which it forms one ninth part. Water exists not only in the ocean and in lakes, rivers, and the atmosphere, but is widely disseminated in a solid, dry state, in many rocks and minerals. The waters of the ocean are sufficient to cover the earth to a uniform depth of more than two miles. This is the principal repository of hydrogen, including which the total quantity is less than one half per cent. of the earth's crust.

10. *Sodium* is next in abundance, and exists chiefly in common salt, in albite and in basalt.

11. Next is *Sulphur*, which exists in a free state in volcanoes, but most abundantly in combination with many metals, as iron, lead, copper, antimony, &c. It also enters largely into the com-

---

What is said of iron? of carbon? of potassium? of hydrogen? of sodium? of sulphur? &c.

bination of gypsum. In minute proportions it is diffused through all soils, and exists in all animal and vegetable bodies.

12. *Manganese* is almost universally disseminated through the rocks and soils, but with an average proportion not exceeding $\frac{3}{10000}$ths of the whole.

13. *Chlorine* exists mostly in common salt, and in the chlorides of magnesium and calcium of the ocean. It is universally disseminated.

14. *Phosphorus* exists in all soils in small proportion, in the bones of vertebrated animals, and in the mineral phosphate of lime.

15. *Fluorine* exists chiefly in fluor spar, combined with calcium; also in small proportion in hornblende, and in many varieties of mica, and consequently in most of the unstratified rocks.

16. *Nitrogen* exists chiefly in the atmosphere and in animal bodies, and is generally diffused through the vegetable kingdom in small proportion.

These simple substances, in the order of their abundance, are. oxygen, about fifty per cent.; silicon, twenty-five per cent.; calcium, seven per cent.; aluminium, five per cent.; magnesium, three per cent.; iron and carbon, each two per cent.; potassium, one per cent. The remaining five per cent. consist mostly of hydrogen, sodium, sulphur, manganese, chlorine, phosphorus, fluorine, and nitrogen.

With the exception of nitrogen, these simple substances exist chiefly or solely in combination. The following are the principal binary compounds in the order of their abundance: silica (quartz), about fifty per cent.; alumina and lime, each ten per cent.; carbonic acid, seven per cent.; magnesia and water, each five per cent.; oxides of iron, three per cent.; potash, more than one per cent. Other binary compounds are, soda, chloride of sodium, sulphuret of iron, oxide of manganese, and sulphuric acid.

Many of these binary compounds unite together, and form salts or ternary compounds. Such are carbonate of lime and carbon

In what quantity do these substances exist? In what state? What are the principal binary compounds? What are ternary compounds?

ate of magnesia, sulphate of lime, the silicates of potash, of soda, of lime, and of other bases.

II. *Mineral Constitution of the Earth.*—As a book is composed of letters of the alphabet, grouped into words and sentences, so is the crust of the earth composed of many distinct kinds of minerals, which, being mixed together, constitute rocks; and many of the rocks, occurring in groups of strata, with certain distinctive marks, forming separate chapters in the history of the globe, constitute what are called formations.

Minerals, then, are the alphabet of Geology. When they occur in homogeneous masses, they are called *simple minerals;* and when two or more simple minerals have been mixed together, they form a *rock.* In a few cases, a simple mineral, as limestone or serpentine, is also called a rock, because it occurs in extensive ledges.

There are, in the crust of the earth, more than 500 kinds of simple minerals; and if a knowledge of all were an indispensable preparation for the study of Geology, most of those into whose hands this work may come might well be discouraged. But *five* only of these minerals constitute about nine tenths of the crust of the earth, and with the addition of two or three more the number will embrace nineteen twentieths of the crust.

1. *Quartz* is the most abundant of all minerals, constituting nearly one half of the crust of the earth. It is one of the harder minerals, scratching glass with facility, although inferior to the diamond. When regularly crystallized, it is called rock crystal, and is more or less transparent. Its numerous varieties occur with all colors. Flint is a variety, with an impalpable structure and conchoidal fracture. The presence of small portions of iron, manganese, chrome, and other foreign substances, produces numerous varieties, some of which are valued as gems, such as jasper, amethyst, agates, cornelian, &c. The sand which is used in making mortar and glass is mostly quartz.

2. *Feldspar* constitutes about one tenth of the crust of the

What is a simple mineral? What is a rock? How many kinds of minerals are known? How many are abundant? What is said of quartz? of feldspar?

earth. It is less glassy in its appearance than quartz, and is not quite so hard. It has a pearly luster, and is more frequently of a grayish white color, although sometimes red, green, &c. When decomposed, it forms an unctuous white clay, called kaolin, which is of great use in the manufacture of fire-bricks, stone-ware, pottery, and porcelain. The undecomposed mineral, pulverized, is used with kaolin in the manufacture of the finest porcelain. Common clay is impure decomposed feldspar. It usually contains a small portion of the protoxide of iron, which by heat is converted into the peroxide of iron, a red substance, which gives the color to common bricks and pottery.

3. *Limestone* (carbonate of lime) forms more than one seventh of the crust of the earth. Its varieties are numerous; those which are crystallized are called calcareous spar. It is much softer than quartz or feldspar, being easily scratched. Varieties which admit of a fine polish are called marble. At a red heat the carbonic acid gas is expelled, and the stone becomes lime. Marl is a pulverulent variety, more or less mixed with particles of clay, and is of great use in agriculture. Calcareous tufa is a deposit of this mineral from mineral springs. Carbonate of lime is easily distinguished from other common minerals by the application of a drop of acid, which will produce effervescence.

4. *Hornblende*, including *augite*, which is now regarded as merely a variety of hornblende, constitutes a large part of the rocks of volcanic origin, and of some of the older slates. It forms from one fifteenth to one twentieth part of the crust of the earth. Some of its varieties are beautiful minerals, as actinolite, amianthus, &c., prized by the mineralogist, but of little economical value. Asbestus is a remarkable variety, consisting of excessively slender silky fibers, that may be picked and wove like cotton into cloth, which will be incombustible. It may also be used for incombustible lampwicks. Compact rocks, of which this mineral forms any considerable portion, although not as hard as quartz rocks, are exceedingly tough.

5. *Mica*, often improperly called isinglass, is about equally

---

What is said of limestone? of hornblende? of mica?

abundant with the preceding. It is not very hard, and usually occurs in thin elastic plates, which are sometimes found containing one or two square feet, but more frequently are very small, like scales, shining, black, brown, or silver colored. The large sheets are used for lanterns, stove-windows, and in the Russian navy for common windows, not being broken by the concussion of a broadside.

6. *Talc* is one of the softest minerals, being very easily cut with a knife. It often resembles mica, but is softer, not elastic, and has an unctuous feel. It contains 30 to 33 per cent. of magnesia. One of its varieties, steatite (soapstone), occurs in extensive beds, and is much used for fire-places, stove-linings, &c. Talc is usually light green.

7. *Chlorite* is generally dark green, and differs but little from talc.

8. *Serpentine* is usually of some shade of green, the varieties of lighter color being called precious serpentine. It is harder than limestone. It receives a high polish, and when free from the cracks and seams with which it usually abounds, is an elegant substitute for marble.

9. *Gypsum* (plaster), *rock salt,* and *coal,* are the only other minerals which form any considerable portion of the earth's crust

III. *General Structure of the Earth.*—By the *crust of the earth* we mean that portion which comes within the reach of observation and legitimate inference; this is much more than is commonly supposed by those who are unacquainted with Geology. The crust is composed essentially of solid rocks—the loam, sand, gravel, clay, and other soft or loose materials, being merely a superficial covering.

There are two kinds of rocks, differing both in structure and origin—the *stratified* and the *unstratified.*

The stratified rocks occur in layers or strata, and were deposited from water. Hence those which lie beneath are most ancient; and where two kinds of stratified rocks occur in junction,

---

What is said of talc? &c. What is meant by the crust of the earth? What are the two kinds of rocks? What is said of stratified rocks?

the one lying under the other, their relative age is obvious. By this simple principle of position, the relative age of most of the rocks has been determined. The various circumstances attending the deposition may also, to a great extent, be inferred from the character of the strata. Thin layers of fine materials, and of very uniform thickness, are deposited from quiet water, and those which consist of coarse gravel and pebbles are the products of agitated waters. The layers of the stratified rocks, originally mud, sand, gravel, shells, coral, &c., have become solid by the agency of heat, pressure, cohesion, crystallization, &c.

The unstratified rocks, on the other hand, usually occur in irregular masses, sometimes overlaying other rocks, or in veins cutting across the layers of stratified rocks, or forming beds interposed between those strata. Modern lavas and ancient granites are alike unstratified and of igneous origin—the various unstratified rocks having been erupted, in a melted condition, at different periods in the history of the earth.

The unstratified rocks which are now erupted from volcanoes have their source beneath all other known rocks; and it is inferred from various data, that those of ancient date were in like manner erupted from beneath the then existing rocks. Going far back in the geological history of the globe, we come to a period when the lowest and oldest stratified rock rested alone upon granite, the oldest of the unstratified. Beneath are doubtless, in immense irregular beds, the reservoirs whence the eruptions of the igneous rocks had their origin, and above we have the successive strata, whose contents reveal the physical history of the earth.

A general idea of the structure of the earth may be obtained from the following figure, which is not intended to represent any given region, but is an ideal exhibition of the positions of the rocks as they would appear in a section through the crust of the earth. A little attention to the explanations will supersede the necessity of a lengthened description.

What is said of the mode in which the unstratified rocks occur? of their source?

Fig. 1.

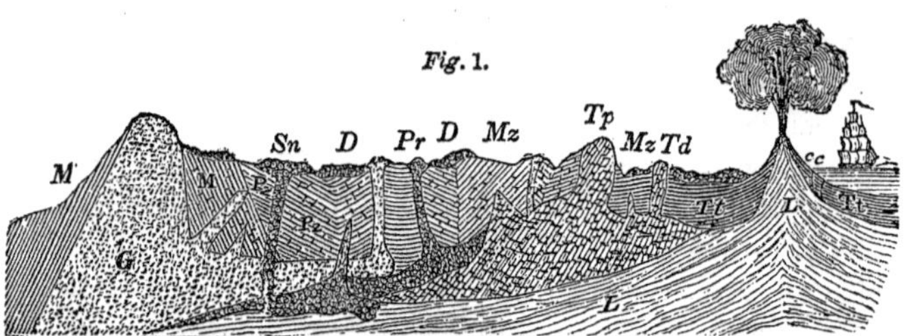

1. *Unstratified Rocks, of igneous Origin.*—G. Granite. Sn. Syenite. Pr. Porphyry. Tp. Trap rocks. Td. Trap dike. L. Lava. *c, c.* Fissures through which the water of the ocean penetrates to the melted lava.

2. *Stratified Rocks, of aqueous Origin*, represented by parallel lines.—M. Metamorphic rocks. Pz. Palæozoic rocks, or the oldest rocks which contain the remains of animals and plants. Mz. Mesozoic rocks. Tt. Tertiary strata D. Superficial covering of sand, gravel, and loose stones.

PART I.

GEOLOGICAL AGENCIES.

WITH a constitutional belief in the constancy of the laws of nature, we look to the operations now in progress, not only on account of their intrinsic interest, but because they alone can enable us to understand the history of the past. When we have seen strata in the process of accumulation beneath the waters, burying within them the present races of animals and plants, marine, fluviatile, or terrestrial, the species of hot or those of cold climates, according to the situation in which these deposits are forming; and when we have seen unstratified rocks resulting from eruptions of lava, we shall be better able to comprehend the origin of ancient deposits, with their imbedded relics of species which have long been extinct, and of those enormous masses of unstratified crystalline rocks which exist where the volcanic fires have long since gone out.

Classification of Geological Agencies. — Geological agencies may, for the most part, be referred to three heads: the *igneous*, comprehending all the effects of heat; the *aqueous*, including the effects of water in all its forms; and the *organic*, or agency of the animal and vegetable kingdoms.

Intensity of Geological Agencies.—Some very able geologists are of the opinion that the course of nature has been, through all the geological epochs, the same as at the present, not only in the nature of the agencies of change, but also in the degree of intensity with which they have acted; that volcanic forces have never been more violent than they are now; and that there have been no other changes of climate than those which are consequent on changes in the relative distribution of land and water.

What are the principal agencies now active in modifying the crust of the earth? On what points do geologists differ?

Many others suppose that volcanic convulsions of the earth's crust have been more violent and on a larger scale than at present; and that the earth's surface was once intensely heated by the internal fires, and after a partial reduction of temperature, was adapted for the support of the dense tropical vegetation and the tropical races of animals, whose remains are now found abundantly in cold climates. Without discussing this question at present—the only elementary question on which geologists differ—we merely hold it up, that it may be seen in the light of the facts which we are briefly to notice.

This difference of opinion, which is gradually diminishing by an approach to a medium, it is important to observe, relates only to the *energy* of the geological agencies of *former* periods. It is admitted by all that these agencies have ever been of the same nature as at present.

CHAPTER I.

AQUEOUS AGENCIES.

The aqueous agencies, which are modifying the surface of the earth, act both chemically and mechanically. Their chemical effects are limited chiefly to the solution of rocks. Mechanically their action is more various and extensive. By penetrating porous rocks, and by circulating in the form of rivers, tides, waves, marine currents, glaciers, and icebergs, water is continually rending the rocks into fragments, and removing the materials to lower levels, or into the ocean. With yet greater facility are materials carried down which have been dissolved in water. *The general tendency of aqueous action is, therefore, to transport the continents into the ocean.*

The ocean covers more than two thirds of the surface of the globe, and the rise of water consequent on the transport of mate-

What is said of the chemical effects of aqueous agencies? of their mechanical effects? of their general tendency?

rials into the ocean tends materially to increase its extent at the expense of the existing continents. There would, therefore, be at length a universal inundation, if the sinking of the bed of the ocean and the elevatory agencies of igneous action did not tend in an equal or greater degree to diminish its area.

Aqueous agencies may be classified as they are or are not *marine.*

Those which are *not marine* are the atmospheric agencies of rain and frost, rivers, lakes, springs, and glaciers.

The *marine agencies* are icebergs, waves, tides, and oceanic currents.

SECTION I.—AQUEOUS AGENCIES NOT MARINE.

I. *Atmospheric Agencies.*—1. *Rain* acts chemically on all calcareous rocks. It is well known that water, when pure, will not dissolve limestone, but that, when charged with carbonic acid gas, it will dissolve calcareous matter with a facility proportionate to the quantity of gas in the water. Falling rain absorbs this gas from the air, and thus acquires the power of slowly dissolving the solid rocks in limestone countries. Calciferous rocks exhibit the effect of this action in the irregular furrows which are worn down their inclined sides, and in their more or less rounded surfaces.

Rain also promotes the union of oxygen and carbonic acid with the iron of iron pyrites, and thus causes rocks which contain this mineral to crumble. In a similar manner it acts on other metals, and on the common alkalies, potash and soda, in such feldspathic rocks as granite and syenite. The chemical part of this process is called *decomposition.* The mechanical effect in the crumbling of rocks is called *disintegration.* Rocks which contain pyrites and alkalies abundantly are therefore most affected in this manner. The effect is much greater when a porous or fissured structure allows the rain to penetrate far into the rock.

Rains also act mechanically by carrying the loose fragments and particles on the surface of the ground into rivers, thus fur-

How are aqueous agencies classified? Describe the action of rain. What is decomposition? What is disintegration?

nishing them not only with their liquid, but also with their solid contents.

This agency is very striking in some countries, where the rock formations are more or less porous like sandstone, or where they have not been first worn and then protected by drift. In Jamaica and Antigua, there is often an imperceptible gradation of coherence from the loose soil of the surface to the solid rock beneath, and the upper portions of the latter may easily be removed with a spade. Heavy rains falling on the steep sides of mountains of such rocks will, therefore, carry off immense quantities of matter, and expose fresh portions of rock.

The removal of mineral matter from higher to lower stations is the greater from the fact that a greater quantity of rain usually falls on high than on low lands. Elevated districts, being cooler, condense a greater portion of the vapor, which is every where present in the atmosphere. This is especially manifest in hot countries, where the intense heat enables the air to contain an immense amount of watery vapor, which is condensed in deluging showers on the sides of the mountains.

2. *Frost* is another atmospheric agency, which in cold climates more or less compensates for the deficiency of rain in the work of destruction. Water penetrating into porous rocks, or entering fissures and expanding by frost with an irresistible force, crumbles the surface, and throws out large blocks of stone. The fragments lie in enormous heaps at the base of precipices, or fall into the beds of mountain torrents, and are removed by freshets. Such an accumulation of angular blocks at the foot of a precipice is called a *talus,* and usually has an inclination of about 40°.

A very compact structure almost wholly preserves a rock from this mechanical agency, and a covering of clay preserves it also from the chemical action of water. In the valley of the Connecticut, porous sandstones have been penetrated and altered to the depth of ten feet. Granite and syenite often have a discolored exterior of only one or a few inches. But on compact varie-

---

Describe the action of frost. Meaning of talus? To what depth have rocks been acted upon by water?

ties of greenstone, the altered exterior is but a small fraction of an inch in thickness, and those striæ are retained which were made by glacial agency in a period long anterior to any human history. In the valley of Lake Champlain, marbles which were polished and striated by the glacial agency, and then covered with clay, are found, when now uncovered, to retain not only the finest striæ, but their brilliancy of polish.

II. *Rivers.*—Rivers co-operate in the work by carrying down limestone in solution, thus furnishing the materials for the solid structures of some of the most extensive and interesting organic agencies which we are to notice. Their most obvious action, however, is in the transport of matter merely by mechanical agency. This is much greater than some would suppose, from the fact that mineral substances lose about three sevenths of their weight in water as compared with their weight in air. A current moving with a velocity of only 300 yards per hour, will tear up fine clay; of 600 yards per hour, will remove fine sand; of two thirds of a mile per hour, will remove coarse sand; and with a velocity of two miles per hour, will transport stones two inches in diameter. The agency of running water is also multiplied by the friction of the transported fragments upon each other and on the bed of the stream.

One of the most magnificent and instructive examples of the denuding agency of rivers is to be seen in the retrocession of the *Niagara Falls*, which have cut an enormous ravine from Queenstown, seven miles back, to their present situation. Soft shales at the base of the falls underlie the harder limestone, which is gradually undermined, and fragments of the overlying rocks are detached from above. In this way the falls are now retrograding at a rate not easily reckoned with precision for the want of historical data, but variously estimated to average from one foot to one yard per year. As the rocks have a small dip backward in the direction of Lake Erie, the water will at length cease to act on the soft shales for the want of sufficient fall below to remove the materials. The process will therefore be arrested long before the falls can have traveled back as far as the lake.

In what manner do rivers act? Describe the action of Niagara River at the falls.

Fig. 2.

VIEW OF NIAGARA FALLS.

In crossing the river just below the falls, the view is justly regarded as one of the most sublime in the natural world. As you look up from this deep ravine, you see at least 20,000,000 cubic feet of water each minute rushing down from a height of 160 feet, and appearing in truth

> "As if God poured it from his 'hollow hand,'
> ———— * * * * and had bid
> Its flood to chronicle the ages back,
> And notch his centuries in the eternal rock."

A remarkable example of rapid erosive action of water is found at the lower falls of the Genesee River, at Portage, N. Y. It is within the recollection of some of the inhabitants that the river flowed over a table rock, and was precipitated 96 feet to the level of the river below the falls. There is, however, now a channel extending back from the falls one eighth of a mile, 80 feet wide and deep, forming a violent rapid, down which the water, bearing along ice and debris, rushes and rapidly wears away the solid rocks. Within five years it is known to have deepened in some places five or six feet (*Fig.* 3).

In 1603 a current of lava flowed down from the highest sum-

What is said of Genesee River?

Fig. 3.

LOWER FALLS OF PORTAGE, N. Y.

mit of Ætna, on the west side, into the valley of the Simeto, and completely blocked up the beds of the river. In the course of 200 years the stream has worn through this lava, which is a com pact blue rock, a passage from fifty to several hundred feet wide, and from forty to fifty feet deep.

The *Ganges* and *Burrampooter*, descending from the Himalaya Mountains, the loftiest on the globe, unite in a vast delta which they have formed. This delta is an extensive alluvial plain, reticulated by an immense number of channels, and is more than half as large as the State of New York. Nothing as coarse as gravel can be found near the Ganges within 400 miles of its mouth. The river has been known to carry away 40 square miles from one district within a few years. Islands of many miles in extent are formed in a short period. Various estimates have been made of the quantity of the solid matter which is carried down by this river; according to the most accurate of which, 35,000 cubic feet of mud pass down every minute during the flood season, or about 3,500,000 tons daily, and the quantity discharged during the 120 days of the flood must therefore amount to 6,000,000,000 of cubic feet. High tides (11 to 16 feet) rapidly disperse this sediment in the Bay of Bengal, whose

What is said of the Simeto? the Ganges and Burrampooter? What amount of matter does the Ganges transport?

waters. 100 fathoms deep at 100 miles out, are gradually shoaled from this distance toward the shore to four fathoms, and for 60 miles are discolored by this turbid stream. The annual discharge of the Ganges would be sufficient to cover a township six miles square with soil to the depth of nearly seven feet.

The delta of the *Nile* is nearly as large as the State of Vermont. Its progress has been arrested in comparatively modern times by an easterly current in the Mediterranean, which carries off much of the sediment that is discharged into the sea, and preys occasionally upon the delta itself. It is very probable that a bay once occupied the site of the delta, and that it must have been of great depth, for while the sea near the shore gradually deepens to 50 fathoms, it then suddenly falls off to 380 fathoms.

The *Amazon* is probably unequaled among all the powerful agents of degradation. The vast amount carried out by its current, which is not entirely lost in the ocean at the distance of 300 miles from land, is furnishing materials, which, instead of forming a delta, become the subjects of oceanic agents.

The *Mississippi*, the father of waters, has formed most of the lower part of Louisiana, and is forming a tongue of land which extends far into the Gulf of Mexico, and which has advanced several leagues since New Orleans was built. The annual discharge of this river is 14,883,360,000,000 cubic feet, equal to 101·1 cubic miles of water. This is about one twelfth part of the quantity of rain which falls in its valley, the remaining eleven twelfths being lost by evaporation. The average amount of sediment is $\frac{1}{528}$th part, making 28,188,000,000 cubic feet, or 2,000,000,000 tons of solid matter. This annual deposit would be sufficient to cover a township six miles square to the depth of 30 feet. The delta comprises an area of 13,000 square miles, with a probable depth of not less than 1000 feet. This amounts to 2700 cubic miles, and would have required 14,000 years for its deposition, if all the sediment had fallen within its area during this time. Since, however, a considerable portion has been more widely distributed in the Gulf of Mexico, the age of the delta must be much greater.

In Massachusetts, the matter carried down by the Merrimac has been estimated, from careful experiments by Dr. S. L. Dana, of Lowell, to be 840,000 tons per annum.

The destructive force of occasional floods and storms is worthy of notice. Oceanic deltas are liable to be flooded not only

What is said of the Nile? the Amazon? the Mississippi? the Merrimac? In what manner do floods increase this action?

by freshets, but by storms from the sea, driving up the tide and current, and when, at rare intervals, these causes all combine, extensive tracts are entirely remodeled, and vegetable and animal life perish on a scale commensurate with the changes in inorganic nature.

Tropical mountainous regions are especially liable to very destructive floods, which pour down in continuous cataracts, sweeping along rocks of many tons' weight, where ordinarily an insignificant brook only is to be seen. Temporary rivers are then formed where, in dry seasons, water is entirely wanting.

Masses of ice co-operate powerfully with freshets, choking up the course of the stream, and forming basins of the accumulated waters, which at length burst their barriers, and rush down, tearing up the loose earth in narrow gorges, like Deerfield River in Massachusetts, and grinding over the solid rocks with the noise of thunder.

The tendency of a river flowing through a plain of unconsolidated materials is to form curves, or bends, as they are usually called. Wherever the current deviates from a straight line, it strikes the opposite bank, wearing it away, while the comparative quietness of the water on the other side promotes the accumulation of sediment, and the degree of curvature is thus continually increasing. At length, in some unusual freshet, the river cuts across the narrowed neck of the bend and forms a new channel. Such bends are numerous on the Mississippi, and are frequently cut off. A few years since, a remarkable bend in the Connecticut, in the beautiful alluvial meadows of Northampton, had a circuit of about three miles, with a neck of eighty rods, when in a freshet the river cut a deep channel across the neck, leaving its former circuit dry except so far as it still receives a small tributary.

Poultney River, in Vermont, affords a very remarkable example of a change of channel near Fairhaven. The change occurred in 1783, during a freshet, and the neighboring inhabitants

---

What phenomena are presented by rivers flowing through plains? What is said of tropical regions? of the effects of ice?

say that it was caused by running a furrow across the neck (*s s* in the figure) of a peninsula.

The accompanying figure is referred to in this description, the present channel being represented by continuous, and the former by dotted lines.

Fig. 4.

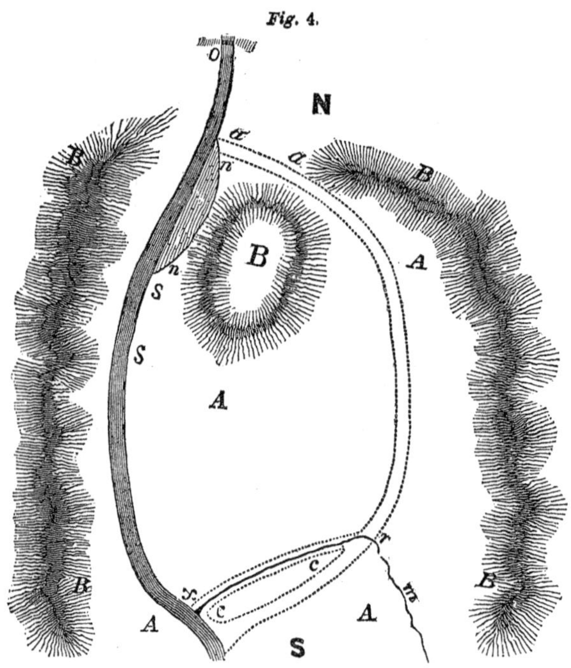

B B B B are hills of taconic slate; A A A A is an alluvial plain, overflowed before 1783, and overlying a thick deposit of fine blue clay.

a a are the Dry Falls, which are about 150 feet high, and fifteen rods long. The water was precipitated over the edges of the strata of soft slate. On the lower parts of the projecting strata are numerous deep furrows, mostly about a foot long, four to five inches wide, and two inches deep, but somewhat deeper in the middle. There are also several pot-holes of various sizes.

n n is recent alluvial deposit. Further up the stream we find high banks of clay, and in the plain A A the river has cut through alluvium and blue clay to a depth of one hundred and fifty feet

Immense slides, on either side of the tortuous stream, cover an area of several square miles with a scene of violent disturbance.

c c was an island in the old channel. *m* is a small rivulet which formerly emptied into the river at *r*, and now runs back through the old channel to *x*, having, since 1783, cut the channel on the north side of the island, *c c*, to a great depth. The old channel is now covered with grass. It was only about ten feet below the banks, the rocks at the falls having been a barrier which prevented a deeper cut.

Numerous rivers, in the lower part of their channels, have probably ceased the work of excavation, and, when confined by embankments, have a tendency to fill up their beds and run at higher levels. The Po and the Adige drain the northern part of Italy, and have caused one hundred miles of coast to encroach twenty miles upon the Adriatic Sea within 2000 years. On these rivers the practice of embankment, which commenced in the thirteenth century, has been carried to a great extent. In consequence, the Po has been filled up so much that the surface of the water is higher than the roofs of the houses in the city of Ferrara. The magnitude of these barriers is a subject of increasing expense and anxiety, it having sometimes been found necessary to give them an additional height of one foot in a single season. The Mississippi is confined by levees for a considerable distance above and below New Orleans, and the future inhabitants of Louisiana may find the river rather unmanageable, should it begin tc fill its bed and to raise its waters. In consequence, however, of the increased evaporation resulting from the clearing of forests about the sources of the river, it is said that the annual discharge has sensibly diminished within fifty years. In a small degree, however, the levees, by preventing the expansion of the river in freshets, diminish evaporation.

Notwithstanding the powerful degrading agency of rivers, they have not, in most cases, formed the valleys through which they flow. These are usually due to agencies which gave the configuration to the surface of the earth long anterior to the historical epoch. Some rivers, as may be seen in *Fig.* 5, turn aside from valleys, through which a moderate elevation would send them,

What is said of the Po? of the Adige? of the Mississippi? of river valleys?

to pass through mountain gorges, which must have been made by other agencies. But the terraces so frequently seen on our rivers were mostly formed by their degrading agency in a former period, in the history of which we shall explain the mode of their formation.

Fig. 5.

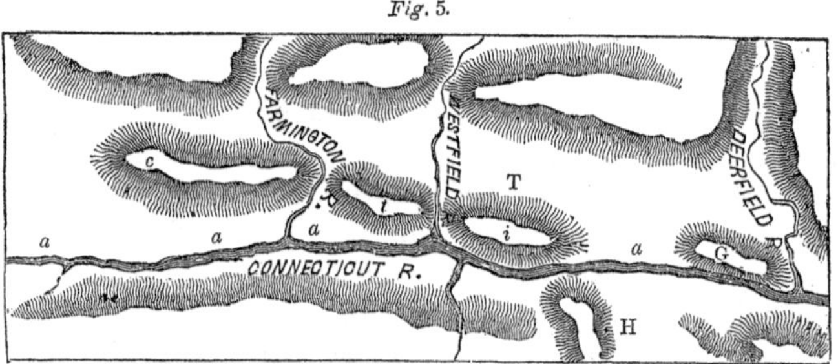

a a a a, a valley, which is crossed by Deerfield, Westfield, and Farmington rivers. *c*, *t*, *i*, and G, are greenstone ridges. H, Mount Holyoke; T, Mount Tom.

III. *Bursting of Lakes.*—The bursting of lakes is an agency which, although occasional and rare, produces powerful effects.

A frightful deluge occurred in 1818 in the valley of Bagnes, in Switzerland. The waters of the Drance were dammed up by the falling of glaciers and avalanches, which formed a barrier 400 feet high and 600 feet wide, above which a lake nearly a mile and a half long accumulated. A bold and persevering engineer tunneled the dike so as to meet the surface of the water of the lake, which flowed through, gradually melting down its channel as the water fell in the lake. In this way 330,000,000 cubic feet of water were carried off in three days without damage, when the dike gave way, and in half an hour 530,000,000 cubic feet of water swept down, running the first thirteen miles in thirty-five minutes, and bearing down 400 houses, with trees, rocks, and earth. Had it not been for the enterprise of the engineer, three times the amount of water might have accumulated before bursting through the dike.

Similar to this was the eruption of Long Pond, in Glover, Vermont. A barrier of fine sand separated this pond from the valley, which extended 20 miles to Lake Memphremagog. Some persons having made, for amusement, a small channel through

---

Describe the effects of the bursting of lakes.

the sand barrier, the running water in a few minutes excavated the channel deep enough entirely to empty the pond. The water rushed into Barton River, the channel of which was much enlarged by the violent inundation, and great numbers of trees were carried down the stream.

IV. *Springs.*—Springs act chiefly by taking up mineral matter at various depths, and afterward depositing it on the surface of the ground. The deposits of greatest magnitude are calcareous or silicious. We have already remarked that water which contains carbonic acid has the property of dissolving limestone. Now the quantity of this gas which water is capable of containing depends upon pressure. Under the pressure of the atmosphere it may contain its own volume; if the pressure be doubled, it will take up double its volume, and so on, and to any additional amount in proportion to the pressure. Consequently, at some distance beneath the surface of the earth, springs may, and especially in limestone countries do, contain a great amount of this gas Hence the subterranean passage of such water through fissures in limestone enlarges those fissures, so that in many cases rivers of considerable size, as in Jamaica (West Indies), after flowing on the surface for many miles, are lost in limestone chasms and flow under ground. Caverns of greater or less size are formed by the same agency, for caves of any considerable extent are almost invariably in limestone districts.

The water, if overcharged with gas and limestone, that is, containing more than the mere pressure of the air will permit, must deposit the excess of limestone when it issues either into an open cavern or upon the surface of the ground. When it drops from the roof of a cavern, *stalactites* are formed, like icicles pendent from the roof, and masses of *stalagmite* on its floor, and sometimes these meet, forming a column, which is continuous from the floor to the roof of the cavern. Sometimes these masses, especially the stalactites, are of a beautiful crystalline structure; the stalagmite is more frequently in thin concentric but irregular layers, a result of the mode of its deposition. Masses of the latter

---

Describe the action of springs; the formation of stalactites.

have been seen rising up, like altars, 10 or 12 feet high, and 15 feet in diameter. Slabs of beautiful calcareous alabaster are obtained from such stalagmite.

The most extensive deposits of this kind are formed where the springs issue on the surface of the ground. At San Filippo, in Italy, the springs have deposited a mass of limestone 250 feet thick, and a mile and a quarter in length. These springs have been known to deposit a solid mass 30 feet thick in 20 years. The High Rock at Saratoga, N. Y., is a calcareous deposit from the spring in its center. The geologist is familiar with numerous cases like this. Frequently the calcareous deposits of springs are more or less filled with irregular pores, and the mass is then called *tufa*. Plants, and any other bodies lying in the water of such springs, are liable to be coated with the deposit, and such cases of mere incrustation are sometimes confounded with petrifaction, which is an entirely different process.

Deposits of silicious matter, often called *silicious sinter*, are the product of hot springs. If water contains an alkali, as soda, it is capable, especially at a high temperature, of dissolving silex, which is deposited when the spring comes to the surface. The basin of the Great Geyser, in Iceland, has been formed in this manner. Silicious incrustations are formed on plants in the same manner as the calcareous incrustations above mentioned. Such deposits are less numerous and extensive than those which are calcareous, but are of much interest, as showing us how water may dissolve rocks of flint.

V. *Landslides.*—Landslides frequently occur on mountains, especially in times of freshets, and sometimes fill up the course of streams and occasion floods. Hills of clay are peculiarly liable to slides, which produce contortions in the flexible strata. Avalanches of snow and ice concur in violently removing rocks and earth from the steep sides of mountains into valleys beneath. The pressure of water in fissures, the undermining process of water passing through soft strata, the action of springs, convert-

---

Mention examples. What is silicious sinter? Describe the Great Geyser What is said of landslides? of their origin?

ing sand into quicksand, the eroding power of streams and torrents, undermining large masses, which, being softened by the water, slide down into the valleys.

Two landslips occurred in Troy, N. Y., in 1836 and 1837, which appear to have resulted from the action of springs of water in fissures of clay beds. The beds of clay and gravel were 227 feet high. A spring of water was obstructed, and filled up a fissure, and, by its pressure, forced off a large mass of clay and earth, the weight of which was estimated to be 200,000 tons, which slid down the declivity, carrying every thing before it, to the second street of the city, a distance of 200 yards. The slide was accompanied by torrents of mud and water. Several buildings were buried in the ruins, and some persons lost their lives.

A few years since, a tract of land at Champlain, Lower Canada, consisting of 207 acres, resting on a steep slope, suddenly slid down 360 yards into the Champlain River, and dammed it up for three fourths of a mile. The slide produced a loud, rumbling noise, and filled the air with a dense, suffocating vapor. One individual was buried to his neck in the moving mass, but finally escaped without injury.

In 1826, a similar slide took place near the notch in the White Mountains, and destroyed a whole family.

VI. *Glaciers.*—The history of *glaciers* has, within a few years, excited much interest, not only on account of their remarkable effects and mode of action, but of their applicability on a grand scale to the explanation of the phenomena of drift.

1. *Origin of Glaciers.*—Under the equator, a perpendicular ascent of three miles brings us to regions where the temperature of the air is below the freezing point, and as we approach the poles this point is reached at a much less elevation, until, in the latitude of 65° or 70°, the soil is frozen most of the year. In most countries there are mountains whose tops extend far above this line of perpetual congelation. Their tops, therefore, become the repositories of eternal frosts and snows. When we ascend such mountains under the equator, we pass through all the climates of the globe, torrid, temperate, and frigid. In temperate climates there will be a broad belt where the snows and ice will

Mention examples. How do glaciers originate?

be extended down a considerable distance during the winter, and retreat much higher up during the summer.

Glaciers and mers de glace (seas of ice) originate in snow which has been partially melted and then frozen. The lower part, by the percolation of water, is converted into solid ice. The upper part is more granular. Increasing by annual layers of snow, the glacier is stratified.

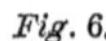

Fig. 6.

GLACIER DU LAUTER-AAR (SWITZERLAND).

In the higher regions of the Alps there are extensive table-lands, which are covered with thick masses of ice, through which the sharp mountain peaks rise to a yet greater height. These icy plains are called *mers de glace*. The glaciers extend down the valleys, until they arrive at a region where the heat arrests their progress. In some parts of the Alps, this limit is met at an elevation of 7000 feet. But some of the glaciers extend down to an elevation of 3000 feet. Their thickness is very unequal, in some places being 600 or 800 feet, while the average is more frequently 100 to 200 feet. Some are ten to fifteen miles long, with a breadth of two or three miles. The upper surface is extremely uneven, and is often covered with needle-shaped masses. Fissures from 20 to 100 feet wide are common, having been pro-

What is said of their thickness, extent, and surface in the Alps?

duced by the contraction of extreme cold in the winter, and enlarged by the melting of the sides in the summer. The slope of glaciers is usually moderate. That of Aar descends 3000 feet in fifteen miles.

2. The rate of *motion* in glaciers is very slow. Professor Hughes, in 1824, built a house on the glacier of the Arvre, which during fifteen years descended at an average rate of eight inches in twenty-four hours.

Several causes co-operate to effect the motion of glaciers. Gravity and the forms of the valleys determine their downward route. Water, freezing in the pores and fissures of the mass, expands with an irresistible force, which is directed by the sides of the valley downward. But the mass is not a perfectly rigid solid; it is flexible, and susceptible of motion like an extremely viscid fluid. It therefore yields to the pressure from above of gravity and expansion, and descends, in some seasons, several miles down the valley, destroying every object in its way.

Fig. 7.

LOWER PART OF THE GLACIER OF THE VIESCH (SWITZ.).

The retreat of a glacier consists merely in the melting of the lower extremity, which varies according to the warmth of the summers. A literal retrograde motion is impossible.

Fig. 8.

UPPER PART OF THE GLACIER OF THE VIESCH (SWITZ.).

3 *Effects of Glaciers.*—In their progress glaciers crowd along all the movable materials. The accumulations in front have the form of rounded hills, and are called terminal moraines. The accumulations along the sides are long ridges, and are called lateral moraines. When glaciers descend from two converging valleys into one, they unite, and the union of the adjacent sides forms a *medial moraine*, which extends down the valley in the middle of the double glacier. *Fig.* 6 shows two moraines which unite in one. Large isolated fragments of rocks are often seen on pedestals of ice (as in *Fig.* 8), that have been protected by these rocks from the melting and evaporation which have removed the general surface. The small stones conduct the diurnal heat through them, so that the icy pedestals are not formed beneath them. On the contrary, the melting and evaporation of the surface expose other stones, which are washed away by torrents of water that descend upon the glacier from the neighboring hills.

Fig. 9.

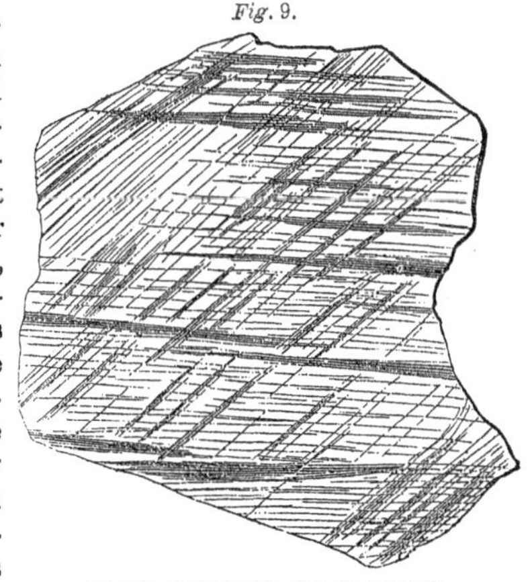

ROCKS STRIATED BY GLACIERS.

Generally the under surface is thickly studded with angular rocks, pebbles, and coarse sand. These projecting fragments, being pressed down by the great weight of the ice on the rocks over which the glacier proceeds, make grooves and scratches, while the finer materials smooth or finely striate the surface. As the glacier retires or advances, new striæ are formed, which slightly vary in direction, although each set are perfectly parallel to each other. This is due to the fact that the materials are firmly frozen into the ice. *Fig.* 9 represents a por-

---

What are moraines, and how are they produced? What is the appearance of the surface? Why do large rocks stand on pedestals of ice? Why do not small stones also?

tion of a rock striated by the Alpine glaciers. Sometimes a stone gets loose and makes in the rock an oblique furrow of unequal depth. The loose materials are crushed, rounded, and ground over the surface, until most of them are reduced to sand and fine mud. At the termination of the glacier there are conical hills, which, as the glacier advances during some seasons further than at others, are crowded upon each other, and present an appearance which resembles the rounded hills along the margin of the river valleys, far from any existing glacial agency. The rocks on the sides of the glacier are striated and embossed in a similar manner with those at the bottom, while those portions of the mountain which are above its influence present a rough and ragged appearance. This effect is seen in the following figure. Em-

Fig. 10.

ROCKS EMBOSSED BY GLACIERS (ALPS, SWITZERLAND).

bossed and striated rocks occur over a large portion of the northern hemisphere as the result of glacial action during a former period of the earth's history.

In colder countries glaciers descend to lower levels. "In Chili, which has the same latitude as the Alps in Switzerland, we have glaciers descending to the sea; but at the Alps they only descend within 3000 feet of the sea level, and this, too, although the Andes are only 7000 feet high, half the height of the

What effects are wrought on the rocks at the bottom and sides of the glacier? Where do glaciers descend to the sea?

Alps in the same latitude. The reason of this singular phenomenon is that to which I have alluded, that the summer heat is less intense in the Alps. In Europe we have to go to latitude 67 degrees before we find a single glacier reaching to the sea. But in the southern hemisphere, in latitude 46 degrees, in Chili, we find this occurring twenty-one degrees nearer the equator; so that there is here an actual generation of icebergs in a region which is almost the limit to where the floating icebergs reach." —*Lyell.* In all latitudes, an open sea terminates glaciers, by melting and breaking off the parts which reach the water.

It will be noticed that large quantities of matter are carried down by glaciers from higher to lower levels. Large blocks of stone are borne along in the glacier, and large hills are formed at its termination. The hills are mostly very finely-pulverized rock. Much of this is taken up by the streams of water and carried onward toward the ocean. It is difficult to estimate the quantity of matter thus transported by glaciers; but when we consider the great extent of surface acted upon not only in the Alps and Andes, but in all high northern and southern latitudes, we must regard them as highly important agents in modifying the surface of the earth.

VII. *Avalanches.*—Avalanches are masses of ice and snow, which are precipitated down the steep declivities of mountains. In Switzerland avalanches are very frequently precipitated into the valleys. In Chili avalanches of ice fall into the sea, and float off as icebergs.

SECTION II.—MARINE AGENCIES.

The materials which are borne along by rivers, if not deposited along their course, are consequently carried into lakes, seas, and oceans, and distributed in sedimentary strata over their beds. We are now to glance at the agencies which regulate and modify this distribution. In lakes, especially those of small extent, the modifying agencies are slight, and the strata will ordinarily

---

What general effects are produced by glaciers? What are avalanches? What becomes of the materials borne down by rivers?

constitute basin-shaped deposits. But in oceans and seas there are several powerful causes which are of great interest not only on account of the effects which they are actually producing, but because the effects are *examples* of agencies which have in former epochs acted in a similar manner to form a large portion of the present surface of continents. They are icebergs, waves, tides, and currents.

1. *Agency of Icebergs.*—Icebergs are masses of fresh-water ice, which are seen floating on the ocean or stranded on shoals both in the northern and southern hemispheres, and with few exceptions in latitudes above 40°.

1. *Origin.*—In their *origin*, icebergs are glaciers formed in the higher latitudes along the coasts and in the bays, in the same manner in which glaciers are formed in the Alps. The new Antarctic continent, which was discovered in 66° south latitude, by the American exploring expedition, was found to be bounded continuously by icy cliffs from one hundred and fifty to two hundred feet in height, without any appearance of rocks. "No break in this icy barrier where a foot could be set on the rocks was observable from aloft." A long range of icebergs was seen stranded in the sea, where bottom could not be reached with a line of nine hundred feet. The margin of the icy barrier was only here and there pierced by deep bays, but otherwise was quite uniform. A few floating icebergs were seen with rocks and earth on them, on one of which a landing was effected and some geological specimens were obtained. One rock was five or six feet in diameter. On the iceberg was a pond of fresh water of an acre in extent. Captain Wilkes describes the icebergs which were seen near their source as distinctly stratified, resulting from successive deposits of snow, which were supposed to fall to the amount of thirty feet per year. By occasional thaws they became more compact, aided not a little by the fogs, which on one occasion formed one fourth of an inch of ice on the rigging in a few hours.

The Astrolabe (of a French expedition) "skirted for sixty

What are icebergs, and where do they occur? How do they originate? Give examples near the Antarctic continent.

miles a perfectly vertical wall of ice, elevated one hundred and twenty to one hundred and thirty feet above the waves. The surface of the ice was perfectly level. Here we have the source of the enormous level icebergs."—*Hayes.*

In other cases, especially in latitudes where the sun has sufficient power in summer to melt most of the snow, icebergs are formed chiefly in narrow valleys at the head of inlets of the sea, where the snow is sheltered from the low summer's sun, and the water flows down on it from the neighboring hills. Thus in South Georgia are formed perpendicular or overhanging cliffs of ice several hundred feet high.

In Sandwich Land, an intelligent navigator observed that "the ice made from the tops of the highest hills down into the sea. In one place in particular, the sea had washed in under the ice as far as we could see, and this huge body of ice, four or five hundred feet in height on its face, and a mile or two in length, hung, not touching the beach by four or five feet, except at the sides of the mountains where it formed. The face next the sea was nearly perpendicular. * * * In Greenland the long narrow bays or fiords, like broad rivers, run far up amid the lofty mountains or table-lands of the interior. The vast plains of the interior abut upon these fiords; hence the greater number are closed by a glacier, close to which the water has a depth of several hundred fathoms. Several of the inlets are now completely filled up, and at others the ice projects far out into the waves, forming a considerable promontory."—*Hayes.*

In the eastern part of Iceland is a region of 3000 square miles almost entirely covered with vast mountains of ice.

Undermined by the waves and ruptured by the frost, immense masses are occasionally detached into the sea, producing by their fall enormous waves, which loosen other masses, and urge on icebergs which have stranded. The noise made by the fall of the enormous masses of ice is compared to thunder, and by the first settlers on the Shetland Isles was mistaken for earthquakes Most of the falling masses, however, are comparatively small fragments, and no one has seen the detachment of the larger ice islands. These are several miles in extent; navigators frequently mention them as being from five to ten miles in length. The

What effect have waves upon ice near the shore? What are ice islands?

French exploring expedition, above mentioned, measured several which were a mile in breadth; and one was 13 miles long, with vertical walls 100 feet high. It must, therefore, have been 600 or 800 feet thick.* Another, seen by the same expedition, was 225 feet high, which would give a depth below the surface of 1200 to 1800 feet. Capt. Ross saw several aground in Baffin's Bay, in water which was 1500 feet deep.

Fig. 11.

ICEBERG SEEN BY CAPTAIN ROSS.

2. *Motion.*—The motions of icebergs are of great importance, not merely for the effects produced at the present time, but for their bearing on the theories of the drift deposits. The most ordinary motion is a uniform slow progress from higher to warmer latitudes, irrespective of wind and waves. This motion is the

* Ice floats with one ninth of its bulk above the surface. Making allowance for any want of compactness, and especially for a greater breadth of the base, the depth of an iceberg may be reckoned at between five and eight times the height.

What is said of the thickness of icebergs? of their motion?

effect of those under currents with which their enormous depth in the water brings them in contact. Immense numbers of them are from this cause often seen to the east of Newfoundland. Floating from the north, they at length, in latitude 43°, come into the warm Gulf Stream, which there has an eastern course. Urged on by the current beneath, they float across the Gulf Stream, and usually disappear before they reach its southern side in latitude 36°. Some of them get aground on the Grand Banks before they reach the Gulf Stream.

One of the most extraordinary examples of size and motion was an iceberg seen by the British steamer, the Acadia, on the 16th of May, 1842, among a hundred others, and which was 400 to 500 feet high, and consequently about 3000 feet deep in the water. It must, therefore, have nearly equaled, from the base to the summit, the highest peaks of the Green Mountains. Having a remarkable resemblance to St. Paul's Cathedral in London, it was named St. Paul's. But the most extraordinary part of the narrative is, that "on the 6th of June the same object was seen, and the immediate exclamation on board was, there is our old friend St. Paul's. In the interim between the two views, the iceberg had drifted about 70 miles." This slow motion, 70 miles in twenty-one days, is worthy of notice. The maximum force of the polar current off Newfoundland is two miles per hour, and, although liable to be retarded, it can hardly be supposed to be reduced to one seventh of a mile per hour for 21 days. It is not improbable that this enormous iceberg was retarded by plowing the bottom of the sea in some parts of its course.

In the southern hemisphere, currents from the polar regions, in the same manner, float the icebergs into warmer latitudes, occasionally as far as the latitude of the Cape of Good Hope.

In their progress into regions of less intense cold, the structure of icebergs changes; the stratification disappears, and the whole becomes a compact mass of translucent blue ice, and the surface presents all conceivable forms, which the imagination easily converts into a city with its spires, domes, and battlements.

Another motion is that of a violent heaving and rolling of the mass when aground. Captain Couthuoy, in August, 1827, saw one stranded on the Grand Bank in about 500 feet of water around which, to the distance of one fourth of a mile, the water was full of mud, stirred up by the violent rolling of the mass.

Icebergs floating into warmer water, and melting more rapidly on some parts than others, sometimes change their center of gravity, and the enormous mass is seen to topple over, producing great commotion in the water.

3. *Dissolution.*—The *dissolution* of icebergs is sometimes effected by a violent explosion, rending the whole into fragments, which soon disappear. Several cases of this kind are recorded, and are supposed to be owing to the expansion of bodies of air confined within the ice at a temperature much below the freezing point, and when the temperature of the ice rises up to this point, the air must expand and the ice explode. But the ordinary process is that of melting in warmer waters.

4. *Effects.*—It is obvious that the foreign materials, rocks and earth, which may be borne along with them, will be dropped in their path on the bed of the ocean. It is, however, rarely that icebergs at a great distance from their original source are seen thus loaded. The one above mentioned, as seen by Captain Couthuoy, was thus loaded, and a few other cases are recorded. But they are rare, and many navigators have seen thousands of icebergs no one of which bore along any foreign materials. On the other hand, such materials have often been seen on them before and soon after they were detached. In many cases, a mass of rocks and earth may be a nucleus around which the ice has accumulated; yet, since these materials must rest on some base, they can not occupy the interior of the ice, and therefore are lost soon after the icebergs are detached. Any materials which adhere to their sides will be dropped near their source.

Icebergs often are stranded, and, being urged along by the force of currents, or turned about by the action of waves, produce important effects on the bottom of the sea.

II. *Waves.* 1. *Size.*—In consequence of the indefinite and imaginative descriptions which are common of waves running "mountains high," those who are not familiar with the open

---

In what manner do icebergs dissolve? What are the effects of icebergs?

ocean seldom have correct conceptions of them. These "mountains" rarely exceed thirty feet in height, although they have been observed in the North Atlantic with a height of forty-five feet. In one example of waves on a scale of unusual magnitude, the height of their summits was, for the most part, not over thirty feet above the bottom of the depressions, and the highest did not exceed thirty-five feet. Although they were suddenly raised by a storm, which had been immediately preceded by another at right angles, in the Gulf Stream, and were, of course, unusually short and narrow, the height was less than one tenth of the width. Enormous as are these masses, which may be half a mile or more in length, the sublimity of such a scene depends more on their *motion* than on their magnitude.

2. The *motion* of waves, which, as is well known, is a motion of the form and not of the substance, is often thirty miles per hour, rapidly rolling past the fleetest ships.

But since the motion is not in the substance of the water, this agency extends but to a moderate depth, and geological effects are produced only when waves are driven on shoals and coasts. Here, on account of the resistance of the bottom, they roll up with a front more and more steep until it becomes perpendicular, and at length fall over and break with enormous force, dashing up the sides of rocky cliffs, or rushing far up the shore in a sheet of foam.

3. *Effects.*—One of the most common *effects* is the wearing of loose stones, originally rough and angular, into smooth oval pebbles. On a sloping shore the loose stones are exposed to continual friction by rolling up and down, and usually the hardest stones are the most perfectly rounded. Where the rocks consist of limestone with a small proportion of flint, the pebbles of flint are more numerous than those of the limestone, most of the latter having been worn out by continual friction. The form of pebbles depends somewhat on the structure of the rock. Slates furnish small much-flattened oval forms, producing a mass of gravel

---

What is the size of waves? What kind of motion have they? Mention the effects of waves on loose stones.

or shingle. Some steep shores of hard rocks are covered deeply with large and well-rounded pebbles, such as are used for paving stones. Storms sometimes pile up at the head of sandy beaches, out of the reach of ordinary tides, enormous ridges of pebbles. An example may be seen near Boston, at Chelsea beach, between which and the marshes within is a ridge consisting mostly of porphyry pebbles four rods wide, ten to fifteen feet high, and two miles long. Stones of several tons' weight are also moved in storms by the force of the waves, and a coast is modified so as to be recognized only in its outline, the minute details of the shore being entirely changed.

On coasts which are fringed with cliffs of loose materials, the waves undermine the cliffs until fragments fall down an easy prey to the next storm. If the cliffs are composed of the drift deposits, the finer materials are washed away, while the shore at its base is covered with large bowlders. But if the cliffs are of solid rock, they will oppose a more effectual resistance. Yet solid rock is not impregnable, for the waves, taking up loose fragments, use them like battering rams to undermine the base of the cliff, while the agency of frost above aids in the work. If the cliffs be of limestone of unequal hardness from the intermixture of silex, peculiar and remarkable effects are produced. The silicious fragments furnish nearly indestructible pebbles, which wear out cavities, and even large caverns are found in such rocks above the present sea level. Smooth concave surfaces within them attest the agency by which they were formed.

Examples of this agency are to be found wherever high lands are in contact with the ocean. The numerous indentations of coasts are mainly due to waves and currents. The following cut, *Fig.* 12, shows the undermining action of the waves at Nahant.

At this place, and along the islands of Boston harbor, we see the harder rocks wearing away very slowly, while the softer rocks and loose materials are rapidly eroded, so as to render sea walls

---

Describe the action of waves on loose stones; on cliffs of loose materials; on cliffs of solid rock; on rocks of unequal hardness. Give examples of the action of waves at Nahant.

Fig. 12.

PULPIT ROCK, NAHANT, MASS.

necessary for the protection of the harbor. During the drift period, this harbor was probably filled up with loose materials, and it has been re-excavated by the joint action of waves and tides.

Remarkable examples of the agency of waves, aided more or less by currents, occur along the shores of Long Island, parts of which have had a much greater extent since the commencement of the historical period. Rocks which were once covered with soil are now naked, and are washed by the waves. The coast from Montauk Point to Nepeague beach, a distance of ten miles, is rapidly wearing away.

At Cape May (Del.), the sea is wearing away the land, in some places, at the average rate of nine feet per annum. During three years, Sullivan's Island (S. C.) has been worn away one quarter of a mile. Much of the eastern coast of England is rapidly crumbling away, and many towns are known only in history, their sites now forming a part of the German Ocean. In the harbor of Sherringham there was, ten years since, depth sufficient to float a frigate, where, forty-eight years before, there was a cliff fifty feet

---

Give examples of the action of waves at Long Island; at Cape May and Sullivan's Island; in England.

Fig. 13.

high with houses on it. Some of the Shetland Isles have been entirely destroyed by the action of the sea; others are now undergoing the process of destruction, and appear like fleets of vessels. The rocks are granite, and the accompanying example, *Fig.* 13, taken from the southern part of Hillswickness, Shetland, gives a good idea of their appearance.

By the erosive power of waves broad channels are excavated. In the thirteenth century, a channel was worn through an isthmus in the northern part of Holland, cutting off the island of Wieringen from the main land. The channel is now 13 miles wide. It is also highly probable that the English Channel was produced by the same erosive power.

Waves in lakes sometimes exert an undermining power upon rocks, especially where their structure is jointed so as to expose them to atmospheric agencies, as in the following example of the cliffs of Cayuga Lake, N. Y.

Fig. 14.

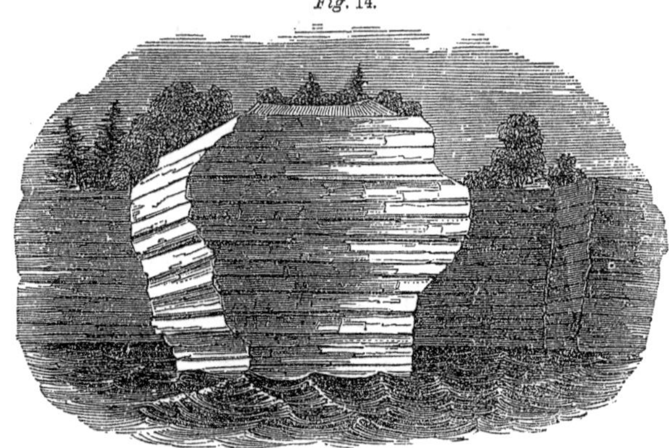
CLIFFS OF CAYUGA LAKE, N. Y.

The action of waves and currents, however, does not always carry the abraded materials out to sea. In some cases it merely

---

Describe the action of waves on the Shetland Isles. What examples of channels worn by waves? Of undermiuing action of waves in lakes?

transports them coastward, and forms shoals, flats, and sand-bars. In other cases, the matter discharged by the rivers and that which is worn down by the waves is thrown back upon the shores, and, when it is fine sand, gives rise to what are termed *dunes* or *downs*. These are formed by the action of the wind upon the sand, blowing it inward, and often destroying the fertility of extensive tracts of country.

III. *Oceanic Currents.*—By oceanic currents the sea is powerfully aided in making extensive depredations on some shores and in building up others.

The materials which are brought down by rivers and removed from the shore by waves are not then left to subside at once to the bottom of the sea. Oceanic currents, some perpetual and fixed in their course, and others intermittent and variable, bear the finer sediment into the deeper part of the ocean. The most remarkable current is the *Gulf Stream,* which flows past the eastern coast of South America, bearing the sediment of the Amazon to the north, and forming vast districts of low land between that river and the Oronoco; then spreading through the Caribbean Sea, it enters the Gulf of Mexico, where, being pent up, it rushes through the Straits of Florida with a velocity of four miles per hour, diminishing to three miles off Cape Hatteras, whence it takes a northeasterly course to the Banks of Newfoundland. There it is met by another current from Baffin's Bay, and deflected toward Iceland, Spitzbergen, and the northern parts of Scotland. In this great river of the ocean there flow about 90,000,000 cubic feet of water per minute, or 2500 times the amount discharged by the Mississippi. The polar current from Baffin's Bay is divided on meeting with the Gulf Stream, one portion being supposed to run under the latter to the south, and the other to flow on the surface between the Gulf Stream and the coast of North America. Another current of great size flows from the Antarctic Ocean along the western coast of South America.

What are the effects when the materials are transported coastwise? What are downs? What are the effects of currents on the shores? on the finer sediment? Describe the Gulf Stream. What is said of other currents?

That Arctic or Antarctic currents flow beneath the surface into the equatorial regions of the Atlantic is proved by the temperature of the Caribbean Sea, which on the surface is 80°, but at the depth of 240 fathoms is 48°, only 1° warmer than at a corresponding depth within some parts of the Arctic circle. One of the under currents has been found at the equator 200 miles broad and 23° colder than the surface water.

These great currents are influenced to some extent by long storms and prevailing winds, and other less and local currents are entirely remodeled by storms.

SECTION III.—GENERAL RESULTS OF AQUEOUS AGENCIES.

I. *Degradation.*—The general tendency of the aqueous agencies, with the comparatively unimportant exception of springs, thus appears to be the removal of soluble and movable materials to lower levels. These levels are formed first at the base of precipices and in the river valleys.

At the base of hills or cliffs there is produced a *talus* of large angular blocks, *a*, *Fig.* 15. These are further acted upon and broken up, as at *b*, *c*, *d*, until the rains or tides remove the finer portions in the form of mud. If these broken materials are pre-

Fig. 15. *Fig.* 16.

cipitated into streams of water, the finer and coarser materials are separated and arranged in successive beds, *Fig.* 16, *a*, *b*, *c*, *d*.

A portion of the finer materials which the rivers transport is deposited along their banks and at their mouths. To show how the mud is deposited by rivers, let *a a*, *Fig.* 17, be a low valley and *b* a stream of water overflowing its banks: the mud will be deposited over the space *a a* in horizontal layers. In the progress of time

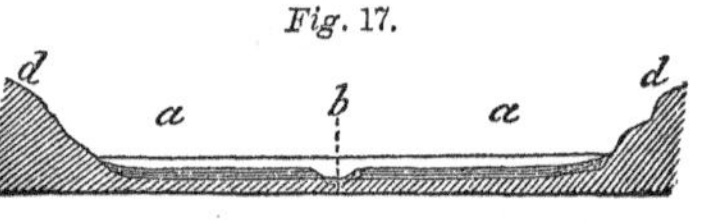

Fig. 17.

How are currents modified? What is the general tendency of aqueous agencies? Describe the deposits at the base of cliffs and in valleys.

the bed of the river will be elevated, as at *b*, *Fig.* 18, and finally the whole valley will be filled with sediment, or dotted with ponds and marshes at a little distance from the raised banks of the stream.

Fig. 18.

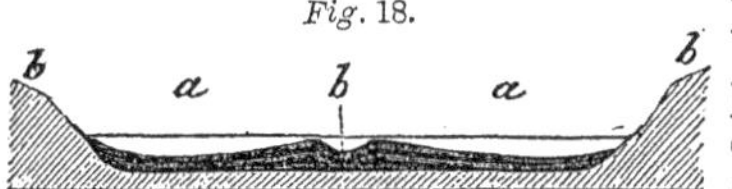

In consequence, however, of the shifting of river channels, the only permanent resting-places for the sediment are in lakes and the ocean.

II. *Salt Lakes.*—In those lakes which have no outlet, there is also a large accumulation of soluble materials. Such are the Salt Lakes in Asia and in Upper California. Since only an extremely minute proportion of salt is universally diffused, the quantity in rivers is not usually appreciable. But the perpetual supply and the perpetual concentration by evaporation in the lakes tend to its accumulation. In this way beds of salt are now in the process of formation.

The ocean, also, is but a great lake without an outlet, continually receiving the discharge of rivers, and accumulating the salt by evaporation. A minute quantity of salt is returned to the land by evaporation, but this is probably less than is carried off by rivers. Complete restoration is effected only when the bed of the ocean becomes dry land.

III. *Basin-shaped Deposits.*—The deposits in lakes, bays, and harbors are more or less basin-shaped. The sediment is in layers, which conform to the general surface of the bottom.

IV. *Formations of Marine Origin.*—By the action of waves, tides, and currents, a large portion of the materials which are brought down by rivers is removed to the open ocean, where it is washed up to form new shores and islands, or distributed in vast sheets over its bed.

So numerous and extensive are the agents of this general distribution, that the ocean may be regarded not as a great lake, but as a mass of broad rivers which skirt the continents, the va-

---

Where is the final resting-place of sediment? What is said of salt lakes? of the origin of salt in the ocean? of basin-shaped deposits? of marine formations?

rying size and complexity of divided, deflected, upper and under currents, present to the geologist a vast and intricate problem, whose details are yet to be solved. If the fine sediment which comes within the influence of these currents subsides at the rate of one foot per hour, it will be carried hundreds or thousands of miles, and in many cases by upper and under currents in different directions, before it will repose on the bed of the ocean. There it will form strata which are destined to be the slate rocks of the present geological epoch. Off the coast of the Southern States, extensive shoals and long low islands occur, which are separated from the main land by large bodies of water, and which are probably the products of oceanic currents. Long Island once constituted several islands, which have been united into one by the action of tides, currents, and waves. Off Massachusetts Bay, far out in the deep sea, but west of the Gulf Stream, are dangerous shoals, which have probably been formed by currents.

Oceanic currents not only distribute the sediment which is mechanically suspended, but they also aid the waves and tides in mingling throughout the ocean, with great uniformity, its saline ingredients, whether they are dissolved primarily by its own waters or by rivers.

A vast quantity of organic bodies, as we shall hereafter show, is enveloped in these deposits of lakes and the ocean. Shell-fish and many other aquatic animals live and die in them. Multitudes of terrestrial animals and plants are carried down by rivers into lakes and the ocean, and are distributed in the strata of mud and sand.

By the powers of cohesion and affinity, which are aided by the pressure of superincumbent waters, and in some regions by igneous agencies, these layers of sediment are consolidated. Should the sea and land change places, these new strata would constitute a formation of stratified rocks, resembling those of the present continents, but, unlike those of former geological epochs, they would contain abundantly the relics of man.

What is the effect of currents on the distribution of the saline ingredients of the ocean? How are organic bodies entombed? How is the sediment consolidated? How may a new formation appear?

CHAPTER II.

IGNEOUS AGENCIES.

We have seen that water is continually corroding the continents and transporting them to the ocean. Its tendency is to level down all inequalities on the earth's surface. Heat, on the contrary, exerts its power in an opposite direction. It tends to elevate, to throw into ridges, and to produce irregularities of the surface. By these opposing forces the general equilibrium of sea and land is preserved.

The crust of the earth has been more or less subjected to the action of heat both from internal and from external sources. The former are the origin of *volcanic action*, using the term in its wider signification, as comprising the kindred phenomena of volcanoes, earthquakes, and thermal springs.

A volcano is an opening in the earth, out of which ashes, stones, and melted lava are ejected. Around the opening which is called the *crater*, a mountain usually rises in the form of a truncated cone.

Extinct volcanoes are those which have not erupted since the commencement of the historical period; and since they therefore belong to the history of antecedent epochs, the consideration of them is reserved for another place.

Active volcanoes are those which have been known to erupt since the existence of man. A very few of these are in *constant* action, as Stromboli and Kilauea, but the greater part are *intermittent*, with intervals of action varying from a few months to many centuries.

Earthquakes are intimately connected with volcanoes, proceeding from the same general cause, and are frequently followed by

---

What is the subject of chapter ii.? What is the general tendency of heat in the earth? What is volcanic action? What is a volcano? What are extinct volcanoes? active volcanoes? earthquakes?

an eruption, by which these convulsive throes of the earth are relieved.

Thermal springs belong to the same class of phenomena, deriving their temperature from past or present volcanic fires, with which they are always associated.

The facts relating to earthquakes, to thermal springs, and to volcanoes, may be classified, as they are *subaërial, submarine,* or *subterranean.* The subaërial phenomena are those which occur on or near the surface of the dry land; the submarine occur in the waters of the ocean; by the subterranean, we mean the deep-seated internal igneous action.

SECTION I.—SUBAËRIAL IGNEOUS AGENCY.

I. *Volcanoes.*—There are about 300 volcanoes, of which one third belong to America, one third to Oceanica, and the remaining third to Europe, Asia, and Africa.

1. *Eruptions.*—Usually the first symptoms of an eruption are heard in rumbling sounds, which seem to travel along for a greater or less distance in the depths of the earth; they are seen in the increased volumes of smoke which arise from the crater, and are felt in tremulous motions of the earth, which assume the violence of earthquakes, and bring in their train the horrors that usually accompany these convulsions. Sulphurous and muriatic vapors fill the air, while electric agencies display their vivid coruscations, accompanied with heavy peals of thunder. Unusual signs of fright are manifested by the brute creation. Showers of stones and cinders fall, sometimes in immense profusion, and the convulsions of the earth become more violent. Masses of rock are ejected from the crater with tremendous explosions, until at length the earthquakes cease, and the imprisoned gases and lava find vent through the crater and sometimes through the sides of the volcanic mountain. A river of molten rock streams down and spreads out into a sea of fire. Sometimes its course is slow

---

What are thermal springs? What are subaërial igneous phenomena? submarine? subterranean? What is the number of volcanoes? Describe an eruption.

sometimes rapid. Glaciers may be encountered and melted, and torrents of boiling water and mud poured down. Showers of cinders again fall, and announce the termination of the eruption to be at hand. The flames, explosions, and ejections of rocks and stones become less violent, and finally nothing but vapors and smoke escape from the crater.

A few examples of eruptions will serve to illustrate the *kind* and *amount* of changes which volcanoes are now effecting, and will furnish the data by which to account for changes which must have been wrought in the crust of the earth during the earlier periods of its history.

2. *Vesuvius.*—Before the Christian era, Vesuvius was in a state of inactivity. History had no records of its eruptions, and a naturalist only, as he observed the volcanic nature of the rocks, would have suspected its real character. Its energies found vent in the neighboring isles of Ischia and Procida, which were shaken by terrific convulsions and desolated by eruptions. But Vesuvius itself was silent; and although Strabo perceived its volcanic character, Pliny omitted it in his list of active volcanoes. In the cone were the remains of an ancient crater which was nearly filled up, and was covered on its interior with wild vines. At the bottom was a sterile plain, on which Spartacus once encamped with his army of 10,000 gladiators, whose descent was a more disastrous eruption than the fiery floods of later years. The mountain was flanked with fruitful fields in a high state of cultivation, and at its base were the luxurious and populous cities of Herculaneum and Pompeii.

In A.D. 63, Vesuvius gave signs of awakening from its repose of unknown ages. From that year to A.D. 79, shocks were frequent, and at length became more violent, when a terrific eruption took place, and buried the cities above mentioned. Since that time its eruptions have been numerous, but those of 1631 and 1822 were most remarkable. In 1631 a stream of lava consumed Resina, which had been built over the site of Herculaneum, and floods of mud were poured down with terrible devasta-

---

What was the condition of Vesuvius before the Christian era? Give the subsequent history of Vesuvius

tion. These floods originated in heavy rains, which are often produced by volcanic action, and which wash down the cinders and dust until they assume the consistency of mud. *Fig.* 19 represents the appearance of an eruption of Vesuvius in 1784, by which Torre del Grecco was overwhelmed, and 400 persons were destroyed.

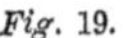

Fig. 19.

VIEW OF VESUVIUS, 1784.

For some time previous to 1822 the crater had been gradually filling up, when it was blown out, with awful explosions, to the depth of 2000 feet, and more than 800 feet of the top were blown off. Since this time eruptions have been frequent, the most remarkable of which occurred in 1849 and 1850.

3. *Ætna.*—Ætna has been observed from the remotest antiquity. It is situated on the island of Sicily, and is about 90 miles in circumference and nearly 11,000 feet high (*Fig.* 20). During the epoch of the eruptions of Vesuvius above described Ætna was occasionally active.

In 1669 an immense quantity of the lava overwhelmed fourteen towns and villages before reaching Catania. Although the walls

Describe Ætna; the eruption of 1669; of 1811 and 1819.

Fig. 20.

VIEW OF ÆTNA, LOOKING UP FROM THE VAL DEL BOVE.

of this city were 60 feet high, the lava accumulated until it gained the top, and then poured over in a fiery cascade and destroyed a portion of the city. The current continued 15 miles further, when it entered the sea with a depth of 40 feet. The progress of this current had been so slow that the surface had time to cool, so that it advanced by breaking through its walls of crust. Since the commencement of the present century, there have been several eruptions through the sides of the mountain. In 1811, seven openings were formed, each at a lower level successively. In 1819, three out of five such openings united in one, and poured an enormous torrent into the "*Val del Bove.*" Arriving at a precipice, it poured over in a cataract of liquid rock, which, cooling in its descent, dashed against the bottom with an inconceivable crash. This current continued to flow for nine months, when it was found to move at the rate of less than five rods per day.*

* A late traveler gives the following description of an eruption of Ætna seen by night:

"It was about half past ten when we reached the foot of the craters, which were both tremendously agitated; the great vent threw up immense columns of fire, mingled with the blackest smoke and sand. Each explosion of fire was preceded by a bellowing of thunder in the mountain. The explosions followed each other so rapidly that we could not count three seconds between them. The stones which were emitted were fourteen seconds in falling back to the crater; consequently, there were always five or six explosions—sometimes more than twenty—in the air at once. These stones were thrown up in the shape of a wide-spreading sheaf, producing the most magnificent effect imaginable. The smallest stones appeared to be of the size of cannon-balls; the greater were like bomb-shells; but others were pieces of rock five or six cubic feet in size, and some of the most enormous dimensions; the latter generally fell on the ridge of the crater and rolled down its sides, splitting into

4. *In Iceland* have occurred some of the most extensive eruptions on record. Most of this island, which is 300 miles long and 150 broad, is covered with volcanoes and vast regions of lava, which is broken into ragged and pointed rocks, or cleft by yawning chasms of many miles in length and of unseen depth. Other parts of the island are filled with innumerable springs spouting forth torrents of boiling water, or with immense mountains of ice.

The convulsions of 1783 appear to have been most violent, and on a scale of extraordinary magnitude. In an eruption of Skaptar Jokul, two streams of lava flowed off in opposite directions, of which one was 40 miles long and 7 broad, and the other was 50 miles long and 12 broad, both containing 70,000,000,000 cubic yards. At least 1300 human beings lost their lives; and not less than 20,000 horses, 7000 horned cattle, and 130,000 sheep were destroyed.

Fig. 21.

5. *On Hawaii* (Sandwich Islands), *Fig.* 21, is an ever-active volcano, *Kilauea*, of unusual form and situation. "The whole surface of the island pertains to the slopes of three lofty volcanic summits, Mount Loa, constituting the southern portion, 13,760 feet in height; Mount Kea, long extinct, covering the north-

fragments as they struck against the hard and cutting masses of cold lava The smoke emitted by the smaller cone was white, and its appearance inconceivably grand and beautiful; but the other crater, though less active, was much more terrible; and the thick blackness of its gigantic volumes of smoke partly concealed the fire which it vomited. Occasionally both burst forth at the same instant and with the most tremendous fury, sometimes mingling their ejected stones.

"If any person could accurately fancy the effect of 500,000 sky-rockets darting up at once to a height of three or four thousand feet, and then falling back in the shape of red-hot balls, shells, and large rocks of fire, he might have an idea of a single explosion of this burning mountain; but it is doubtful whether any imagination can conceive the effect of one hundred of such explosions in the space of five minutes, or of twelve hundred or more in the course of an hour, as we saw them!"—*Mantell.*

Describe Iceland; the eruption of 1783. What is said of Hawaii?

ern portion, 13,950 feet in height; and Hualalai, toward the western shores, estimated at 10,000 feet."—*Dana.*

The crater of Kilauea, *Fig.* 22, is not in a truncated cone of a

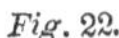

Fig. 22.

CRATER OF KILAUEA.

mountain, but in an upland country, near the base of Mount Loa. The crater is an immense chasm, in the middle of a plain, which is surrounded by a precipice from 200 to 400 feet high, and is 15 miles in circumference; or rather this plain is itself a chasm, and the crater is a chasm within a chasm. Mr. Stewart, formerly missionary at these islands, gives an interesting account of its appearance:

"Standing at an elevation of 1500 feet, we looked into the horrid gulf, not less than eight miles in circumference, directly beneath us. The hideous immensity itself, independent of the many frightful images embraced in it, almost caused an involuntary closing of the eyes against it. But to the sight is added the effect of the various unnatural and frightful noises, the every agonized struggling of the mighty action within. This gulf contains 50 or 60 conical craters, many of which are in constant action. About half way down the perpendicular side of the chasm is a ledge or piazza of lava from a few feet to several yards in width, which extends all around. Below this, all was of a dismal black color, except two or three of the conical craters at the bottom, which were covered with sulphur of various shades of yellow and green. The cliffs above the piazza were red on the north and west sides; on the east the bank was less precipitous, and consisted of entire banks of sulphur of a delicate and bright yellow.

"As the darkness of the night gathered around us, fire after fire began to glimmer on the eye, appearing in rapid succession. Two or three small craters were in full action, every moment casting out stones, ashes, and lava, with heavy detonations, while the flames glared over the surrounding obscu-

---

What is said of the crater of Kilauea?

rity, richly illuminating the more distant volumes of smoke. The great seat of action, however, seemed to be in the southern and western end, where an exhibition of ever-varying fire-works was presented. Rivers of fire were seen rolling in splendid coruscations among the laboring craters. During the second night, the noises were redoubled, rolling from one end of the vast chasm to the other with inconceivable velocity, and the flames burst from a large cone, which in the morning appeared to have been long inactive. Red-hot stones, cinders, and ashes were propelled to an immense height, and soon the lava boiled over in two curved streams glittering with indescribable brilliancy.

"A whole lake of fire opened in a more distant part, two miles in circumference. Its surface had all the agitation of an ocean; billow after billow tossed its monstrous bosom in the air, and occasionally met with such violent concussion as to dash the fiery spray 40 or 50 feet high. This spray is blown out by the winds into delicate threads, as melted glass may be drawn out, and accumulates on the sides in masses which resemble bunches of tow."

An eruption, which had its source in this volcano, in June, 1840, was one of the most extraordinary of modern times. It has been described by Mr. Coan, an American missionary in the Sandwich Islands, who, soon after the eruption, discovered the place where the current broke forth, and traced it to the sea. We give his narrative somewhat abridged, as furnishing one of the most instructive examples of this class of geological agencies:

"For several years past the great crater of Kilauea has been rapidly filling up. The great basin below the black ledge has been computed to be from three hundred to five hundred feet deep. Silent eruptions had occurred at intervals, until the black ledge was repeatedly overflowed, each forming a new layer from two feet thick and upward, until the whole area of the crater was filled up, at least fifty feet above the original black ledge. This process of filling up continued till the latter part of May, 1840, when, as many natives testify, the whole area of the crater became one entire sea of ignifluous matter, raging like the ocean when lashed into fury by a tempest. For several days the fires raged with fearful intensity, exhibiting a scene awfully terrific. The infuriated waves sent up infernal sounds, and dashed with such maddening energy against the sides of the awful caldron as to shake the solid earth above, and to detach huge masses of overhanging rocks, which, leaving their ancient beds, plunged into the fiery gulf below. Every thing within the caldron is new. Not a particle of the lava remains as it was when I last visited it. All has been melted down and recast. The whole appears like a raging sea, whose waves have been suddenly solidified while in the most violent agitation.

"On the 30th of May, the people of Puna observed the appearance of smoke and fire in the interior, a mountainous and desolate region of that district. Thinking that the fire might be the burning of some jungle, they took little notice of it until the next day, when the sudden and grand exhibitions of fire left them no room to doubt the cause of the phenomenon. The fire augmented during the day and night, but it did not seem to flow off rapidly in any direction. All were in consternation, as it was expected that the molten flood would pour itself down from its height of four thousand feet to the coast. On Monday, June 1st, the stream began to flow off in a northeasterly direction (from *n*, *Fig.* 21); and on June 3d, at evening, the river reached the sea (at

Nanawale, which it destroyed), having averaged about half a mile an hour in its progress.

"The source of the eruption is in a forest, and in the bottom of an ancient wooded crater (*a*, *Fig.* 21, p. 54), about four hundred feet deep, and probably eight miles east from Kilauea. From Kilauea to this place the lava flows in a subterranean gallery, probably at the depth of a thousand feet, but its course can be distinctly traced all the way by the rending of the crust of the earth into innumerable fissures, and by the emission of smoke, steam, and gases. The eruption in this old crater is small, and from this place the stream disappears again for the distance of a mile or two, when the lava again gushes up and spreads over an area of about fifty acres (*c*). Again it passes under ground for a few miles, when it reappears in another old wooded crater (*m*), consuming the forests, and partly filling up the basin. Once more it disappears, and, flowing in a subterranean channel, cracks and breaks the earth, opening fissures from six inches to ten or twelve feet in width, and sometimes splitting the trunk of a tree so exactly that its legs stand astride at the fissure. After flowing under ground several miles, it again broke out, like an overwhelming flood (*n*), and, sweeping forest, hamlet, plantation, and every thing before it, rolled down with resistless energy to the sea, where, leaping a precipice of forty or fifty feet, it poured itself in one vast cataract of fire into the deep below, with loud detonations, fearful hissings, and a thousand unearthly sounds. Imagine to yourself a river of fused minerals, of the breadth of Niagara, and of a gory red, falling, in one emblazoned sheet, one raging torrent, into the ocean! The atmosphere in all directions was filled with ashes, spray, gases, etc., while the burning lava, as it fell into the water, was shivered into millions of minute particles, and, being thrown back into the air, fell in showers of sand on all the surrounding country. The coast was extended into the sea for a quarter of a mile, and a sand-beach and a new cape were formed. Three hills of scoria and sand were also formed in the sea, the lowest about two hundred, and the highest about three hundred feet high.

"For three weeks this terrific river continued to disgorge itself into the sea with little abatement. Multitudes of fishes were killed, and the waters of the ocean were heated for twenty miles along the coast. The breadth of the stream, where it fell into the sea, is about half a mile, but inland it varies from one to four or five miles in width, conforming itself, like a river, to the face of the country over which it flowed.

"The depth of the stream probably varies from ten to two hundred feet, according to the inequalities of the surface over which it passed. During the flow, night was converted into day on all eastern Hawaii. The light rose and spread like the morning upon the mountains, and its glare was seen on the opposite side of the island. It was also distinctly visible for more than one hundred miles at sea, and at the distance of forty miles fine print could be read at midnight.

"The whole course of the stream from Kilauea to the sea is about 40 miles. Its mouth is about 25 miles from Hilo station. The ground over which it flowed descends at the rate of one hundred feet to the mile. The crust is now cooled, and may be traversed with ease, although scalding steam, pungent gases, and smoke are still emitted in many places.

"Hills have been melted down like wax; various and deep valleys have been filled; and majestic forests have disappeared like a feather in the flames. In some places the molten stream parted and flowed in separate channels for a great distance, and then, reuniting, formed islands of various sizes, from one to fifty acres, with trees still standing, but seared and blighted by the intense heat. On the outer edges of the lava, where the stream was more shallow, and the heat less vehement, and where, of course, the liquid mass cooled soonest, the trees were mowed down like grass before the scythe, and left charred, smouldering, and only half consumed. As the lava flowed around the trunks of large trees on the outskirts of the stream, the melted mass stiffened and consolidated before the trunk was consumed, and when this was effected, the top of the tree fell, and lay unconsumed on the crust, while the hole which marked the place of the trunk remains almost as smooth and perfect as the caliber of a cannon. These holes are innumerable, and measure from ten to forty feet deep, but they are in the more shallow parts of the lava, the trees being entirely consumed where it was deeper. During the flow of this eruption, the great crater of Kilauea sunk about three hundred feet, and her fires became nearly extinct, one lake only out of many being left active in this mighty caldron. This, with other facts which have been named, demonstrates that the eruption was the disgorgement of the fires of Kilauea. The open lake in the old crater is at present intensely active, and the fires are increasing, as is evident from the glare visible at our station, and from the testimony of visitors.

"While the stream was flowing, it might be approached within a few yards on the windward side, while at the leeward no one could live within the distance of many miles, on account of the smoke, the impregnation of the atmosphere with pungent and deadly gases, and the fiery showers which were constantly descending and destroying vegetable life. During the progress of the descending stream, it would often fall into some fissure, and forcing itself into apertures and under massive rocks, and even hillocks and extended plats of ground, and lifting them from their ancient beds, bear them with all their superincumbent mass of soil, trees, etc., on its viscous and livid bosom, like a raft on the water. When the fused mass was sluggish, it had a gory appearance, like clotted blood mingled and thrown into violent agitation.

"Sometimes the flowing lava would find a subterranean gallery, diverging at right angles from the main channel, and, pressing into it, would flow off unobserved, till meeting with some obstruction in its dark passage, when, by its expansive force, it would raise the crust of the earth into a dome-like hill of fifteen or twenty feet in height, and then, bursting this shell, pour itself out in a fiery torrent around."—*Coan, Sept.* 25, 1840.

In this eruption the quantity of matter which was ejected amounted to 6,000,000,000 cubic feet. The old crater of Kilauea lost 15,400,000,000 cubic feet of its contents, the greater part of which therefore must have been drawn off into subterranean fissures.

6. Another illustration we quote from Lyell's Principles of Geology:

"In April, 1815, one of the most frightful eruptions recorded in history occurred in the mountain *Tomboro*, in the island of Sumbawa. It began on the 5th of April, and was most violent on the 11th and 12th, and did not entirely cease till July. The sound of the explosions was heard in Sumatra, at the distance of 970 geographical miles in a direct line, and at Ternate, in an opposite direction, at the distance of 720 miles. Out of a population of twelve thousand, only twenty-six individuals survived on

---

What is said of the quantity of lava lost by Kilauea? Describe the eruption of Tomboro.

the island. Violent whirlwinds carried up men, horses, cattle, and whatever else came within their influence, into the air; tore up the largest trees by their roots, and covered the whole sea with floating timber. Great tracts of land were covered with lava, several streams of which, issuing from the crater of the Tomboro Mountains, reached the sea. So heavy was the fall of ashes, that they broke into the president's house at Birna, forty miles east of the volcano, and rendered it, as well as many other dwellings in town, uninhabitable. On the side of Java, the ashes were carried to the distance of 300 miles, and 217 toward Celebes, in sufficient quantity to darken the air.

The floating cinders to the windward of Sumatra, formed, on the 12th of April, a mass two feet thick and several mile in extent, through which ships with difficulty forced their way.

The darkness occasioned in the daytime by the ashes in Java was so profound, that nothing equal to it was ever witnessed in the darkest night.

The area over which the tremulous noises and other volcanic effects extended was one thousand English miles in circumference, including the whole of the Molucca Islands, Java, and a considerable portion of Celebes."—*Lyell.*

7. *Static Pressure in Volcanoes.*—The enormous pressure of the liquid lava against the interior of volcanoes, when it is raised to a great height, is the cause of the frequent eruptions through their sides. The force requisite to raise lava to the edge of the crater of Ætna, from the level of the base of the mountain, exceeds five tons per square inch. In Cotopaxi, which is 19,000 feet in height, the force required is ten tons per square inch: yet this volcano has projected matter 6000 feet above its summit, and once threw a stone weighing 200 tons to the distance of nine miles. Aconcagua, in Chili, the highest of all volcanoes (23,000 feet), would require at its base, for an eruption from the summit, a force of twelve tons per square inch.

8. *Characters of Lava.*—The melted matter ejected from volcanoes is composed chiefly of feldspar and augite. When the former prevails, the lava is said to be feldspathic or trachytic, and when the latter predominates, the lava is said to be augitic.

What is said of the static pressure in volcanoes? Of what minerals is lava composed?

When lava is cooled near the surface of the mass, it is usually light and porous, having been inflated with bubbles of gas. In some cases these bubbles of gas are of great size, and the cavities constitute large caverns. Sometimes the elasticity of the gas explodes the bubbles, and throws fragments of lava into the air. When melted feldspathic lava comes into contact with water, the water is converted into steam, which causes the lava to froth up, and converts it into *pumice.* At the time of the eruption of Skaptar Jokul, in 1783, there was an island thrown up seventy miles from land, and the ocean was covered with vast quantities of floating pumice for a distance of 150 miles. Dr. C. T. Jackson has made pumice from the slags of furnaces by throwing water in the path of the running slag. It is, indeed, a common form of the slag of furnaces.

When lava cools under the pressure of a superincumbent mass, it is generally as compact and solid as the older rocks. Some of the products of volcanoes consist mostly of silex, and resemble impure glass or the slag of furnaces, and frequently the artificial can not be distinguished from the natural product.

II. *Earthquakes.*—1. One of the most remarkable earthquakes was that which began at *Lisbon*, on the 1st of November, 1755. A sound like thunder under ground was heard, and in six minutes the greater part of the city was thrown down, and 60,000 persons had perished. The sea retired, and then rolled back 50 feet higher than its ordinary level. A new marble quay was suddenly swallowed up with a great concourse of people, and the vortex drew down many boats and small vessels. The highest mountains in Portugal were shaken from their foundations; their summits were split and rent, and portions of them were thrown down into the valleys. The shock was felt at sea, and a ship 120 miles west of St. Vincent experienced a violent concussion, which was probably occasioned by a wave of translation, that is, a motion of the water itself.

This earthquake was felt over a large part of the earth. All

When is lava porous? What is pumice? When is lava compact? Describe the Lisbon earthquake.

Europe, even to Norway, felt the shock. The waters of Loch Lomond (in Scotland) rose two feet and four inches. Terrible eruptions and earthquakes occurred in Iceland. Springs throughout Europe and Great Britain became hot, and some of them turbid. The north of Africa was violently shaken, and a village near Morocco was ingulfed with 8000 or 10,000 persons in a fissure, which opened and closed over them. The shock was felt in the West Indies; and the waters of Lake Ontario were violently agitated.

2. The *West Indies* have been more or less subject to earthquakes since their discovery. In 1692 a great earthquake ingulfed three quarters of the then large and opulent city of Port Royal, in Jamaica; the Blue Mountains, 8000 feet high, were rent and shattered, and the surface of the island swelled and heaved like a rolling sea. In 1843, Guadaloupe and several towns were suddenly destroyed.

3. The *motion* of the ground in earthquakes is said to be like that of waves pursuing each other with great velocity, in many cases of 20 miles per minute. Sometimes the force acts in an upward direction, as in the earthquake of Calabria in 1783, when loose masses bounded into the air to the height of several yards, and in some of the towns the pavements were thrown up. During the same earthquake, however, other places were affected with a force acting horizontally. This year there were 349 shocks in Italy, the first of which, February 5th, "threw down, in the space of two minutes, the greater part of the houses in all the cities, towns, and villages, from the western flanks of the Apennines in Calabria Ultra, to Messina in Sicily, and convulsed the whole country."—*Lyell.* On March 28th, another shock occurred. The extent of country convulsed was very great. Around the city of Oppido as a center, all the towns and villages were destroyed for twenty-two miles in every direction. The surface of the country heaved like the waves of the sea, and the inhabitants experienced the sensation of sea-sickness. The earth open-

---

What is said of the earthquake of 1692? of the motion of the ground in the earthquake of Calabria?

ed in many places; large masses were shaken down from the mountains and cliffs of the sea; deep lakes and fissures were produced, valleys were excavated, rocks and trees were thrown into the air, the courses of rivers were changed, sand-hills and ridges were thrown up in some places, and depressions were produced in others.

The *billowy* motion has been remarked by all who have described the phenomena of earthquakes. It is a cause of ridges, arches, and faults in the rocky strata. In Darwin's travels in South America, it is stated that an engineer who was following *up* an ancient river bed, found himself, on one occasion, suddenly going *down* hill.

4. The effects of earthquakes, most important in a geological point of view, are the *subsidence* and the *elevation* of land. In the year 1772, Papandayang, one of the loftiest volcanoes of Java, was in eruption, when the greater part of the mountain, for a space 15 miles long and 6 wide, fell in and disappeared with its inhabitants. On the other hand, the bed of the sea, in the harbor of Conception (Chili), was raised, in 1750, to the amount of 25 feet, and in 1822 the coast of Chili was permanently elevated through an area of 100,000 square miles. January 16th, 1819, an earthquake occurred at Cutch, on the delta of the Indus. The fort and village of Sindree were submerged, the tops only of the houses appearing above the water. 2000 square miles of land were converted into a lake. Immediately after the shock, a mound 50 miles long, 15 miles wide, and 10 feet high, arose within a few miles of the place of subsidence.

III. *Thermal Springs.*—Geysers, or springs of hot water, occur in volcanic countries. The most celebrated are those of Iceland, in the southwest part of which is an assemblage of perforations in the earth, through which are emitted jets of boiling water.

The Great Geyser is an opening from 8 to 10 feet in diameter, in a basin-shaped mound, *Fig.* 23, through which a column of

---

What is said of the billowy motion? What are the most important geological effects? What is said of Papandayang? of Chili? of the Cutch? Where are the most celebrated geysers? Describe the Great Geyser.

Fig. 23.

VIEW OF THE GREAT GEYSER.

water is forced up to the height of 150 or 200 feet. The water in its ascent is divided into innumerable jets, and descends in showers of spray, as represented in the figure. The action of the geyser is not constant, but intermittent, there being alternate periods of activity and repose.

Hot springs are also common in countries which are quite remote from any existing volcanic agency, but which at no very remote geological period have been the theater of volcanic action. In Arkansas and in Oregon are examples.

In some cases the heat is that of boiling water, but there are

What is said of the temperature of thermal springs?

all degrees of temperature, and the thermal character of some springs is so slight as to be detected only by continued thermometrical observations.

In most, if not in all cases, thermal springs, which are distant from active volcanoes, are situated near ancient volcanic rocks, or near chains of mountains which were broken up and elevated by the igneous agency of former epochs. The latter is the case with the hot springs of Virginia and many others in the United States.

SECTION II.—SUBMARINE IGNEOUS AGENCIES.

I. *Volcanic Islands.*—Many islands of considerable size have been seen to rise from the sea, and many others composed of lava are doubtless of volcanic origin. Authentic narratives of such events in remote ages are limited to the Mediterranean Sea. In the Grecian Archipelago there is a group of islands, of which *Santorini* is the chief, several of which have been thrown up at different times.

In July, 1831, a volcano rose up through the sea off the coast of Sicily, and was called *Graham's Island. Fig.* 24 represents its

Fig. 24.

GRAHAM'S ISLAND, 18TH OF JULY, 1831.

appearance on the 18th of July, and *Fig.* 25 on the 4th of August,

What is said of the situation of thermal springs? What is said of volcanic islands? Describe Graham's Island.

Fig. 25.

GRAHAM'S ISLAND, 4TH OF AUGUST, 1831.

as seen by Corrao. In August it was 180 feet high and $1\frac{1}{3}$ miles in circumference; but the part above the water, being composed of loose materials, disappeared in two or three years, leaving a rocky shoal. The Azores and the Aleutian Isles have been repeatedly increased by new volcanic islands. In 1814, near Unalaschka, an island was raised with a peak 3000 feet high. Many new islands have been thrown up near the shores of Iceland, and some of them have since subsided.

II. *Submarine earthquakes* not unfrequently furnish proofs of their occurrence in sudden and extraordinary waves. A few years since, at the Sandwich Islands, a series of such waves swept away a village near the shore. Perhaps a sudden rise and fall of one or two feet, which was observed in the harbor of Nantucket, in June, 1836, may be referred to such a cause.

Volcanic forces may be active in the lowest depths of the ocean. At the depth of five miles, the pressure of the superincumbent mass of water is so great that the most powerful volcanic agency would act silently and be unfelt at the surface. With that pressure the water would become red hot without bursting into steam, and currents of incandescent lava might flow quietly down the

---

What are the effects of submarine earthquakes? What must be the character of eruptions at great depths under the ocean?

sides of submarine mountains. But the effects on the chemical constitution, and on the crystalline structure of the melted and cooling mass, would be great. Doubtless many of the older igneous rocks had such an origin.

SECTION III.—SUBTERRANEAN IGNEOUS AGENCIES.

The history of igneous agency on the surface of the earth naturally leads us to look within for its source. We shall briefly proceed from the less to the more general conclusions, until we arrive at a theory sufficiently comprehensive to embrace all the phenomena of igneous action.

I. *Theory of Geysers.*—The most probable *theory of geysers* has been well illustrated by the following figure. (Lyell's Principles, Am. ed., i., 472.)

Fig. 26,

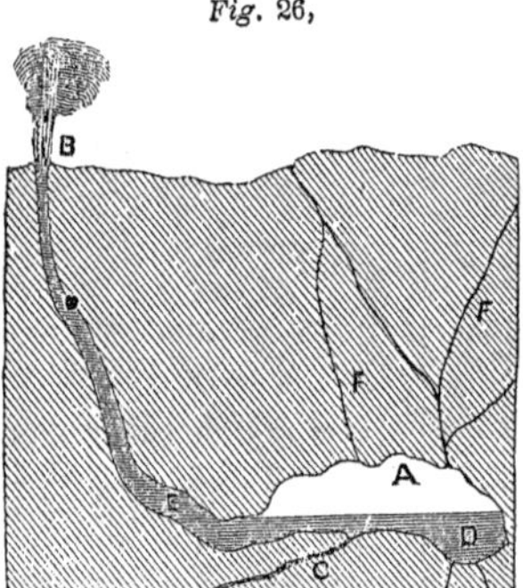

"Suppose water percolating from the surface of the earth to penetrate into the subterranean cavity A D by the fissures F F, while at the same time steam at an extremely high temperature, such as is commonly given out from the vents of lava currents during congelation, emanates from the fissures C.

"A portion of the steam is at first condensed into water, while the temperature of the water is elevated by the latent heat thus evolved, till at last the lower part of the cavity is filled with boiling water, and the upper part with steam under high pressure. The expansive force of the steam becomes at length so great that the water is forced up the fissure E B, and runs over the rim of the basin. When the pressure is thus diminished, the steam in the upper part of the cavity A expands, until all the water [in the upper part of A D] is driven into the pipe; and when this happens the steam also rushes up with great velocity."

II. *Theory of Volcanoes.*—1. The same theory may be applied to volcanoes, if we suppose that the seat of operations is on a

---

Where is the source of igneous agency? Describe the theory of geysers. How may this theory be applied to volcanoes?

mass of molten lava at the depth of several miles. Water entering is converted into steam, which forces up the lava in the crater of some neighboring volcano or through the crust of the earth. That water is an important agent in volcanic eruptions would seem to be proved by the great quantities of steam which escape, and by the hydrochloric acid and salt emitted, which the salt water may have furnished. In South America, even fishes have been discharged.

It may perhaps be objected to any theory which shall make the expansive force of steam the sole agency, by which lava is elevated many thousand feet, that the force may not be sufficient. When water is converted into steam at a temperature of 800°, nearly a red heat, it merely doubles its volume, and has a pressure of only two tons per square inch. But since the pressure of lava in most volcanoes exceeds this amount, it would seem that the ordinary source of the great power of steam by increase of heat, viz., the increased quantity in a given space, would avail little. The resistance of the superincumbent matter would be such that the water would not double its bulk. The universal law of expansion by increments of heat would hold good; but, although the application of the law under such circumstances may not be obvious, yet it may be conjectured that the water would merely be heated red hot, without a very considerable expansion.

2. The geographical distribution of volcanoes tends to the same conclusion. On inspection of the following map, *Fig.* 27, on which the volcanic regions are indicated by the shaded parts, it will be seen that volcanoes are, with a few exceptions, situated near the ocean. About two thirds of the number are on islands, and of the remaining third nearly all are within a short distance of the coast. In Mexico and in Central Asia there are a few exceptions to this general statement. It will be seen, also, that most volcanoes are arranged in lines or zones of great extent.

3. The history of volcanoes and earthquakes has shown us that

What objection to this theory of volcanoes? What is said of their geographical distribution?

Fig. 27.

the source of their action is very deeply seated in the earth. This is especially manifest from the *quantity of matter erupted.*

The single eruption of Skaptar Jokul, in 1783, protruded a mass of lava exceeding the bulk of the entire mountain. In 1660, it was computed that the ejections of Mount Ætna, if collected, would form a mass twenty times the size of the original mountain; yet the subsequent eruptions do not indicate any exhaustion of the source. It was, therefore, a correct idea of Seneca, who says that the volcanic mountain does not supply the fire, but merely affords it a passage. Volcanoes are only chimneys over the subterranean fires.

4. But we may go further, and affirm that these fires are sufficiently extensive to furnish a communication between distant volcanoes. This conclusion unavoidably follows from their *alternate action*, from their relations to earthquakes, and from the geographical extent of the latter. From the numerous facts which sustain this conclusion, we can present only a few illustrations.

In 1797, the volcano of Pasto, in the western part of Columbia, emitted a volume of smoke for three months, which ceased at the moment when a violent earthquake, with an eruption of mud and water, of vapors and flames, occurred at the distance of 180 miles. Vesuvius and Ischia have alternated with each other. Eruptions of the latter were frequent and violent in the earlier

---

What is proved by the quantity of matter erupted? Give examples. What is said of the alternate action of volcanoes?

ages, while Vesuvius was quiet until A.D. 79, when it overwhelmed Herculaneum and Pompeii. Ischia has since had its period of repose, and Vesuvius of activity.

It has been observed that, from the commencement of the thirteenth to the latter half of the seventeenth century, earthquakes in Syria and Judea almost entirely ceased, while Southern Italy suffered extraordinary convulsions. A comparison of their history leads to the conclusion that both are not violently affected at the same time, whence it has been inferred that a subterranean connection exists between regions nearly fifteen hundred miles distant.

In 1811, South Carolina was convulsed with earthquakes, which continued until Laguira and Caraccas (South America) were destroyed. At the same time the valley of the Mississippi was convulsed, especially under New Madrid. New lakes, 20 miles in extent, were formed, and others were drained; the earth undulated like the ocean, and split into frightful chasms. These chasms were in a direction from southwest to northeast, and the people, soon observing it, felled trees at right angles to this direction, and placed themselves upon them. Many were thus saved from being swallowed up.

The directions of these chasms were that which would have been produced by an undulating force, propagated in a northwest and southeast line; and it is worthy of notice that New Madrid, in the valley of the Mississippi, South Carolina, and Caraccas, are nearly in this direction. This circumstance, and especially the coincidence in time, indicate a subterranean communication between these places, whose extremes are distant 2500 miles.

The extent to which the earthquake in 1755 was felt has been mentioned (p. 61), and indicates a seat of subterranean action under a very considerable portion of the crust of the earth.

The influence of earthquakes may be considerably extended by the transmission of vibrations through the solid rocks. But the extent of country affected in this manner can not be much larger than that which is directly acted upon. A mere transmitted vibration could not affect the *quantity* and *temperature* of the waters in thermal springs and render them turbid, which was remarkably the case throughout Europe in the Lisbon earthquake. Says Baron Humboldt, in the view of such facts: " They demon-

---

Describe the earthquakes of 1811. What is said of the vibratory influence of earthquakes? What is Humboldt's opinion?

strate that these forces act, not superficially in the outward crust of the earth, but at immense depths in the interior of our planet."

III. *Theory of Internal Heat.*—Thus far, most geologists are agreed; but if we extend our inquiries further, and require a more definite theory of the source of volcanic fires, we then find some difference of opinion.

Let us consider the theory which is most in favor, and which supposes that the whole interior of the earth is in a state of fusion, within a crust of 50 to 100 miles in thickness. According to this theory, earthquakes are the effects of the heaving of this melted interior, occasioned, perhaps, by steam or chemical action, and volcanoes are the safety-valves through which earthquakes are relieved.

In support of this theory, it is urged that it affords a satisfactory explanation of the great extent of earthquakes and of the alternation between distant volcanoes, of the great quantity of matter ejected through the latter, and of their being associated together along lines, or in groups where the crust of the earth may be supposed to be of less thickness. The undulatory motion of earthquakes would be merely the waving of the crust.

The theory of internal heat, however, reposes chiefly on experiments made upon the temperature of mines and Artesian wells, which concur in the interesting result that in all places the temperature of the earth increases after passing below the stratum of surface temperature, in a ratio, which varies in different places from 1° Fahr. for 25 feet of descent, to 1° for 70 feet, but which is quite constant in the same place. From the very numerous observations which have been made, we select a few examples:

"At the Dolcoath mine of Cornwall (England), the mean annual temperature at the surface of the ground is 50°; of a spring 1440 feet below, 82°; and of the rock three feet three inches within its surface, and 1381 feet deep, 76·6°. At the silver mine of Guanaxuato the mean annual temperature at the surface is 68·8°, and the temperature of a spring 1713 feet below is 98·3°. A single experiment in the deepest coal mine in Great Britain,

What is the theory of internal heat? How does it explain the igneous agencies? What is the direct argument for this theory?

near Sunderland, gave the following results: depth of the place of observation, 1584 feet; below the level of the sea, 1500 feet Mean annual temperature at the surface, 47·6°; temperature on the day of observation (Nov. 15, 1834), 49°; do. of the air at the bottom of the pit, 64°; close to the coal, 68°; do. of water collected at bottom, 67°; do. of salt water issuing from a hole made the same day, 70·1°; do. also of gas rising through the water, 72·7°; do. of the front of the coal, 68°. Hence the heat increases at the rate of about a degree for every 60 feet."—*Hk.*

At Berlin (Prussia), the mean temperature is 47·1°; the temperature of the water of an Artesian well at the depth of 675 feet is 67·6°. At Paris, the mean temperature is 51·1°; the temperature of an Artesian well 1800 feet deep (the deepest in the world), is 83°. In Wurtemburg, the water obtained in Artesian wells is used with success to prevent the stopping of machinery by the freezing of the water which carries it, and also for warming a paper manufactory.*

As similar results have been obtained in all cases, it is inferred that the heat would be found to continue to increase, if we were able to penetrate much deeper, until, at the depth of one or two miles, we should reach the temperature of boiling water.† At the same ratio of increase, the heat at a depth of forty miles would be sufficient for the fusion of all known rocks.

It has been objected to this theory, that the diffusion of heat through the liquid interior should melt also the crust of the earth. If it were supposed that the heat continued to increase through all the melted nucleus to the center, in the same ratio in which it increases through the solid crust, the objection would be fatal to any *such* theory. But we can not see any more difficulty in conceiving that the earth has had a crust 50 miles in thickness congealed in the lapse of ages, than that a body of water should

* Warm springs which rise from deep sources, have been used for warming conservatories and irrigating gardens. It is said that a piece of ground at Erfurt (Germany), which is thus kept at an equable and high temperature, yields to its proprietor an annual profit of $60,000, derived from raising salad.

† On account of the great pressure, it is not to be inferred that water *would* boil at this or at any other depth. See p. 67.

What is said of Artesian wells? of the heat at the depth of one or two miles? at the depth of forty miles? What objection to the theory? What reply is made?

be incrusted with ice by a similar process. It might as well be urged that it is impossible for a pond of water to be frozen over, because the diffusion of heat through the water would melt the ice.

Another theory (of Sir Charles Lyell) supposes that there are, beneath the crust of the earth, immense reservoirs and subterranean channels of lava, whose action is the source of eruptions and of earthquakes. The ground of preference for this theory on the part of its advocates is, that it is *sufficient* to account for all the phenomena of earthquakes and volcanoes, and for the temperature of mines and of Artesian wells. If, however, these reservoirs and channels are thus numerous, this theory is obviously exposed to the objection which has been urged against the former; and the existence of partitions protected from the colder temperature of the surface by miles of non-conducting materials, but exposed to the action of these adjacent lakes of fire, is inconsistent with the laws of the equilibrium of heat in liquids.

SECTION IV.—GENERAL RESULTS OF IGNEOUS AGENCY.

The general effects of igneous agency may be thus summed up:

1. The transfer of matter from the interior to the surface of the earth.
2. The discharge of gases and of pent-up vapors, which gives relief to earthquakes.
3. The production of fissures, which, being filled with lava, form *dikes*.
4. The formation of peculiar kinds of stratified rocks by dis charges into the sea.
5. The elevation of parts of the earth's crust.
6. The depression of parts of the earth's crust.
7. The lifting up of strata out of a horizontal position.
8. The burial of animals, of plants, and of works of art, sometimes without entirely destroying them.

State Sir C. Lyell's theory; the argument for it and against it. What is the first-mentioned effect of igneous agency? the second? the third? &c

9. The heating of thermal springs.

10. An increase of the erosive action of water. This effect appears in the floods produced by the sudden melting of snow and ice in volcanoes; but especially in the waves of translation, which are produced by earthquakes. It is well known that ordinary waves have but little onward motion, except when meeting a shore. On the contrary, in waves of translation, there is a horizontal motion of the water, which greatly increases its erosive power.

CHAPTER III.

JOINT EFFECTS OF AQUEOUS AND IGNEOUS AGENCIES.

The two great powers which we have thus considered, fire and water, are to some extent in their geological effects, as they are in their nature, antagonist forces. Running water is constantly removing the land by slow degrees from higher to lower levels, and into the sea, while volcanoes are raising, from within the crust of the earth, to a greater or less height, their floods of melted rocks. Earthquakes are also engaged in lifting whole countries by violent shocks, while some unknown power, perhaps *the plication of the crust of the earth*, is more gradually effecting the same result. As, however, water in some cases reverses its usual mode of operation, and in springs raises its soluble contents from deep places and deposits them on the surface of the ground, so, on the other hand, the igneous agencies not unfrequently reverse the more common process of elevation, and we see earthquakes swallowing down large portions of land. It is obvious that rivers carry into the sea an amount of sediment generally exceeding the eruptions of volcanoes, and if the elevatory effects of earthquakes did not exceed those of depression,

---

In what way are heat and water antagonist forces? What is the effect of springs? What is the average result of these agencies?

the general tendency would be to reduce the earth to a uniform level. But if, as is stated, the great earthquake of Chile raised an extent of 100,000 square miles three feet in height, here was the amount of two and three fourths millions of millions of cubic feet raised at once above the sea level. Similar elevations may occur in places which are at a distance from the sea, and therefore show no obvious proof of elevation.

Elevation of land in some cases is going on gradually, without any apparent earthquake shocks. This is remarkably the case with the country around the Baltic Sea, which is rising at the rate of two or three feet per century. This is shown by the marks made at various periods to designate the high-water level, and is confirmed by the marine shells of species now living on the shore, which are found from 10 to 200 feet above the present level of the water, and 50 miles inland.

The coast of Greenland appears to be undergoing a gradual depression. In this country the Northern States appear to have been rising during a geological period not very remote, and perhaps the process is yet slowly going on.

The great problem, therefore, of the general tendency of existing forces is as complicated as it is grand in its numerous and conflicting elements.

CHAPTER IV.

INORGANIC AGENCIES NEITHER AQUEOUS NOR IGNEOUS.

There are some other physical agencies, of more or less importance, which we have not introduced in our classification, because they are of limited extent. Rocks are sometimes split by lightning, and terrestrial magnetism or rather galvanism has been supposed to exert an important influence, especially in producing

Give an example of a great amount of elevation; of gradual elevation; of gradual depression. What effects are produced by lightning and by terrestrial magnetism?

metallic veins. In districts of fine sand, winds exert an important agency in their distribution. It is well known that many of the remains of the ancient Egyptian architecture are more or less buried in impalpable dust. We are familiar with the accounts of caravans on the great deserts of Northern Africa, which have been overtaken and buried by moving sand. In Europe some villages have been destroyed by the same cause. A few examples of such sands in this country are known, but have not attracted much attention.

CHAPTER V.

ORGANIC AGENCIES.

The operation of organic agencies is of great interest to the geologist, because it shows how extensive rock formations are and have been formed. It is a circumstance truly remarkable, that animals the most minute, nearly or quite invisible to the naked eye, and among the lowest in the scale of beings, are those only which are forming solid rocks or deposits of sufficient magnitude to engage the attention of the geologist.

SECTION I.—CORALS.

The most important organic agency undoubtedly is to be found in these well-known structures.

I. *The Animals.*—The animals which construct corals belong to the class *polypi*, or polypes, and consist of a homogeneous fleshy bag, open at one extremity, which is fringed with tubercles. The polypi are, however, very minute, seldom exceeding the size of a pea, more frequently less than a pin's head, or are even invisible. They have the remarkable property of living united in one common mass, so that whatever is swallowed and digested by

---

What is the agency of winds? What is the rank of those animals which produce important geological effects? What is said of the animals that construct corals?

each, contributes to the nourishment of the community by a common circulation. The solid coral is not a structure voluntarily made by them, but a secretion in which they have as little design as the oyster in secreting his shell, or a quadruped in the growth of its bones.

II. *Temperature in which Corals grow.*—One interesting result of the recent Exploring Expedition has been the knowledge of the *temperature* requisite for the existence of these rock-building animals. In seas whose temperature is below 66° they do not flourish, and hence along the western coast of South America, where the water is cooled by an Antarctic current, coral reefs do not exist, while in the same latitude on the eastern coast they are very extensive. Nor do they build at great depths, as was once supposed, but only in water not exceeding a few fathoms. Most of the deep soundings in tropical seas indicate a temperature far too low for their existence. In colder waters, many species of polypes occur, but they do not construct extensive reefs.

III. *Varieties of Coral Reefs.*—Coral reefs are, for the most part, either fringing or barrier reefs, or they inclose only a smooth body of water called a *lagoon.*

1. *Fringing reefs* are those which occur along the shore. In their origin coral reefs may all have been fringing.

2. *Barrier reefs* are more or less parallel with the shore, but are separated from it by a broad and deep channel. When an island, which is fringed with living corals, subsides beneath the waters very slowly, the reef is built up to the surface of the water as fast as the land sinks. After a certain period, the island is contracted by its submergence, and the reef becomes a barrier reef. *a a*, *Fig.* 28, are barrier reefs, and *b b*, the water between the island and the reef.

Fig. 28.

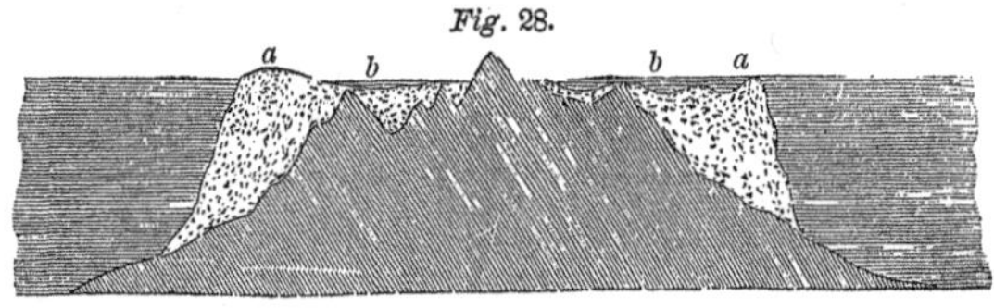

What is the origin of coral? At what temperature do coral reefs grow? At what depths? What are fringing reefs? What barrier reefs?

3. *Lagoon reefs* are formed in the last stage of this process. When the island has disappeared, we have a circular reef inclosing a *lagoon*. *Fig.* 29 represents a section across an island inclosing a lagoon, *b b*. Such circular reefs

Fig. 29.

with a diameter from one to thirty miles, are common in the Pacific, and were formerly supposed to be on the summits of volcanoes. The following is a view of Whitsunday Island, *Fig.* 30, in which the circular reef has become habitable.

Fig. 30.

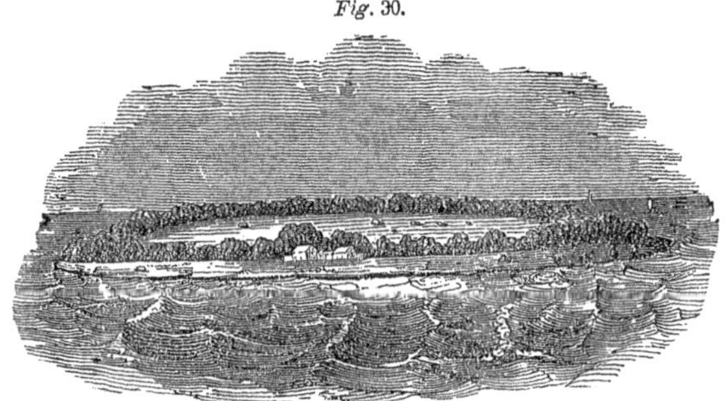

VIEW OF WHITSUNDAY ISLE.

It is thus easy to determine whether the movement of islands is that of emergence or subsidence. Since it is not probable that many islands are stationary through periods which are geologically long, it may be correctly inferred, from the existence of fringing reefs on islands, that the process is that of emergence; a conclusion which is sustained by the independent evidence of ancient sea-lines and corals within the island. If, however, the reefs are separated from the land by broad and deep channels, the island is in the process of subsidence. Near such reefs dead corals are found several hundred feet below the depth at which they could have been formed.

IV. *Extent of Coral Reefs.*—Coral reefs are very numerous in many parts of the Indian and Pacific Oceans, and constitute one

---

What are lagoon reefs? What may be inferred from fringing reefs? What from barrier reefs? What is said of the extent of coral reefs?

of the greatest dangers of navigation. Notwithstanding the minuteness of the animals, reefs of vast extent are described.

"On the eastern coast of New Holland is a reef 350 miles long. Disappointment Islands and Duff's Group are connected by 500 miles of coral reefs, over which the natives can travel from one island to another. Between New Holland and New Guinea is a line of reefs 700 miles long, interrupted in no place by channels more than thirty miles wide. A chain of islets, 480 geographical miles long, has long been known by the name of the Maldivas. Some groups in the Pacific, known as the Dangerous Archipelago, are from 1100 to 1200 miles long, and from 300 to 400 miles broad."—*Hk.*

V. *Rate of Increase.*—It has been supposed, on account of the existence of shells nearly buried in the substance of corals, while the mollusca were yet living, that the growth of corals was rapid. But Mr. Couthuoy (zoologist of the Exploring Expedition) finds that in such cases the animals in the shells have excavated the cavity in the reef. On the other hand, the growth is so slow as not to present any very satisfactory data for determining its rate. Thus the most stupendous effects are produced not only by the feeblest agents, but by an imperceptible progress.*

* "I saw the living pile ascend,
The mausoleum of its architects,
Still dying upward as their labors closed;
Slime the materials, but the slime was turned
To adamant by their petrific touch.

"Frail were their frames, ephemeral their lives,
Their masonry imperishable. All
Life's needful functions, food, exertion, rest,
By nice economy of Providence,
Were overruled, to carry on the process
Which out of water brought forth solid rock.
Atom by atom, thus the mountain grew
A coral island, stretching east and west;
Steep were the flanks, with precipices sharp,
Descending to their base in ocean gloom.
* * * * * * * *

"Compared with this amazing edifice,
Raised by the weakest creatures in existence,
What are the works of intellectual man
His temples, palaces, and sepulchers?
Dust in the balance, atoms in the gale.
Compared with these achievements in the deep,

What is said of the growth of coral reefs?

SECTION II.—ANIMALCULES.

The least of all the animal kingdom rank next to polypes in importance as geological agents. Some animalcules are visible to the naked eye, but others are only one twenty-four thousandth of an inch in diameter, and a number equal to the entire population of the globe might sport freely in a single drop of water; yet their remains constitute strata which are many feet in thickness. The part most important in this connection is the *shell*, with which a majority of the species are furnished. Fragile as such shells might be supposed to be, yet, as they consist of silex, they retain their original form and structure for indefinite periods of time. Silicious marl, an impalpable pulverulent deposit, which is found usually beneath the beds of peat in Massachusetts and other places, chiefly consists of these shells to the depth of several feet.

The vast quantity of these deposits is less astonishing when we consider the fecundity of animalcules, some of which will have millions of descendants in a few days. Thus one individual of the species *Hydatina senta* in twelve days may produce 16,000,000, and one of another species in four days 170,000,000,000; and Ehrenberg, to whom science is indebted more than to all others on this subject, affirms, that if the price of tripoli (rotten stone or polishing powder, which is composed of their shells) should rise materially, he could raise animalcules, and collect the shells, and sell them at a profit for polishing powder. He has actually obtained several pounds by rearing the animalcules. We may have occasion to notice again deposits of these shells in Germany

Were all the monuments of olden time.
Egypt's gray piles of hieroglyphic grandeur,
That have survived the language which they speak,
Preserving its dead emblems to the eye,
Yet hiding from the mind what these reveal;
Her pyramids would be mere pinnacles,
Her giant statues, wrought from rocks of granite,
But puny ornaments for such a pile
As this stupendous mound of catacombs."—MONTGOMERY.

What is said of animalcules? what of their shells? What is silicious marl? What is said of the fecundity of animalcules? What is the thickness of their deposits in Germany?

of the thickness of fourteen and twenty-eight feet, and to describe some as existing in this country.

On account of the great extent of silicious deposits known to have originated from this source, some geologists have been inclined to believe that most of the silicious strata in the crust of the earth once constituted the shells of animalcules.

Other species have shells composed of iron rust, and much of the iron scum, which may be seen floating on water, consists of animalcules.

This class of animals is not limited to fresh-water deposits, but many kinds exist in the sea, and fossil species abound in some marine deposits, as in the chalk of England.

SECTION III.—SHELLS.

Another organic source of rocky strata is to be found in the vast quantities of the shells of mollusca and of the crustaceous coverings of other marine animals, which are most abundant in some parts of the tropical seas. In such regions the sand of beaches is often seen to be composed of the comminuted calcareous fragments. So in the ancient palæozoic rocks of Lake Champlain, we find entire ledges composed of pieces of corals and shells. It is probable that many beds of crystalline marble were originally composed of such materials, and have been subsequently altered by igneous agency.

SECTION IV.—PLANTS.

Plants are well known to form deposits of vegetable matter. From their complete decay results vegetable mold, the particular consideration of which belongs to agricultural geology. While many quickly perish, others are more enduring, and form accumulations of muck, peat, and drift-wood. Peat results chiefly from the partial decay of mosses, especially Sphagnum, which constantly grows as the older portions decay. Since, however, in hot climates the decay of plants is rapid and complete, when

---

What is said of iron scum? What kinds of water are inhabited by animalcules? What is said of the agency of mollusca? What is the origin of vegetable mold? of peat?

exposed to atmospheric agency, beds of peat are found only in cold countries.

Immense rafts of drift-wood are well known to accumulate in the lower part of the rivers of this continent.

The Atchafalaya is supposed to have been formerly the channel of Red River, but is now an arm of the Mississippi. Some obstruction occurred, and a mass of timber accumulated, which was called "the raft." After 38 years, in 1816, it was 10 miles long, 220 yards wide, and 8 feet thick. The Washita was described in 1804 as covered for 50 miles with a raft, on which all the plants of the forest, except large trees, were growing. Immense deposits of drift-wood occur at the mouth of the Mississippi.

In a few cases, as in Maine and Louisiana, the process by which drift-wood may be converted into coal has been seen in its incipient stages. Facts of this kind are of great interest as illustrating the origin of the immense beds of coal in the carboniferous formations.

Describe the raft in the Atchafalaya; in the Washita. **What do these facts illustrate?**

PART II.

PRINCIPLES OF PALÆONTOLOGY.

PALÆONTOLOGY treats of the remains of animals and plants, which have been buried in the strata during former periods of the earth's history.

CHAPTER I.

PRINCIPLES OF ZOOLOGY WHICH RELATE TO FOSSILS.

A MINUTE practical knowledge of the natural history and anatomy of the numerous tribes of animals and plants is essential in the investigations of this subject; but it requires the devotion of a lifetime to attain this knowledge in any one of the numerous sciences which treat of animals and plants. The geologist, therefore, after becoming familiar with the outlines of these sciences, contents himself with depending on the aid of zoologists and botanists in determining the character and habits of extinct species.

SECTION I.—CLASSIFICATION OF THE ANIMAL KINGDOM.

We subjoin a brief outline of the *classification* of the animal kingdom, as indispensable even to the mere reader of geology. About 150,000 species of animals (of which two thirds are insects) and 100,000 species of plants are known to naturalists. The object of classification is not merely to aid the memory in the knowledge of such a multitude of objects, but also to exhibit

---

Of what does Palæontology treat? How does the geologist ascertain the character and habits of extinct species? What is the number of known species of animals and of plants? What is the basis of the classification of animals?

the natural relations of affinity, which constitute the Divine plan of their creation.

The animal kingdom is divided into the following five grand divisions, which are characterized primarily by peculiarities of the nervous system. This part is the most important and distinguishing feature of animals, and is therefore the most suitable basis of primary divisions.

I. *Vertebrata.*—This division comprises the highest orders of animals, with man at the head. The nervous system consists of a *brain* and *spinal cord*, from which branches are given off ramifying indefinitely through the body. There is also a *sympathetic system* of nerves, which are the source of the involuntary motions, as the beating of the heart, and which originate from several centers in the body. The vertebrata, also, are characterized by the possession of a *skeleton*, consisting of bones, which, unlike the hard parts of other animals, grow and are continually nourished by the circulation of the blood. The organs of the five senses are manifest in all of them, and four of the senses have a distinct apparatus placed in the cavities of the face. Examples of this division are the warm-blooded quadrupeds, birds, reptiles, and fishes. The name is derived from the vertebræ or spine. Their fossil remains, although far less common than shells in the strata, lead to very interesting and important conclusions respecting the condition of the surface of the earth during the periods of their existence.

II. *Articulata.*—The nervous system in this division consists of nervous centers arranged in two parallel lines along the length of the body, the anterior one in the head being the larger. These centers send out nerves, and may be regarded as so many brains, the number of which is inversely proportioned to the activity and intelligence of the animals. In this division we find an external frame-work of jointed rings. Insects, spiders, crabs, barnacles,

---

What is said of the nervous system of the vertebrata? of their skeletons? of their senses? What examples are mentioned? What is said of the nervous system of the articulata? of their frame-work? What examples are mentioned?

and worms are examples of this division. The name is derived from *articulus*, a joint. The astonishing number of individuals, their direct or indirect agency on vegetation and on other animals, render the study of them one of the most important branches of Zoology. But, with the exception of the crustacea, comparatively few relics exist in a fossil state.

III. *Mollusca.*—In the mollusca the nervous system consists of several irregularly-scattered masses, of which one around the throat is the largest. Most of these animals are covered with shells. The study of their *shells* belongs to the science of *conchology*, a subject which is of the highest importance to the geologist, for the great majority of the fossils, by means of which the age of the geological formations is ascertained, are shells. The study of shells becomes, therefore, indirectly of great economical importance. Thus, inspection of the shells in the rocks of New York shows the geologist that the expectation of finding coal in them, or in any rocks associated with them, is utterly futile. Univalve or bivalve shells are those which are composed, the former of one, and the latter of two principal pieces. Of the univalve shells, some are divided by partitions into several chambers, which are air-tight, and serve as a float to the animal. At the present time these are not numerous, but in former geological epochs they were very abundant. They belong to the class of cephalopods (see table opposite). The multivalve shells (those which are composed of more than two pieces), for the most part, do not, however, belong to this division of animals, but to the cirrhopods, a class of articulated animals.

IV. *Nematoneura.*—In this division the nervous system is less fully developed. It consists of filaments of nervous matter, with indistinct masses or centers in a few of them. Examples are sea-stars, many intestinal worms, and some animalcules, and a few minute marine animals allied to the coral polypes. The

---

What is said of their numbers? What is said of the nervous system of mollusca? of their shells? What is said of the importance of conchology? What are univalve and bivalve shells? What is said of chambered shells? What is said of multivalve shells? What is said of the nematoneura?

Classes of the Animal Kingdom.

				CLASSES.	EXAMPLES.
VERTEBRATA.	Viviparous, suckling their young.	Respiration aerial, with lungs.	Warm blood.	*Mammalia.*	
	Oviparous.			*Birds.*	
			Cold blood.	*Reptiles.*	
		Respiration aquatic, with gills.		*Fishes.*	
ARTICULATA.	White blood.	Terrestrial.	With 6 legs.	*Insects.*	Beetles. Flies.
			24 or more legs.	*Myriapods.*	Centipedes.
			8 legs.	*Arachnidans.*	Spiders. Scorpions.
		Aquatic.	10 or 14 legs.	*Crustaceans.*	Lobsters. Crabs.
			No legs.	*Cirrhopods.*	Barnacles.
	Red blood.	Aquatic or Terrestrial.		*Annelidans.*	Earthworms Leeches.
MOLLUSCA.	With a distinct head.	Naked, or with a univalve shell.		*Cephalopods.*	Nautilus. Cuttlefish.
				Pteropods.	Hyalæa.
				Gasteropods.	Slugs. Snails. Cowries.
	Without a distinct head.	With a bivalve shell.		*Brachiopods.*	Terebratula.
				Conchifers.	Clams. Oysters.
		Naked.		*Tunicata.*	Ascidia.
NEMATONEURA.	Marine.			*Echinodermata.*	Sea stars.
	Parasitic and Aquatic.			*Cœlelmintha.*	Cavitary intestinal worms.
				Epizoa.	Parasites on fishes and on crabs.
	Aquatic.			*Rotifera.*	Wheel animalcules.
	Marine.			*Bryozoa.*	Flustra.
ACRITA.	Parasitic.			*Sterelmintha.*	Parenchymatous intestinal worms.
	Marine.			*Acalephæ.*	Sunfish and Portuguese man-of-war.
	Aquatic.			*Polygastrica.*	Animalcules with many stomachs.
	Marine.			*Polypi.*	Animals which form corals.

name is derived from the Greek words νημα and νευρα, *thread nerve.*

V. *Acrita.*—In these animals no nervous system has been recognized, and their bodies consist of fleshy or gelatinous masses, without distinct organs of digestion or motion. They may be cut in pieces with impunity, each part becoming an entire animal, and they naturally increase by spontaneous division, as well as by offshoots and by eggs. Examples are the animals which secrete the well-known coral structures, some animalcules, sunfishes, &c. The name is derived from two Greek words, *α* and χρινω, *indiscernible,* because no nerves have yet been found in them.

SECTION II.—ORIGIN AND NATURE OF SPECIES.

This subject involves two questions—*When* and *how* does Creative Power introduce new species?

I. *When are species introduced?* Geology shows us that they have been introduced from time to time during long periods. Some suppose that the introduction of species forms a part of the present order of things. But if it would be difficult to learn the precise time when species die out, much more is it to determine the year or the century of their introduction. If some are now introduced from time to time, it is probable that such an event is not more frequent than the extinction of others, and as it would be more common among the smaller and inferior tribes, on account of their numerical preponderance, it might occur frequently, without being demonstrable. When species before unknown are discovered, we know not whether some of them have been created within a few years, or whether they have been overlooked. It is only when the view is extended through periods in which centuries dwindle, like terrestrial distances seen from the planetary spaces, that we may speak with confidence of the *gradual* introduction and extinction of species.

What is said of the acrita? What questions relate to the origin of species? What does geology show of the time of their introduction? What is said of the question of their introduction during modern times?

II. *How are species introduced?* This question has been much discussed, and some of the theories have been supposed to conflict with the great truth of religion, that species are created by a Being of infinite intelligence and goodness. Some writers, therefore, with more zeal than knowledge, have made upon these theories attacks which are too absurd to engage the attention of sober science. Whether species are introduced by a special act of creation, or whether the actions of the Supreme Being, in this as in other events of nature, have such a uniformity and such fixed principles that we may call them laws of nature, are questions which should disturb no man's religious belief. Our limits, however, will not permit us to discuss the theories, and we can merely give a synopsis of them without adducing any of the facts with which they are connected.

1. One theory supposes that species are introduced by a direct *creative* agency, which is of the nature of a miracle; that is, can not be referred to any uniform course of nature.

2. Another theory is that of *transmutation*, which supposes that beings of the most simple organization having somehow come into existence, the more complex and the higher orders of animals have originated in them by a gradual increase in the complexity of their structure. Of this theory there have been various modifications. It was supposed by Buffon, originally, that there were elementary particles of living matter, viz., animalcules, whose fortuitous aggregation formed larger animals, which are therefore only heaps of animalcules. Lamark supposed that animalcules are elementary particles of living matter, but that the larger animals were formed from them by the process of appetency, which we shall presently notice. The basis of this theory and the basis and superstructure of Buffon's theory, were overthrown by those naturalists who discovered that animalcules have a very complicated organization. The doctrine of appetency supposed that new organs are formed in animals before destitute

---

What is said of discussion on the manner of introduction? What is the first theory? What is the theory of transmutation? What was Buffon's opinion? Lamark's? What is said of the doctrine of appetency?

of them, by the existence of desires in the animal constantly tending in a given direction. Thus the desire of masticating food produced teeth, the desire of handling made hands and fingers grow, &c. This seems sufficiently absurd, and those who have adopted the theory of transmutation have generally detached it from this absurdity, and not attempted to explain *how* the process of transmutation goes on.

It was never supposed that the process of transmutation was so rapid that any perceptible progress could be made within the limits of human observation. But the long periods of geology were supposed to afford ample scope ; and the idea formerly prevailing, that species which first existed were of the simplest forms, and that the more complex were gradually introduced, seemed to corroborate the theory. The geologist, however, finds that some of the higher orders were among the early inhabitants of the globe, and that while in some portions of the animal kingdom there has been a gradual introduction of the more complex forms, in others the process has been retrograde. This theory appears to have been suggested by the fact that in many groups of species there is a gradual passage from one to another, without any distinct line of division between them. But gradation is not progress, and transmutation, therefore, is not logically inferred.

3. Another theory is that of *equivocal generation*, which is not exclusively applied to the introduction of species, but which supposes that individuals of species, which may or may not have preexisted, may be generated by circumstances supposed to be favorable to such a process. It once had a wide application, and bees, flies, snakes, weeds, mushrooms, &c., were supposed to be formed in this manner. The substitution of the rigid and exact observations which characterize modern science, in place of the idle speculations of former days, and especially the use of the microscope, have shown the relation of parent and offspring to exist in most of these cases, and the unexplored portions of nature to which this theory is applied are greatly reduced. The

---

What is said for and against the theory of transmutation? What is said of the theory of equivocal generation? of the vitality of seeds and of animalcules?

inquiries on this subject, without confirming this theory, do, how ever, extend our views in respect to the vitality of seeds, and of the inferior forms of animal life. Seeds which have been dormant for centuries have been made to grow, and every one knows that the clearing of wood-land is often followed by a different growth of timber, whose seeds have been for centuries in the shade of the forest. Animalcules are known to exist for years without manifestations of life, in a state of dry powder, the sport of the winds, and on application of moisture have revived.

We conclude, therefore, that the first hypothesis is correct, viz., that each species is introduced by the special direct agency of the Creator.

4. We have yet to inquire into the question of the unity or plurality of the origin of species. A variety of considerations, which are drawn from the geographical distribution of the species, and which are more properly described in zoological works, lead to the conclusion that each species was originally created by the introduction of many individuals, in some cases in distant regions, but more frequently in one region of greater or less extent, and that they were, with few exceptions, created in the same countries which they now inhabit.

Of the nature of the process by which the originals of the species were created, science is entirely ignorant.

III. *Nature of Species.*—The original individuals of a species appear to have had the same peculiarities which now characterize races and varieties. The gradation from one species to another was original in their first creation. Species are, therefore, of the same nature as genera; that is, they can be defined only as groups collected around types, and they are often destitute of well-defined limits. Gradation is thus more logically accounted for by original creation than by transmutation.

What is said on the question of unity or plurality of origin? What is said of the process of creation? of the nature of species?

SECTION III.—DURATION OF SPECIES.

I. *Means of their Preservation.*—Nature has not left the individuals of species without many safeguards against the dangers to which they are exposed. Yet multitudes of seeds perish without germination, and the greater portion of the animal kingdom perish by violence long before reaching the limit of which their life is susceptible, so that a "natural death" is rather the exception than the rule in nature. But the life of a species, if we may so speak, is far more carefully guarded. To every species is given a degree of fecundity far beyond that which would be required for its perpetuity, if not exposed to occasional accidents and disastrous seasons; while, for the excessive multiplication consequent on a favorable season, checks are provided in the limited supply of nourishment, and in the enemies which increase by this increase of their prey. Thus individuals may perish in countless myriads, but the species endure from age to age, and not a few of the feeblest and most minute species have existed many times longer than man himself.

II. *Causes of extreme Longevity.*—Without the aid of geology, zoology can give us no information on the duration of species, for, in the rare cases where the species has become extinct within the historical period, it teaches nothing definitely respecting their origin. Happily, however, the case of some species now in existence, and whose remains are found in some of the tertiary strata, is full of instruction. Of the twelve hundred and thirty-eight species of shells which existed in the earliest of the tertiary periods, forty-two are supposed to be identical with living species. If, now, we can satisfactorily account for their longevity, or even discover any condition connected with it, we shall obtain a view of principles which are highly instructive on this subject.

Some of the species which have thus survived the changes on the earth's surface, that have proved fatal to most of their former cotemporaries, have a wide geographical distribution. This "indicates a capacity of enduring a variety of external circumstan-

---

What is said of the means of their preservation? What examples of extreme longevity of species? of the distribution of such species?

ces, and may enable a species to survive considerable changes of climate, and other revolutions of the earth's surface." — *Lyell*. *Lucina divaricata*, which was introduced in the tertiary periods, now inhabits all our coast south of Long Island, the West Indies, and the Mediterranean. *Saxicava rugosa*, now living on the shores of Europe and America, is found in the pleistocene deposits of both countries, and as far back as the eocene in Europe.

If thus a species is capable of enduring extremes of climate in a wide geographical range, the probability that it is a long-lived species may be inferred. Among fossil species, therefore, those which occur in several successive formations are the more likely to be found in the strata of distant countries, or conversely, those which are known to have this wide geographical range are less likely to be restricted to a single formation.

Others which have also survived the three great tertiary periods, or, in the language of Sir Charles Lyell, "have, like Nestor, survived three generations," are species which, living in deep water, find a uniformity of temperature and of other conditions that do not exist at the surface. Such frequently have, from this cause rather than from any innate capacity of endurance, a wide geographical range.

Others, again, are species which have gradually migrated into warmer latitudes, as the climate has gradually become colder. Several species of shells, which are now found only in a fossil state in Middle Europe, are living within the tropics. Hence we should derive a caution in inferring the cotemporaneousness of distant formations from the identity of a few species, for it is possible that some of these formations may have had the same climate at periods which, although not far distant on the geological scale, may yet have been by no means cotemporaneous.

III. *Comparative Longevity of different classes of Animals.*—The species of some classes are much more enduring than of others. Of the mammals no living species can be traced further back

---

What inferences may be made from their wide distribution? What is said of species that live in deep water? of the migration of species? What is said of the longevity of species of mammals?

than the most recent of the tertiary periods. The fossil remains of birds are perhaps too few to warrant any positive generalizations, but those which have been found belonging to former periods are distinct from any now living. Of fishes which have been found in greater or less numbers in nearly all the fossiliferous rocks, the same species is scarcely ever found in two successive formations. They are, therefore, the most serviceable in determining the cotemporaneousness of distant formations.

Most, perhaps all of the species of fresh-water mollusca, which were introduced in the United States during the pleistocene period, continue at the present time. But the mastodon, megatherium, and other quadrupeds of that period, are all extinct, and their places have been filled by other species. In general, the species of mollusca have a greater longevity than the vertebrata.

SECTION IV.—EXTINCTION OF SPECIES.

Since the vast majority of the species, and many entire genera and families, whose remains are found in a fossil state, have long since become extinct, any facts of a similar character within the historical period become invested with extraordinary interest. Of so recent a date, however, is zoological science, to which alone we can look for any information respecting the countless host of small species which people the air and the waters, that the facts are few and relate only to the larger species. The extinction of the fossil species was owing primarily to the constitution of the species, and partly to gradual changes of climate, aided by mutations in the distribution of land and water. Since the existence of man, his agency has had a marked effect in the more or less complete extermination of some species, and in the increased development of others, to subserve his purposes. Many quadrupeds, as the beaver, wolf, bear, &c., are now extinct in Great Britain. The perseverance with which the fur trade is carried on is rapidly tending to the extermination of many species.

"Immediately after South Georgia was explored by Captain

What is said of birds? of fishes? of mollusca as compared with vertebrata? To what was the extinction of the fossil species owing?

Cook, in 1771, the Americans commenced carrying seal-skins to China, where they obtained most exorbitant prices. *One million two hundred thousand skins* have been taken from that island alone since that period, and nearly an equal number from the island of Desolation! The numbers of the fur-seals killed in the South Shetland Isles, in 1821 and 1822, amounted to three hundred and twenty thousand. This valuable animal is now almost extinct in all these islands."

"An extraordinary bird, a native of New Zealand, of which but few living individuals are known to naturalists, appears to be on the point of extinction; it is the *Apteryx australis*." "This bird is of a grayish brown color, and has neither wings nor tail." "The Dodo was a bird of the gallinaceous tribe, larger than a turkey, which abounded in the Mauritius and adjacent islands, when those countries were first colonized by the Dutch, about two centuries ago. This bird formed the principal food of the inhabitants, but was found to be incapable of domestication, and its numbers therefore soon became sensibly diminished. Stuffed specimens were preserved in the museums of Europe, and paintings of the living animal are still extant in the Ashmolean Museum at Oxford, and in the British Museum. But the Dodo is now extinct—it is no longer to be found in the isles where it once flourished; and all the stuffed specimens are destroyed. The only relics that remain are the head and foot of an individual in the Ashmolean, and the leg of another in the British Museum. To render this illustration complete, the bones of the Dodo have been found in a tufaceous deposit beneath a bed of lava, in the Isle of France; so that if the remains of the recent bird alluded to had not been preserved, these fossil relics would have constituted the only proof that such a creature had ever existed on our planet."—*Mantell.*

In Ireland a large species of elk became extinct probably not long after the island was inhabited by man.

"Its remains commonly occur in the beds of marl beneath the peat-bogs." "In Curragh, immense quantities of the bones of the elk lie within a small space, as if the animals had assembled in a herd; the skeletons appear to be entire, and the nose is elevated, the antlers being thrown back on the shoulders, as if the creatures had sunk in a morass, and been suffocated." "The skeleton is upward of ten feet high from the ground to the high-

---

What is said of human agency in the extinction of species? What is said of the Irish elk?

est point of the antlers, which are palmated, and from ten to fourteen from one extremity to the other." "In the county of Cork, a human body was exhumed from a wet and marshy soil, beneath a bed of peat eleven feet thick; the body was in good preservation, and enveloped in a skin covered with hair, which there was every reason to conclude was that of the elk. A rib of the elk has also been found, in which there was a perforation that evidently had been formed by a pointed instrument while the animal was alive, for there is an effusion of callus or new bony matter, which could not have resulted from something remaining in the wound for a considerable period; such an effect, indeed, as would be produced by the head of an arrow or spear. There is, therefore, presumptive evidence that the race was extirpated by the hunter-tribes who first took possession of the British Islands." —*Mantell.*

But the most remarkable of all the species which have become extinct within the historical period are those of *Dinornis*, a genus of birds, some of which were of enormous size, far exceeding the ostrich. Their bones are found in recent deposits in New Zealand, and the natives have a tradition of the former existence of this bird, which they called *Moa.* It was allied to the Apteryx in its characters, being without wings. The thigh bone of the larger species was upward of two feet in length. Being the only land animal in that country of sufficient size to furnish food to the people, it was probably exterminated by the early inhabitants.

These are the only animals which are known to have become extinct since the commencement of man's existence. Many which flourished during the periods immediately preceding have since become extinct, but so slow are the changes in nature and so long is an entire period, the unit in geological computation, that it is not easy to determine whether any such were in existence in the earliest days of our race.

SECTION V.—GEOGRAPHICAL DISTRIBUTION.

The *geographical distribution* of animals and plants is a subject of great importance in its application to geology. Although the subject is so copious in its details that it can never be ex-

---

What is said of the genus Dinornis?

hausted, a few outlines may be sketched as sufficient to give some general view. The common observer, who is unaccustomed to careful discrimination between the species, especially of small animals and plants, which nearly resemble each other, is not aware of the great difference between the Fauna and Flora of different countries. Thus of the species of shells in the West Indies, probably less than one in one hundred exists in the New England States. The native quadrupeds of the United States are specifically distinct from those of Europe. Of the countless hosts of insects, few, if any, are common to both sides of the ocean, except those which have been transported by human agency.

The distribution of the species, apart from the influence of human agency, depends on the places of original creation, on climate, and on station or the kind of place which their habits require. Climate and station have influenced distribution in two ways; by determining, in the Divine plan, the place of original creation, and by restraining subsequent dispersion.

I. *Places of Original Creation.*—A few species appear to have been originally introduced into different countries. Thus a small snail, Helix pulchella, is aboriginal in both Europe and the United States. Fresh-water fishes of the same species inhabit distant rivers. But more frequently the places of original creation of the individuals of a species are adjacent.

II. *The connection between climate and distribution* is very obvious. Let any one travel a thousand miles of latitude, and he will find that he has left behind him a great majority of the familiar species. Let him go two thousand miles, and the appearance of a single species which he has been accustomed to see will be an era in his wanderings. Not only so, but there will be a striking difference in the more comprehensive groups: whole genera, families, or even orders have disappeared, and new ones

What is said of the differences in the animals of different countries? On what does the distribution of species depend? In what ways have climate and station influenced distribution? What is said of the places of original creation? of the connection between climate and distribution?

surround him. On the other hand, an equal change of longitude will present fewer changes in the more comprehensive groups, and in the species, which changes will be found chiefly dependent on intervening barriers of mountains, water, &c.

In some cases whole tribes of animals and plants, and in others the individuals of particular species only in a genus, are equally affected by climate.

Thus the whole tribe of palm-trees and many families of shells, either exclusively or with the exception of a few of the smaller and less characteristic species, such as the cowry family, with their richly-colored and highly-polished shells, are found in great profusion only in hot climates. The oaks, if found within the tropics, are on elevated regions, where they have the climate proper to a higher latitude. Again, there are some genera which exist in all climates, as Helix (snails), but are represented in each climate by peculiar species. In most cases there exist peculiarities by which the experienced naturalist can distinguish the species of one climate from those of another. Rarely the same species exists in very different climates. The *Lucina divaricata* mentioned on page 91, and a few others, are exceptions to the general rule. A peculiar class of examples of this kind are the migrating animals, which find a greater uniformity of climate by changing their abode as the season changes. From these facts we should expect, if the climate had ever been uniform over the surface of the earth, that species would have had a much wider geographical range than at present, and this appears to have been the fact in the earliest periods.

III. The circumstances of *station* restrain the dispersion of species. Bodies of water interpose an obstacle to the distribution of terrestrial animals; fresh water to the marine, and salt water to the fresh water species. Deserts and mountains separate zoological as well as political provinces. Islands have their peculiar Fauna and Flora. Some species can exist only on certain geological formations; thus the larger and heavier land-shells of some tropical countries exist only in limestone regions, and the land-shells of this country are far more abundant in such districts, although not exclusively confined to them. Some re-

---

What examples are mentioned? In what ways do the circumstances of station restrain the dispersion of species?

quire dry and others wet land; some pure, and others impure water; some live on sand, others in mud; some on high land, others on low land; some in deep water, others on shore. A few species, again, can accommodate themselves to a variety of stations. All these facts it is the business of the zoologist to in vestigate, and on him, consequently, the geologist is dependent for the means of drawing conclusions.

Sometimes there is a peculiar grouping of species; as when a river flowing into an estuary mixes terrestrial, fresh water, and marine species in the same deposit; or when beasts of prey drag into caves the carcasses of various animals; or when the hermit crabs,* so abundant in hot climates, and with amphibious habits, occasionally mingle terrestrial and marine shells on the open coast without fluviatile agency.

In the application of these general facts to the case of fossils of extinct species in older formations, no *universal* rule can be given by which to determine whether a given species inhabited a hot climate or otherwise, but the decision must depend on its zoological affinities. If the species belongs to a group which is found exclusively in the tropics, the probability that it was tropical will be very strong; and if a large number of such species are found in the same formation without any which are proper to colder climates, the conclusion may be considered certain. If, however, the extinct species belong to a group which is universally distributed, and if it has none of the distinctive marks of climate on it, it will fail of establishing any conclusion respecting the climate.

From the peculiarities of station which characterize not only species, but more comprehensive groups, the geologist is usually able to make very interesting and important inferences respecting

* Crabs, whose instinct leads them invariably to enter the empty shells of dead molluscs for the protection of their bodies, that are not sufficiently protected by the thin crust with which they are covered, are called "hermit crabs."

What peculiar groupings of species are mentioned? What is said of inferences respecting the climates inhabited by the fossil species? What inferen ces respecting the geography of ancient periods?

the geography of a region, as it was at the time when the species now extinct was in existence.

Fossil shells show whether the deposit was made in fresh or salt water; or if species of both fresh and salt water are mingled, it will be generally inferred that the deposit, especially if the layers form a basin-shaped depression, was made in an estuary into which a river emptied. The depth of the water may often be inferred from the genus to which the fossil belongs. The character of the shells will also show whether it was a sandy or a muddy shore; in this particular, however, merely confirming the inferences made from the mineral characters of the strata. Fossil plants show not only the temperature of the climate, but its humidity, and the general features of the surface. The geological student will find in the copious details of zoology, with which we are not here concerned, that the geological inferences are as full of interest as are the zoological premises from which they are made.

Terrestrial animals are, for the most part, distributed in the following manner. Commencing with the Arctic regions, the same species generally occur in both hemispheres. In the temperate regions, most of the species of North America, Europe, and Asia are peculiar to each of these regions, but are very similar to each other. In the tropical regions the differences are greater. In the southern extremities of the great bodies of land, New Holland, South Africa, and Patagonia, the differences are greatest.

The details of the subject of geographical distribution vary in different classes, and even in different families and species of the same class. Some are dispersed over regions of several thousand miles in diameter, while others of the same genus or family are restricted to a few square miles.

SECTION VI.—DETERMINATION OF ORGANIC REMAINS

Were we permitted to go into details, it might easily be shown in what manner important inferences may be drawn from very

How may fresh-water deposits be recognized? How estuary deposits? What other information may be derived from fossil shells? What from fossil plants? How are terrestrial animals distributed? What is said of the difference in the extent of distribution of different species?

imperfect remains of animals and plants. Often only a tooth, or a few scattered fragments of bones, or even the tracks, may be all the relics which remain to testify to the character of the extinct species. The determination of the character of organic remains, therefore, becomes a problem of the highest importance, and often taxes to the utmost the resources of science. Nor is it a question of scientific interest only; for, since the real character and the age of the rock is to be determined usually by the nature of the organic remains, it becomes a question of great practical interest.

Not only zoology and botany contribute from their boundless resources to the aid of the geologist, but comparative and vegetable anatomy render no less efficient aid. Every tribe of animals has some peculiarities, of which some are necessarily connected, and others are invariably associated without any obvious connection.

An example of the necessary relations between parts may be seen in the relation of the claws of the tiger, or of any kindred species, to the other parts of the animal. The sharp, curved, retractile claws are fitted for seizing and tearing its prey, and all the other parts appear to be necessarily associated with them. The teeth must be sharp, and adapted for cutting flesh rather than for bruising grain; the food being flesh, the digestive apparatus must be less complicated, and the intestines not a quarter as long as in herbivorous animals; and the general form and structure must be consistent with agility in order to seize the prey. All these and many more details would be inferred from finding a single retractile claw, whether it should belong to any known or unknown species, and with the aid of zoology all the general characters of the tribe of carnivorous mammals, with their anatomy and physiology in their numerous details, would also be inferred.

Of the characters which are *invariably associated* without ap-

---

Why are imperfect fossils often important to science? Why are organic remains of economical importance? What sciences aid in determining fossils? Describe the example of necessary relations of parts. What is said of associted characters?

parent reason, we have a striking example in the ruminants, for every cloven-footed quadruped chews the cud, and thus the number of stomachs is infallibly inferred from the characters of the foot. The naturalist is familiar with numberless examples of this curious principle in zoology. It extends even to color and clothing as well as to form and structure, so that certain colors only will be found in certain species or groups of species.

Both of these principles are familiar facts in reference to well-known animals. A single tooth of a horse will show not only to what kind of animal the tooth belonged, but even his age; and every one knows that a green horse, a blue dog, or a red elephant are as unlikely to exist as a ruminating quadruped with four toes or with the solid hoof. Now those who are as familiar with the various tribes of organic nature as the husbandman is with his cattle, are able to make similar inferences respecting the multitude of species with which the mass of mankind are unacquainted. Thus, when a single tooth of the iguanodon was shown to Cuvier before the discovery of the skeleton, he laid down the characters and habits of that herbivorous reptile with an accuracy which the subsequent discoveries have only confirmed. A single fish scale was found in the intestines of an ichthyosaurus, and shown to Professor Agassiz, who recognized it as belonging to an extinct species known to him, and he was able to point out the part of the fish to which it was originally attached.

In the vegetable kingdom, although many general inferences may be made from parts, yet details can not often be deduced in this way, and the geologist finds fragments of plants without being able to decide whether they belonged to the same or to different species.

What familiar examples are mentioned? What is said of Cuvier? of Agassiz? of the utility of fragments of fossil plants?

CHAPTER II.

FOSSILIZATION.

THE proportion of the various tribes of animals which are found in a fossil state must depend not only on their number, but on the facility with which they may be preserved. It becomes, therefore, of great importance to consider this subject, in order to avoid erroneous conclusions respecting the Fauna and Flora of former epochs. Thus two thirds of all the known species of organic remains are shells, but it is not to be inferred that molluscs constituted two thirds of all the organic beings in existence. The durability of the parts, the agents of burial, and the nature and process of petrifaction, are the principal topics.

SECTION I.—DURABILITY OF ORGANIC BODIES.

The durability of parts of the bodies of the various tribes of animals and plants differs extremely, and is found to correspond to a considerable extent with the relative quantity of their remains which have been found in a fossil state. It is therefore important to consider this subject a little in detail.

I. *If Vertebrated* animals are buried so deep beneath a mass of sand or mud that their putrefying bodies can not rise to the surface, their skeletons will be preserved entire. If, however, they are not thus buried, they will, soon after the commencement of putrefaction, from the formation of gases, become lighter than the water, and float on the surface; decomposition will be more rapid, and the bones will fall scattered to the bottom, and be gradually covered by the deposits of mud or sand which may be going on. Birds, however, are always buoyed up by their feathers, and hence only scattered bones are likely to be preserved.

What is said of the proportions of various tribes of fossil remains? What are the three principal topics relating to fossilization? What is said of the durability of the parts? What is said of vertebrated animals? of birds?

Their bones also being tubular, are more likely to be crushed than those of other vertebrata. The skeletons of some fishes, as the sharks, being cartilaginous, are more subject to decay, but their teeth, being very hard, are preserved. Accordingly, the bones of birds are extremely rare; and while the skeletons of many kinds of fishes are common, it is seldom that any thing more than the teeth of sharks is found. Birds, however, frequently leave their tracks in fine sand and mud to be buried by additional layers of sediment, and although little search has been made by laying open strata on the banks of rivers and on the shores of the sea, a number of examples have been found. It will be seen in the sequel that the tracks of birds are far more frequently found in the rocks than are any other relics of them.

II. *Of the Articulata;* insects, although far more numerous than any or all other classes, except the animalcules, are very rarely preserved. The naturalist, in searching for objects of natural history, rarely finds dead insects. Multitudes of them, with a ferocious activity, prey upon each other. On the water or on the land they are the favorite food of numerous tribes of vertebrata; bats, lizards, frogs, fishes, and birds devour myriads. Permeated by the innumerable air tubes of respiration, they after death speedily decay by atmospheric agency. Being from the same cause lighter than the water, they are rarely buried, but, like birds, float on the surface, without, however, eventually dropping any solid parts, but either decaying or becoming the prey of fishes. Rarely, therefore, either in the recent or older rocks, are their remains found. Yet as it would obviously be incorrect to infer, from the paucity of their remains in alluvial deposits, that few exist at the present day, so it would be equally erroneous to infer positively that they were not more numerous in former periods than the number of their remains would indicate. The remains which have been found are either those of insects that frequent water, and are therefore more likely to be buried

What is said of cartilaginous fishes? of bird tracks? of insects? How is the scarcity of fossil insects accounted for? What kinds of fossil insects are found?

in the mud, or the elytra of beetles, which are the most indestructible parts.

Of the *Myriapods* the same remark may be made: covered with crustaceous rings of about the consistence of the harder parts of beetles, they are about equally durable, and are found about as frequently in their haunts.

Of the *Arachnida* a majority are extremely frail, but a few, especially the scorpions, are as durable as any of the beetles, being covered with a hard crust, and a few have been found in a fossil state. The most durable species of this class are, however, not aquatic.

Most of the *Crustaceans* are covered with a crust much harder than any of the preceding articulata, and parts of some of them have the solidity of bones. They are mostly marine; some, especially of the minute species, inhabit fresh water, and a few only are capable of living on dry land. They are, therefore, much more frequently preserved.

Of the *Cirrhopods*, many are covered with true shells of solid carbonate of lime, and as they are marine, there can be no reason why they should not be preserved in as large proportion as any tribes of organic beings. From their deficiency in the geological formations, it may quite safely be inferred that the class is of comparatively recent introduction.

Annelidans are less durable than the myriapods, which they resemble in their general form. But many of them are aquatic, and, having the habit of burying themselves in the mud or sand, are not very unlikely to be preserved, notwithstanding the exceeding frailty of their structure. It is quite remarkable that some of the oldest animals yet described as existing in any geological formation are marine worms with very slender and perishable bodies. Their remains, although distinct, usually resemble faint impressions. The tracks of earth-worms, which are so abundantly seen immediately after a heavy rain has driven them out of their holes, are often in a situation to be soon covered with

---

What is said of myriapods? of the arachnida? of crustaceans? of cirrhopods? of annelidans? of their fossil remains?

a layer of mud, and are quite likely to be preserved. Impressions on some of the solid rocks probably have a similar origin.

III. *Of the Mollusca,* those which are destitute of shells are very perishable. Hence their remains are rare.

But a great majority of the mollusca are furnished with durable calcareous shells. Of the several classes, the shells of the gastropods and conchifers are most solid and durable, and the latter most abound in species whose habit is to live buried in sand or mud. Yet in the older formations they are either very rare or entirely wanting, and the shells of the cephalopods and brachiopods are far more abundant, although at the present time they are extremely rare: whence it may be safely inferred that the proportions of these great classes have been totally reversed.

Of land-shells, the actual number of living species is probably nearly as great as of marine species. But since the species, for the most part, have a much more restricted distribution, they are comparatively less known to naturalists. The same cause may be an obstacle to our knowledge of extinct species. Terrestrial shells are also much less likely to be imbedded and preserved.

IV. *Of the Nematoneura,* the echinodermata are covered with shells, which are densely crowded with calcareous portions, only less solid than the shells of molluscs and cirrhopods, and, from their marine habits and great numbers, they are likely to be preserved. Their remains are common in many of the formations.

The rotifera and bryozoa are not uncommon in a fossil state, as they are aquatic, and have hard parts which are silicious or calcareous. Of the parasitic classes of nematoneura, it would perhaps be premature to say much of the probability of their existence in a fossil state; less solid, however, than most of the other classes of this division, they are less likely to have been preserved, if they existed.

V. *Acrita.*—The acalephæ, although marine, being mere gelatinous masses, with but a few grains of solid matter, are very un

---

What is said of the mollusca which have no shells? What shells are the most durable? What are most abundant in a fossil state? What inference? What is said of land-shells? of echinoderms? of the acalephæ?

likely to be preserved. The absence of the remains of these animals from geological formations affords, therefore, little proof that they did not exist. Of the polygastric animalcules, the many which had shells of flint leave their shells in extensive deposits, as before remarked.

Of the polypi a few species are naked, and although marine, their soft gelatinous bodies could scarcely be preserved. But the great number of coraliferous species are engaged in the involuntary labors which we have before described, and their durability is proved by the remains of corals in nearly all the fossiliferous rocks.

Sponges have sometimes been regarded as animal and sometimes as vegetable bodies. Their situation is favorable to fossilization. So fine is their texture and permeable to mineralizing agents, that the microscopic investigations of late years have shown that many flint nodules in chalk consist of silicious petrifactions of sponges. They are not numerous in the older formations, and, had they existed abundantly in the most ancient periods, it is probable that their remains would have been abundantly preserved.

SECTION II.—BURIAL OF ORGANIC BODIES.

I. *Human agency* has exerted a powerful effect in depositing in the earth the relics of man himself, both human skeletons and objects of art.

During the war between England and France, in the earlier part of the present century, 125 British ships of the line and frigates, and an immense number of smaller vessels, went to the bottom. The other European powers suffered a much greater loss. During the year 1849, there were 566 British vessels wrecked. Probably more than 3000 European and American vessels are annually lost. Coins in immense quantities, seals and ornaments consisting of the hardest minerals, and often engraved, and durable articles of glass, earthenware, &c., are thus buried in the marine strata which are now in the process of formation.

What is said of polypes? of sponges? What are the effects of human agency? What is said of the loss of vessels? of the burial of works of art in marine strata?

Human bones are as durable as those of other animals. Cuvier says, that "in ancient fields of battle the bones of men have suffered as little decomposition as those of horses which were buried in the same grave." They have been found in a fossil state, and even in solid rocks. The most remarkable instance is that of the skeleton found in a fragmentary rock in Guadaloupe. But as this rock is daily increasing by the minute fragments of shells and corals which are united by a calcareous cement, no remote antiquity can be ascribed to these remains. In short, *all* the remains of man are limited to deposits which are of recent origin.

II. *Natural Agencies.*—A large majority of the organic bodies which are imbedded in the strata now forming are buried by aqueous agencies. In a few instances, bodies have been preserved without the agency of water. Thus bodies of men, and remains of birds and eggs, have been found in guano; and the moving sands of deserts bury various objects which may be preserved indefinitely. Animals are often buried in caves and fissures by inundations, of which numerous examples have been found in Europe. Peat bogs also preserve the bodies of animals which are mired in them. Peat also accumulates over and preserves prostrate forests. The only vestiges of the forests described by Julius Cæsar, along the great Roman road in Britain, are the trunks of trees in peat. In frigid climates, animals are sometimes entombed for ages in ice.

The ejections of volcanoes, as we have seen, may bury even entire cities, and various organic bodies may be preserved in the same manner.

Floods and storms often bury immense numbers of organic beings. In 1787, on the coast of Coromandel, there was a flood occasioned by a hurricane which drove the waters of the sea inland 20 miles. This flood covered the country with mud, in which were the bodies of 10,000 inhabitants and 100,000 domestic animals. When, however, animals are buried permanently

---

What is said of the durability of human bones? What is the character of most natural agencies? Give examples of agencies not aquatic. What is said of floods and storms on sea-coasts?

beneath the water, the probability of their preservation is much greater Such may have been the effect of the earthquake wave which in 1780 rushed over the city Savanna la Mar, in Jamaica, and swept away the whole town, leaving not a vestige of man, beast, or habitation on the surface.

But the most efficient agents are the floods of rivers, by which plants and animals are borne into deep water and often into the sea, and permanently submerged. In these cases the bodies may be buried at once beneath a heavy mass of sand and stones. Even marine animals are often destroyed by the mass of materials swept down in floods.

"We are informed by Humboldt that, during the periodical swellings of the large rivers in South America, great numbers of quadrupeds are annually drowned. Of the wild horses, for example, which graze in immense troops in the savannah, thousands are said to perish when the River Apure is swollen, before they have time to reach the rising ground of the llanos. The mares, during the season of high water, may be seen, with their colts, swimming about and feeding on grass, of which the top alone waves above the waters." * * * "In Scotland, in August, 1829, a fertile district on the east coast became a scene of dreadful desolation, and a vast number of animals and plants were washed from the land, and found scattered around the mouths of the principal rivers. An eye-witness thus describes the scene which presented itself at the mouth of the Spey, in Morayshire: For several miles along the beach, crowds were employed in endeavoring to save the wood and other wreck with which the heavy rolling tide was loaded; while the margin of the sea was strewed with the carcasses of domestic animals, and with millions of dead hares and rabbits.' "—*Lyell.*

The solid parts of marine animals, as bones of fishes, and the shells of molluscs, of crustacea, and of echinodermata, are of course often, at the death of the animal, in places favorable for

---

What is said of the floods of rivers? What example in South America? Describe the example in Scotland. What is said of the situation of the solid parts of fishes and other marine animals?

preservation, or are swept into such places. Thus, to the east of the Faroë Islands, a bed of sand and mud, full of broken and entire shells, has been traced for twenty miles; and for the space of three and a half miles in length, the mud is so full of fish bones, that the sounding lead is seldom drawn up without some vertebræ attached. Between Gibraltar and Ceuta, fragments of shells have been found on a gravelly bottom at the depth of 4800 feet, carried thither by a current. Fishes are also buried by submarine eruptions of lava or mud.

Most of the genera of mollusca have aquatic habits, and exist in great numbers in places favorable for their preservation. Not a few live buried in mud or sand, where after death only the soft parts perish. The proportion, therefore, of those which are preserved should by far exceed that of any other division of the animal kingdom. We are, therefore, not surprised that fossil shells should constitute the greater portion of many strata of the fossiliferous rocks of all ages.

SECTION III.—MODES AND DEGREES OF PRESERVATION.

I. *Petrifaction* consists in the substitution of particles of mineral matter in the place of the particles of vegetable or animal matter, and consequently preserves the structure of the original body. In some cases this process is known to take place at the present time, as when bones are enveloped in clay containing sulphuret of iron. Sticks, nuts, &c., in a place where bog iron ore is accumulating, are found to have been converted into ore, probably within a few years. Leaves have been artificially baked in clay, and then resembled ancient petrifactions. But little, however, is known of the *process* of petrifaction, for the chemical conditions favorable to it are more likely to exist under a pressure of superincumbent materials excluded from the air.

II. In some cases the space occupied by the organic body is left empty, and only a *mold* remains. When this is filled with mineral matter, we have a *cast* of the body, which differs from a

---

What is said of the habits of most mollusca? What is petrifaction? What modern examples of petrifaction are mentioned? What are molds? *casts?*

petrifaction in the absence of an organic structure. The interior of shells is usually filled with sand or mud, and when the shell subsequently decays without petrifaction, a cast of the interior only remains. Frequently we find casts of the shells themselves containing the casts of their interior.

III. The calcareous shells of molluscs and of echinoderms, and the bones of vertebrated animals, are often preserved through several geological periods without other change than the loss of the animal matter. Multitudes are found in this condition in the tertiary and cretaceous formations. It is obviously inaccurate to call such fossils petrifactions.

In Siberia a mammoth and a rhinoceros have been preserved since the pleistocene period in ice, with their flesh entire. Human bodies have been preserved in peat bogs for more than a thousand years.

IV. Various bodies are incrusted with depositions of carbonate of lime from calcareous springs, and are often, but very erroneously, called petrifactions. They are not fossils unless they belong to a remote period.

SECTION IV.—COMPARISON OF THE NUMBER OF FOSSIL AND LIVING SPECIES.

It may be interesting to show, in a tabular form, the comparative number of species known to be living, and of those which have been found in a fossil state.

Classes or divisions.	Living species.	Extinct species.
Mammalia	1,500	275
Birds	8,000	20
Reptiles	1,500	120
Fishes	8,000	1,500
Insects	100,000	250
Articulata (Other than insects.)	3,000 ?	500
Molluscs	20,000	6,000
Polypi	1,000 ?	900
Echinodermata	200	700
Animalcules	800	100
Plants	100,000	1,800
	244,000	12,165

What examples are mentioned of preservation without petrifaction? What is said of incrustations? What is the number of known species of existing animals and plants? of extinct species?

In this comparison, the first column comprises the species of one period, and the second those of hundreds of periods. The comparative deficiency in the number of extinct species is due to three causes: first, the liability of many tribes to perish without being fossilized; second, the small portion of the fossil remains which appear on the surface, the great majority being concealed within the strata, while the living species are on the surface, and exposed to notice also, in the case of many animals, by their habits of activity; third, the greater number of observers and collectors of the existing species of animals and plants. Of so much consequence are these three principal circumstances, that, with some exceptions, it would be mere speculation to make a definite comparison of the actual numbers of living and of extinct species.

Why are so few extinct species known?

PART III.

HISTORY OF THE EARTH.

The first and second parts of our subject may be regarded as preliminary to the third, in which we come directly to the history of the earth. In this history the geologist describes the condition of the earth in successive periods, in respect of the distribution of land and water, the features of the land, the climate, the sedimentary deposits and igneous rocks, and the successive generations of animals and of plants.

CHAPTER I.

SOURCES OF KNOWLEDGE.

The direct sources of knowledge are to be found in the rocks, which have been formed during the successive periods of the past.

SECTION I.—RELATIVE AGE OF THE STRATIFIED ROCKS.

The basis of the whole history is a knowledge of the relative age of the rocks.

I This is ascertained primarily from the *position* of the strata of the sedimentary rocks. It is obvious that the vertical order in a series of the strata is the order of time; that the lowest stratum was first deposited, and that the uppermost one is the most recent. In this way we learn the relative age of those stratified rocks which are found in junction.

What does Geology describe in the history of the earth? What is the basis of the history? What may be learned from the order of the strata?

II. But all the strata do not occur in any one place in junction, and their relative age is usually ascertained in the following manner: If the actual chronological order be represented by the order of the letters of the alphabet, we may have in one place the series A, B, C, E, G; in another, C, D, E, F, G; in another, A, B, M, N, P, &c. Thus the parts of a complete series occur in different places. These parts may be arranged with ease and accuracy in the order of time, provided that we are able to identify synchronous strata; or, to recur to our illustration, when the formations A, B, C, E, G occur in one place, and C, D, E, F, G in another, we must be able to identify C and E in the second series with C and E in the first series. This is done by the *fossils*. Each formation is characterized by peculiar fossils; and, having become acquainted with the species in the formations in each local series, we find that those of C and E are the same in both series. We then infer the synchronism of these formations.

But since the species of animals and plants differ widely on distant parts of the earth's surface, we can not expect thus to de termine an exact synchronism between the formations of very distant countries. This difficulty is found to be less in the older formations than would have been anticipated, because it appears that there was, during their deposition, a much greater uniformity in the animal and vegetable kingdom than at the present time. The differences also between the Faunæ and Floræ of periods, which are in geological time quite remote from each other, are greater than between those of a given period, which are geographically distant.

III. Another means of directly proving the relative age of rocks occurs when we find the fragments of one stratified rock in another. The inference is then unavoidable, that the first was not only deposited, but consolidated before the latter was formed.

IV. The synchronism of deposits in distant regions may also

Is the series of strata complete in any one place? How is the fact illustrated? How can distinct formations of the same age be identified? What difficulty is mentioned? What is the third method of ascertaining the relative age of rocks? the fourth method?

be approximately established, in the case of the later geological periods, by *a comparison of their fossils with existing species.* In receding from the present time, the proportion of extinct species will be found to increase, until we arrive at a period anterior to the existence of any of the present races. If the introduction of new species and the extinction of old ones had taken place with a uniform rate in all times and places, we should have a means of measuring time with a degree of precision proportioned to the completeness of the collections of fossils. Although the rate of change may not have been characterized by exact uniformity, this principle is of the highest utility in geological investigations. This method may be illustrated by taking A, B, C for a series of successive deposits in one region, and *a*, *b*, *c* for a similar series in another place. If all the fossils in A and *a* belong to existing species, we may refer the deposits to the present period. If in B and *b* ten per cent. belong to extinct species, they are synchronous, and are referred to a previous period; if in C and *c* thirty per cent. are extinct, they are synchronous, and more ancient than B and *b*.

The same method may be extended to a comparison of any formations which immediately succeed each other in position. If many of the fossils are common to both, we may infer that they are not remote in geological time. The converse, however, evidently may not be true when two successive deposits contain the remains of animals of very different habits, as marine species in one and fresh water species in another.

V. More or less aid is sometimes derived from the lithological characters of the strata. The lithological characters comprise the mineral constitution and the structure. They depend partly on original deposition and partly on subsequent changes. Their peculiarities are dependent much more on local circumstances than on the period of time during which the deposit was formed. Thus, during the same period, chalk beds were formed in England and strata of sand in New Jersey. The lithological char-

How is this method illustrated? How may it be applied to ancient formations? What is said of the lithological characters?

acters are used, in determining the age of rocks, with much caution and to a limited extent.

VI. The minerals which are disseminated through, or associated with certain strata, may also be of some service in ascertaining their age. Some minerals appear to have been formed more abundantly during certain periods.

SECTION II.—RELATIVE AGE OF THE UNSTRATIFIED ROCKS.

The age of unstratified rocks may often be more or less accurately determined by their association with stratified rocks. It is obvious that an igneous rock must be more recent than the strata through which it has been erupted. It is not always correct to infer that strata which lie over an igneous rock are therefore more recent, since the igneous rocks may have failed to penetrate all the overlying strata. Frequently, however, it can be determined by local circumstances whether the overlying strata were there at the time of eruption.

SECTION III.—HISTORY OF EACH PERIOD

After having fixed the relative chronology of the formations, we may learn from their organic remains the characters and habits of the plants and animals of each period. From these data we infer the peculiarities of climate, and from the same data and the lithological characters of the strata we may learn much respecting the distribution of land and water, and the features and extent of the land. From the strata themselves, and from the unstratified rocks associated with them, we learn much respecting the aqueous and igneous agencies of the period.

Since, also, some thick deposits thin out in certain directions, it may be inferred that the materials of which they are composed were derived from the opposite direction; in other words, that there was an island or continent in that direction, the size of which and of its rivers must have been in some measure proportionate to that of the formation derived from it.

What is said of the mineral contents of the strata? How is the age of the unstratified rocks ascertained? What may be learned from the organic remains of the strata? What from differences of thickness?

From the alternation, in some cases, of fresh water with marine formations, it is also obvious that some portions of the earth have been subject to both elevation and subsidence repeatedly.

It appears to have been the general fact that the palæozoic formations, with their fossils, were uniformly spread over much more extensive areas than the secondary and tertiary formations; the change from the former to the latter is not, however, abrupt, but the deposits, for the most part, are more limited, as they are more recent.

With the exception of some metamorphic districts, and some very limited fresh-water formations, the greater part of existing continents is covered with strata which abound with marine fossils. These fossils, with scarcely any exceptions, are the remains of animals and plants which lived and died in the places where they are now found. This is attested by the preservation, in numerous instances, of parts so delicate that they would have been destroyed by transport, and also by the nature of the stratum in which they are found, which corresponds to the known station of kindred species now living.

It is obvious, therefore, that the existing continents have been under the ocean since the commencement of the palæozoic period.

CHAPTER II.

MEANS OF OBSERVATION.

THE means by which the rocks are exposed to observation are both artificial and natural.

SECTION I.—ARTIFICIAL EXCAVATIONS.

Excavations for *roads* and *canals* often expose the rocks and the superficial deposits to a considerable depth. Fresh excava-

What may be learned from the alternation of marine and fresh-water strata? What is said of the extent of the older formations? What proof that the existing continents have been under the ocean since the palæozoic periods? What is said of artificial excavations?

tions are especially useful in exhibiting the order and structure of the unconsolidated strata. *Artesian wells* (p. 72) often furnish valuable information respecting the order of the strata, as well as concerning the internal temperature.

Mines are yet more useful to geologists. The deepest is that of the Eselschacht in Bohemia, which is 3778 feet deep. The greatest depth below the sea-level is that of some coal mines in Newcastle, England, which are from 1500 to 2000 feet below this level. Since many deep mines are carried to a great horizontal extent in following metallic veins or beds of coal, they have furnished a great amount of knowledge on those subjects.

SECTION II.—NATURAL MEANS.

I. *Excavations by Water.*—Among the natural means are the *excavations by rivers.* The section of the rocks which is exhibited in the channel of the Niagara is seven miles long, and from 150 to 300 feet deep. The structure of an uneven country is often best exhibited by the river-courses and the *lines of sea-coast.*

II. *Uplifts.*—Yet more instructive are those *uplifts* where the strata on one side of a fracture have been lifted up, forming a precipice. The order of several successive formations is often thus exhibited, and the geologist is able with great facility to ascertain, also, their characters, contents, and thickness.

A beautiful example of this kind is found in Snake Mountain, in the valley of Lake Champlain. Here most of the lower part of the Silurian system is shown within the distance of less than half a mile, as in the following figure:

Fig. 31.

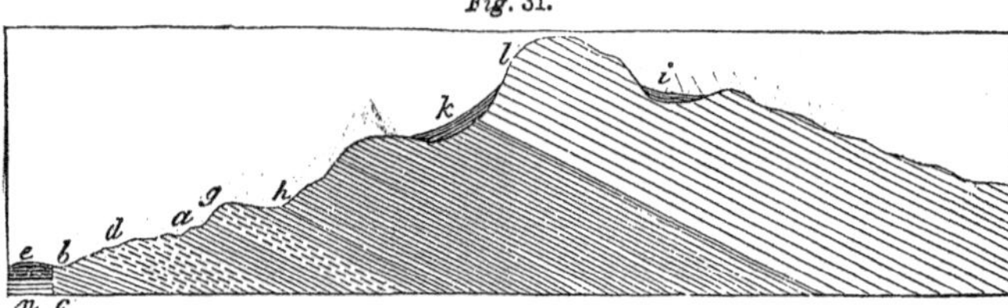

SECTION OF SNAKE MOUNTAIN, ADDISON COUNTY, VT.

c. Fracture.

What is said of mines? of excavations by water? of uplifts? Describe the example.

e. Brown clay, extending six miles west to Lake Champlain, where it covers (*n*) the Trenton and Isle la Motte limestones.
b. Calciferous sandrock; dip* 10°.
d. Isle la Motte limestone, dip increasing from 10° to 20°.
a. Trenton limestone; dip 25°.
g. Utica slate; dip 38°.
h. Hudson River shales; mostly covered with drift and debris.
k. Debris, from
l. Red sandrock; dip 20°.
i. Cranberry meadow, over liquid peat.

This example has been espe ially useful in deciding in the negative an important question which had arisen in American Geology, as to the existence of a Taconic System older than the Si lurian System.

III. *Inclination of the Strata.*—But the limits of geological observation would be very narrow if it were restricted ordinarily to a thickness of a few hundred feet, with some rare opportunities in high mountains and deep mines. By means of the *inclination of the strata*, we become acquainted with the structure of the earth to the depth of many miles. The convulsions of ancient times have more or less broken and lifted up the layers of the stratified rocks, so that few of them now lie in their original position. The layers must once have been nearly or quite horizontal. If Lake Erie should be drained, extensive layers of sediment would appear, and if these should be consolidated, fractured, and inclined by igneous agency, their whole structure and contents would be exhibited.

Most of the older strata have been repeatedly subjected to igneous agencies, while various aqueous agencies have more or less modified the parts which have been thus exposed. But the natural and artificial sections of canals, rail-roads, rivers, and sea-coasts more or less perfectly exhibit the upturned edges of the ancient strata. It is obvious that when we pass over the surface

* Dip is the inclination below the horizon.

What can we learn from the inclination of the strata? What must have been their original position? How are their edges exhibited? What can we learn by traveling across their edges?

of the ground directly across the edges of the layers, we obtain the same knowledge of their order, structure, and contents which we should have obtained by a perpendicular descent had the strata retained their original horizontal position. If, for example, we travel six miles across the edges of strata, which have a dip of 45°, their thickness will be found, by a trigonometrical calculation, to be a little more than four miles. It is quite unnecessary, therefore, for the geologist to descend into the earth to learn the structure of its crust. Probably all the formations, and consequently all the materials for a chronological history of the earth, have been exposed to the light of the sun.

CHAPTER III.

CLASSIFICATION OF THE STRATIFIED ROCKS.

SECTION I.—CLASSES OF STRATA.

I. THERE are five classes of stratified rocks. The first and oldest are called *primary strata,* and are characterized by their position beneath the fossiliferous rocks. Many of the strata which were once referred to this class of rocks have been proved to have been originally fossiliferous and are now *metamorphic;* that is, the fossils have been obliterated, and their structure has been rendered crystalline by heat. Such rocks belong mostly to the next two classes. Probably very few of the supposed primary strata will eventually remain in this class.

II. The second class includes those strata which contain the remains of the earliest organic beings, and a long series of subsequent formations. They are called *Palæozoic rocks* (from two Greek words, which signify ancient animals), because they contain the remains of the earliest animals known to have existed. Most of them have also been called *Transition* rocks, because

---

What are the primary strata? What are metamorphic rocks? What is said of the palæozoic rocks?

they were once supposed to have been formed during a transition of the earth's surface from an uninhabitable to a habitable state.

III. The third class contains another long series of formations, commencing with the New Red Sandstone, in which occur the extraordinary bird tracks discovered and described by President Hitchcock, and terminating with the chalk formation. They have been called *Secondary Rocks*, and also *Mesozoic Rocks*—the word mesozoic signifying animals of the middle periods.

All the animals of the palæozoic and mesozoic periods have long since become extinct.

IV. The fourth class, called the *Tertiary or Cainozoic strata*, consists of regular and nearly horizontal strata of limestone, sand, and clay; and, with many extinct species of animals and plants, they contain some which still exist. The word *cainozoic* signifies new or recent animals.

V. The fifth class, called the *Quaternary strata*, comprises the unstratified masses of sand, gravel, pebbles, &c., termed drift, and all the subsequent deposits up to the present time.

What is said of the mesozoic rocks? of the tertiary strata? of the quaternary strata?

SECTION II.—TABULAR VIEW OF THE FORMATIONS.

The following table exhibits the classes, orders, systems, divisions, and formations of all the stratified rocks, the average thickness of the formations, and the countries where they are most freely developed.

Classes.	Orders.	Systems.	Divisions.	Formations.	Thickness.	Countries.
Quaternary.				Alluvium.	Variable.	
				Newer Pleistocene.		Southern States,
				Older Pleistocene.	200	Verm't., Canada.
				Drift.	Variable.	Northern States, British Am., and North'n Europe.
Tertiary. Cainozoic.				Pleiocene.	2,000	England, Italy, France, Southern States.
				Meiocene.	500	
				Eocene.	1,000	
Secondary. Mesozoic.	Newer Sec'd. " Mesozoic.	Cretaceous System.	Chalk.	Maestricht beds.	1,000	Southwestern States, New Jersey, England, Western Europe.
				Chalk with flints.		
				Chalk without flints.		
			Green Sand.	Upper Green Sand.	480	
				Gault.		
				Lower Green Sand.		
	Middle Secondary. Middle Mesozoic.	Wealden.		Weald Clay.	900	England.
				Hasting's Sand.		
				Purbeck Strata.		
		Oolitic System.	Upper Oolite.	Portland Stone.	500	
				Portland Sand.		
				Kimmeridge Clay.		
			Middle Oolite.	Upper Calcareous Grit.	450	
				Coral Rag.		
				Lower Calcareous Grit.		
				Oxford Clay.		
				Kelloway's Rock.		
			Lower Oolite.	Cornbrash.	450	
				Forest Marble.		
				Great Oolite.		
				Carbonaceous and Stonesfield Slate.		
				Inferior Oolite.		
				Calcareo-silicious Sand.		
	Older Secondary. Older Mesozoic.	Liassic System.		Upper Lias.	700	
				Marlstone.		
				Lower Lias Shale.		
				Lower Lias Limestone.		
		Triassic System.		Variegated Marls or Keuper.	900	Germany, Massachusetts, and Connecticut.
				Muschelkalk.		
				Bunter Sandstone.		

Classes.	Orders.	Systems.	Divisions.	Formations.	Thickness.	Countries.
Palæozoic.	Newer Palæozoic.	Permian System.	Magnesian Limestone.	Zechstein.	500	England. Germany.
				Bituminous and Argillaceous Schist. Lower New Red.	300	
		Carboniferous System.	Seral Series.	New Shales. New Coal Measures. Older Shales. Older Coal Measures.	5,700	New Brunswick, Virginia, Ohio, Michigan, Pennsylvania, England, Illinois, Kentucky.
				Silicious Conglomerate.	1,400	
				Shales and Sandstone.		
			Vespertine Series.	Carboniferous Limestone. Lower Red Shale	3,000	Western States. Virginia, Pennsylvania.
	Middle Palæozoic.	Old Red System. Devonian System.	Upper Division.	Yellow Quartzose Sandstone. Impure Limestone. Gritty Red Sandstone.	10,000	Russia, Scotland, Pennsylvania. Ohio, Michigan.
			Middle Division.	Gray Fossil Sandstone.		
			Lower Division.	Variegated Sandstone. Bituminous Schists. Coarse Sandstones. Great Conglomerate.		
	Older Palæozoic.	Upper Silurian.	Erie Division.	Chemung Group.	1,500	Western States, New York, Great Britain.
				Portage Group.	1,000	
				Genesee Slate.	150	
				Tully Limestone.	16	
				Hamilton Group.	1,000	
				Marcellus Slate.	50	
		New York System.	Helderberg Series.	Corniferous Limestone.	70	
				Onondaga Limestone.	14	
				Schoharie Grit.		
				Cauda Galli Grit.	70	
				Oriskany Sandstone.	700	
				Upper Pentamerus Limestone.		
				Encrinal Limestone.		
				Delthyris Shaly Limestone.	100	
				Pentamerus Limestone.	80	
				Water Lime Group.	100	
				Onondaga Salt Group.	700	
			Ontario Division.	Niagara Group.	164	
				Clinton Group.	80	
				Medina Sandstone.	350	
				Oneida Conglomerate.	500	
		Lower Silurian.	Champlain Division.	Red Sandrock.	400	Vermont, New York.
				Hudson River Group.	600	
				Utica Slate.	100	
				Trenton Limestone.	400	
				Isle la Motte Marble.	25	
				Bird's-eye Limestone.	30	
				Calciferous Sandrock.	300	
				Potsdam Sandstone.	300	New York.

Primary strata.

We proceed now with a hasty sketch of the geological history of the earth. It would seem more natural in a historical science to commence with the most remote period of antiquity, and to follow down the course of time. But the condition of the earth in its earliest ages was so unlike the present, in its geography, its climate, and all its features, that there is little in common between the most ancient and the present period, except that the same material atoms and the same laws of nature remain. He who for the first time becomes acquainted with this extraordinary history, is lost in the strange scenes of those earliest days, unless, proceeding step by step from the present to the past, he shall have become gradually accustomed to the change.

We shall therefore trace up the stream of time into the remote regions of the past. Each system will be the subject of a distinct chapter, commencing with the most recent.

A strictly chronological order would require that all the events of the geological history should be narrated in the precise order of their occurrence; that the characters derived by the strata from agencies acting on them subsequently to their deposition should be described as belonging to the time when those agencies acted. So much, however, remains to be discovered, in order to render geology complete in this respect, that it is impossible to observe strictly a chronological order. We shall therefore describe the phenomena of the strata in the reversed order of their original deposition. The time is probably distant when geologists will narrate events, as well as classify deposits, in a strictly chronological order.

What order of time will be followed in this work? What order would be desirable?

CHAPTER IV.

QUATERNARY SYSTEM.

THE dependence of the formations of this system, each on the preceding, is both so obvious and so intimate, that we shall commence with the first in the order of time, when the features of the earth were quite similar to those of the present day.

SECTION I.—DRIFT.

The name of this deposit, Drift, is derived from the fact that all its materials have evidently been drifted in a common or similar movement. All its phenomena either are consistent with or directly prove this fact. The most important facts may be referred to the following heads:

1. Lithological characters.
2. The geographical limits of the drift.
3. The transport of the materials.
4. The wearing down of the solid rocks.
5. Streams of stones.
6. The absence of fossils from unaltered drift.
7. The crushing of slate hills.

I. *Lithological Characters.*—The deposits of this period consist of sand, gravel, hard-pan, pebbles, and bowlders variously mingled, and sometimes have a very irregular and confused stratification.

The pebbles and bowlders, especially the latter, are distinctly scratched. But a large portion of the materials which were originally accumulated during this period has been remodeled, without much change of place, during a subsequent period. The original *unaltered drift* is recognized by the striæ on the pebbles, and by a more dense structure. It is rarely, if ever, found

---

Why is the drift so called? What are the principal classes of facts relating to it? What are its lithological characters?

in North America on the surface at less than 400 feet above the present sea level.

II. *The Geographical Limits of Drift.*—The deposits of this period cover this continent north of 40° north latitude, and in the valleys of the Delaware, Susquehanna, and Mississippi they extend a few degrees further to the south.

In Europe they occupy the northwestern part, reaching to the ocean on the north; commencing a little east of the White Sea, their boundary extends in a southeasterly direction nearly to the Ural Mountains, in latitude 61°, then southwesterly through the central region of Russia nearly to latitude 51°, and thence westerly in a very irregular zigzag line through the southern parts of Poland and Prussia. The great irregularities of the southern boundary are due, as in the United States, to southerly elongation in valleys.

Drift is also said to occur in high latitudes in the southern hemisphere, as in Patagonia and in the Falkland Islands. It is wanting within the tropics. The few accounts which have been given of its existence in the West Indies are based on the deposits of rivers which formerly flowed at higher levels and were subjected to violent freshets.

III. *Transport of Drift.*—1. In the regions of unaltered drift, the greater accumulations appear to have been on mountains. On the lower parts of the Green Mountains of Vermont, at an elevation of 1000 to 2000 feet, drift so uniformly and abundantly covers the surface, that the solid rocks are wholly concealed throughout extensive districts. The extreme height of drift in North America is about 5000 feet above the sea level. The summit of Mount Washington is the only part of New England which was above its reach. Within its geographical limits and below a certain height, the drift appears to have been distributed without much regard to levels, and is universal. In this respect it differs from all aqueous deposits.

What are the limits of drift on this continent? what in Europe? What is said of drift in southern latitudes? in the West Indies? What is said of the accumulations of drift?

2. The *direction* of the transport is ascertained by tracing the materials of the drift to the source from which they were derived. In almost every place there are local peculiarities of the rocks which the practiced eye of the geologist easily recognizes, so that the fragments which have been removed may be identified with their original source. In North America, the direction of transport was generally to the south or a little east of south, rarely varying on account of local causes from ten degrees west of south to a little south of east.

In Europe the phenomena are quite different. The materials have been dispersed, in nearly straight lines, from the mountains of Scandinavia as a center.

3. The *distance* of transport differs much in different places, and especially in respect of different parts of the materials. The great mass of the materials is derived from ledges which occur within a few miles. It is rarely the fact that the greater portion of the gravel or pebbles, or even of the bowlders, have traveled more than five or ten miles. If in some places the bowlders which have come from a greater distance are more numerous than those of the neighboring rocks, it is because the latter are more perishable. Such cases are therefore no real exception to the general statement.

The distance to which isolated *bowlders* have been transported is sometimes very great.

In Vermont, heavy iron ore has been carried 30 or 40 miles. On Long Island, Nantucket, and Martha's Vineyard, bowlders are found which were derived from the continent. It is thought that some bowlders in New England have been removed 100 miles. In Ohio and other Western States, bowlders are found which were derived from the rocks north of the Great Lakes, and which must have been transported several hundred miles.

4. In general, there is a very close *correspondence* between the *durability* of the rocks and the *abundance* of bowlders and pebbles derived from them and contained in the drift. Durability

---

What is said of the manner of ascertaining the direction? of the direction in North America? in Europe? of the distance? of the durability and abundance of bowlders?

depends on the power of resisting either chemical or mechanical agencies. Very soft slate rocks are unaffected by the solvent powers of rain, but are often impressible by the finger nail. Yet such rocks are found to retain now the finest striæ, which were imprinted on them by the mechanical agency of the drift period. But their fragments enter very sparingly into the composition of the drift. Limestone is not only soft, but decays also by solution. It is therefore deficient in the drift even of limestone regions.

On the other hand, a narrow strip of quartz rock, on the western flank of the Green Mountains, has filled the drift with pebbles and bowlders, whose ellipsoidal form has procured the provincial appellation of "hard-heads."

It is scarcely necessary to remark that the detachment of the fragments and the rounding of their angles are not to be exclusively ascribed to the drift agency. The ordinary agencies of rivers and waves, and of the atmosphere, are ever producing these results. If, as there is good reason to believe, extensive regions, which are covered with drift, were above the ocean during many long periods anterior to the drift, these agencies were probably producing bowlders, pebbles, and other loose materials. Then came the drift agency, adding more or less to these materials, but operating eminently as an agent of dispersion. Since the lapse of time would alike accumulate the durable and consume the more perishable fragments, we find here one cause of the deficiency of the latter, in comparison with the extent of the parent rocks. The long-continued violence of the drift agency itself was, however, an additional cause of this deficiency.

5. Hills and mountains, not exceeding 1000 or 2000 feet in height, do not appear to have presented any insurmountable obstacles to the passage of the materials. Gravel and bowlders are often more abundantly accumulated on the north sides of mountains. But examples are very common of large bowlders having been carried up steep acclivities to great heights.

How were many of the bowlders originally rounded? What are the two causes of greater abundance of more durable bowlders? What is said of the obstacles presented to the drift by mountains?

On Hoosac Mountain, in the town of Adams, Massachusetts, is a bowlder weighing six hundred or seven hundred tons, which has been transported across a valley, and one thousand feet up the mountain. On the summits of Mount Holyoke range are large fragments of sandstone, which have been carried up an almost perpendicular precipice several hundred feet from the valley beneath.

6. The *size* of some of the transported *bowlders* is enormous. One in Fall River, Massachusetts, weighs 5400 tons. The pedestal for the statue of Peter the Great was hewn from a bowlder which weighed 1500 tons.

IV. *The Wearing down of Solid Rocks.*—1. The transport of the drift over the solid rocks acted more or less to wear them down, producing well-rounded and smoothed surfaces, on which are *striæ* and *furrows.* There being no exception to the general fact of coincidence between the direction of the striæ and that of the movement of the drift materials, the one may be taken as an index of the other. But from the great facility and precision with which the direction of these striæ may be measured, they afford the best means of ascertaining the direction of the drift movement.

In some cases, on the tops of mountains of the tougher rocks, as on Mount Holyoke, in Massachusetts, furrows are found which are a foot wide and two inches deep. In Dorchester, Massachusetts, the sides of an angle between two portions of a ledge of hard quartzose conglomerate, which were nearly in the direction of the drift, were rendered concave and smooth, as if the moving bodies were forced through with great friction.

More frequently the furrows are but an inch to three inches in width and about a third as deep. Still more frequently scratches occur which do not exceed one eighth or one fourth of an inch in width, and others are as fine as the stroke of a pen.

Sometimes the entire surface of the rock is smoothed down, rounded, and polished, as in the case of the Dorchester rocks above mentioned, where the surface of the hard quartz pebbles

What is said of the size of bowlders? of the direction of the striæ and furrows? of the size of the furrows and striæ? of polished surfaces?

is a little in relief, like the silicious specks in a piece of marble artificially polished.

Perhaps no part of North America has furnished, within an equal space, so many beautiful examples of polished and striated rocks as have been found in Vermont. In that state more than three hundred examples (reckoning as distinct cases those only which are at least one quarter of a mile distant) have been found.

Among the districts most remarkable for their abundance is the valley of the Winooski. There the perfectly rounded surfaces of the talcose slate, the broad furrows, and delicate scratches were protected by a fluviatile deposit of fine sand, during the long subsequent period of the older pleistocene, and therefore have a freshness which adds much to the beauty and interest of these results of the drift agency. In the northwest part of this state few ledges of rocks can be found which are not rounded and furrowed. In this region, many furrows are from 12 to 20 inches wide, and 3 or 4 inches deep. The recent removal of clay often exposes a striated surface of marble, which retains a polish nearly equal to that of artificially polished marble.

The limestone of Addison county abounds with well-smoothed surfaces covered with scratches, which have been protected from decay by a covering of brown clay, and have within a few years been exposed in the gulleys at the road side. *Fig.* 32 represents the drift striæ in black marble, near Lake Champlain, in Shoreham, Vt.

Fig. 32.

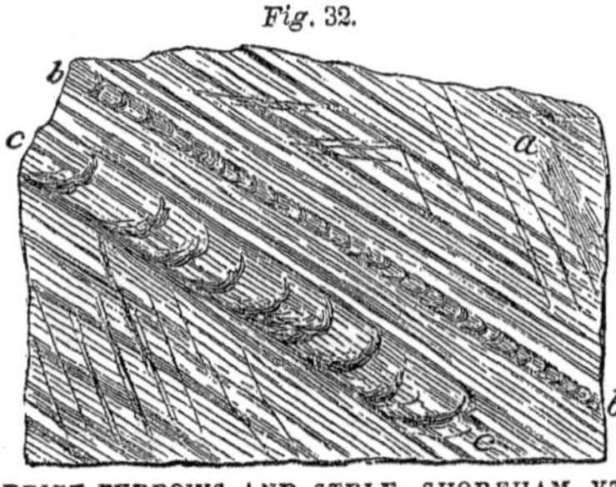

DRIFT FURROWS AND STRIÆ, SHOREHAM, VT.

a is the abrupt commencement of several coarse striæ.

b b is a furrow which appears to have been the effect of unequal pressure, perhaps the rolling of a loose pebble.

c c is a broad furrow in the bottom of which pieces were broken out, probably by the rolling of a very angular stone.

Furrows and striæ are wanting on surfaces of limestone which

What is said of examples in Vermont? Describe the specimen represented n *Fig.* 32. What is said of limestone surfaces?

have been long exposed to atmospheric agencies. The surfaces of rocks which are more or less feldspathic, as the granite and gneiss regions, and which decay by yielding up their potassa, have rarely retained the furrows. The same is true of surfaces which crumble by the admission of water into a porous or loosely laminated structure. The *parallelism* of the striæ is one of their most obvious characters. On very convex, and even on hemispherical surfaces, there is scarcely any perceptible deviation from parallelism. Glacial furrows and striæ differ from those which are made by fluviatile agency. The former are straight, or sometimes slightly and regularly curved; the latter are very irregular and tortuous. Both kinds of furrows are represented in the accompanying figure.

Fig. 33.

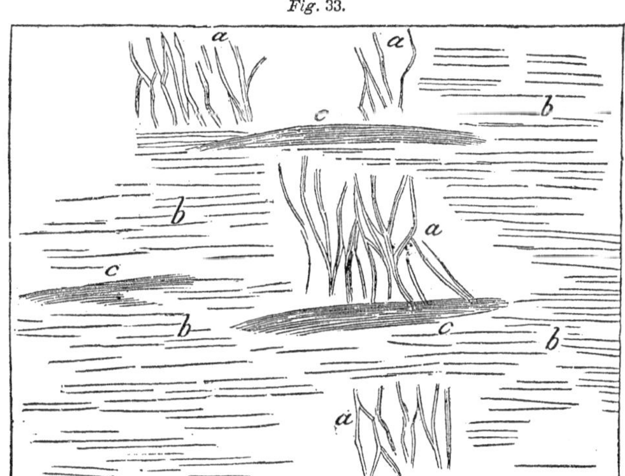

CASCADE OF HÖNNE VOSS, NORWAY.

a a. Furrows made by the river. *b b.* Glacial striæ. *c c.* Glacial furrows.

Some furrows and impressed lines on the surfaces of rocks resemble drift striæ and furrows, but are due to structure. They

What is said of the parallelism of striæ? of the difference between fluviatile and glacial furrows? What is said of furrows owing to structure?

are produced by the unequal weathering of the strata, and may be very easily distinguished from those which have resulted from an external mechanical force. Furrows are also produced by the solvent power of rain on limestone slopes, but they are too irregular to resemble drift furrows.

2. Furrows and striæ are wanting only where the rock is undergoing disintegration. It would not, therefore, be an unreasonable inference, that the entire surface of the rocky floor of the northern parts of this continent was smoothed into cavities and convexities, and covered with furrows and scratches by the drift agency.

3. The greatest elevation at which drift furrows exist is an interesting question which has received some attention. They exist abundantly on Jay Peak, at the northern extremity of the Green Mountain range. The height of this summit is 4025 feet above the ocean. From the size of the furrows, we can hardly avoid the conclusion that such agency was nearly or quite as energetic as at lower levels.

On the White Mountains of New Hampshire are drift furrows at an elevation of 5000 feet above the sea. This is the greatest height at which they have been found in North America. Above 6000 feet, the summit of Mount Washington entirely escaped the effects of the drift agency.

On the other hand, the lowest level reached by the striæ is unknown. They are seen descending beneath the sea, where it is impossible to trace them any further.

4. The minor inequalities of surface, such as hills and isolated rocks, did not perceptibly influence the direction of the current. But the general direction was essentially modified by the prominent features and outlines of the surface. In Vermont, the Green Mountain range has a direction nearly north and south, and is

---

What is said of furrows caused by rain? of the universality of the polishing and striating agency? of the height at which furrows exist on the Green Mountains? on the White Mountains? of their lowest level? What is said of the influence of the minor inequalities of surface? of mountain ranges and valleys?

crossed by the valley of the Winooski. On either side of the valley the mountains rise to a height of 3000 and 4000 feet. Here the current was deflected so as to ascend the valley in nearly an easterly direction. The same is true of the valley of the Lamoille, a few miles further north. But on the tops of the adjacent mountains the current held its usual north and south direction. But the proximity of longitudinal valleys appears to have deflected the current at greater elevations. Thus, in Putney, Vermont, striæ on the west side of a mountain have a direction to the southwest.

5. The north and west sides of the ledges of rocks, having been most exposed to the drift agency, are often very perfectly rounded, forming the *roches moutounees* of European writers; but the opposite sides have generally escaped the influence of this agency. To the former or struck side has been applied the epithet *stoss* (Swedish for struck), and the latter is called the *lee* side.

An interesting example is found on the south side of the Winooski, in Bolton, Vermont. In this example there is a peculiarity, that the rock is shelving on the stoss side, as below *a* in *Fig.* 34. *a e* is the stoss surface covered with furrows and striæ, and *c e* is the lee side. *a m* and *c c* are entirely destitute of any marks of drift agency. It is very obvious that the surface *a m* proves that the drift agency was that of a solid. Any bodies moving in a liquid, like bowlders and pebbles in water, would have exerted their greatest force against this shelving surface.

Fig. 34.

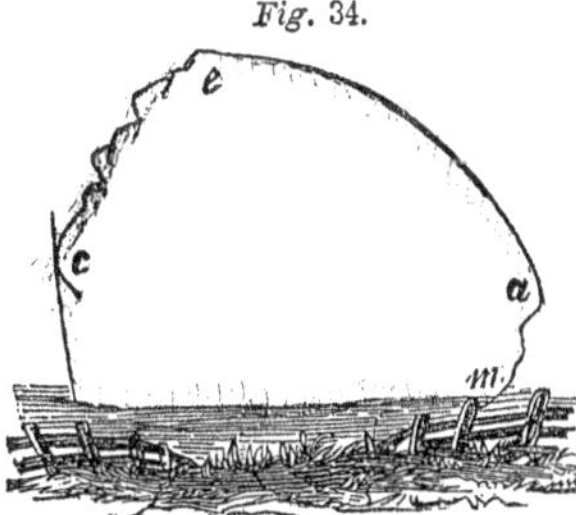

The same phenomena appear on a larger scale in the stoss and lee sides of mountains. The effects, however, were modified by the original form and structure of the mountains. Thus the lee sides of many of the Green Mountains are on the east, although the direction of the drift agency was but little east of south. Those mountains which are most isolated on the north and west sides were most modified in their outlines. Thus Cam-

What is said of the north and west side of rocks? Describe the example represented in *Fig.* 34. What is said of the stoss and lee sides of mountains?

el's Hump, in Vermont, owes its present form almost entirely to the drift agency, the steep and irregular lee side being on the south, and the well-rounded outline on the north, for this mountain, rising up suddenly on the south side of the Winooski, unprotected by any considerable elevation to the north or west, received the full force of the drift agency.

6. In numerous examples of striated rocks, two or more distinct sets of striæ may be recognized with different directions, but those of each set are parallel. At Hill's quarry, on the Isle la Motte, in Lake Champlain, there are eight sets;* but it is not usual to find more than two or three sets at one locality. It is probable, however, that the earlier striæ of other directions were entirely obliterated by the long continuance of the drift agency.

It is often possible to determine the relative age of the striæ. But much discrimination is requisite, for one which is shallow may appear to have been cut off by another which is narrow and deep, and yet the latter may or may not be the older; or, if the striæ are of nearly equal depth, it may not be possible to arrive at any conclusion. But when a small one is continued across the bottom of a somewhat deeper and much broader stria, and when several such examples occur on the same surface, there can be no doubt of their relative age. We have succeeded in determining the relative age of two sets of striæ, in about thirty examples in Vermont, and find no exceptions to the general conclusion that those which had the more westerly origin are the more recent.†

7. In the northwestern countries of Europe the drift radiates from the Scandinavian Mountains. In other respects, as to the character of the striæ and furrows, the rounded and polished

* These directions are, south 8° west, south 3° west, south 10° east, south 25° east, south 43° east, south 45° east, south 47° east, and south 65° east. In Vermont, those striæ which have a direction of south 10° east are most abundant; next are those which run south 20° to 30° east.

† In one case, that of Hill's quarry above mentioned, the age of three sets was determined as follows, beginning with the oldest, south 10° east, south 8° west, and south 47° east.

What is said of different sets of striæ? of the relative age of such striæ? of the striæ and polished rocks in Scandinavia?

rocks, and the stoss and lee sides of rocks and mountains, the drift phenomena in Norway and Sweden are similar to those in North America. The greatest height at which striæ occur in Scandinavia is nearly 6000 feet. They descend beneath the sea to unknown depths. In other parts of the northwestern half of Europe they are less common.

8. That the striæ and furrows belong to the same period with the drift, being indeed the effect of the transport of the latter, has been regarded as sufficiently proved by the coincidence of direction, and by the want of any deposits intervening between the striated rocks and overlying drift. It necessarily follows that they are as ancient as the drift accumulations. It can also be shown that they immediately preceded the deposition of the pleistocene, which is more recent than the drift; for, wherever a surface has been recently uncovered, the most delicate striæ are found to have been preserved by a few inches of clay. It follows, then, that the furrows and striæ could not have been long exposed to atmospheric agencies which would have obliterated them. It would not, of course, follow that the striating agency was limited to this point of time, which indicates its close only. It had undoubtedly been continued through a long series of ages.

V. *Streams of Stones.*—1. The type of this class of phenomena was discovered in Richmond, Massachusetts, by President Hitchcock, and has since elicited much attention. These streams are arranged in long lines, and consist of an enormous quantity of angular fragments of rock, overlying the common drift, and extending from a ledge of the same kind of rock in the direction of the drift current.

From their position it is inferred that they were formed at or near the close of the drift period. It is not, however, to be supposed that this effect was peculiar to the end of the period, for whatever similar streams may have been strewed along previously would subsequently have been either rearranged or buried

---

What is said of their greatest height? What proves the striæ to be as ancient as the drift? What proof that they immediately preceded the pleistocene? What are streams of stones? What is said of their age?

beneath other drift. A considerable portion of erratic blocks may have been distributed in the same manner; but well-characterized examples are rare, and difficult of recognition.

2. A very interesting example occurs in Huntington, Vermont.

Fig. 35.

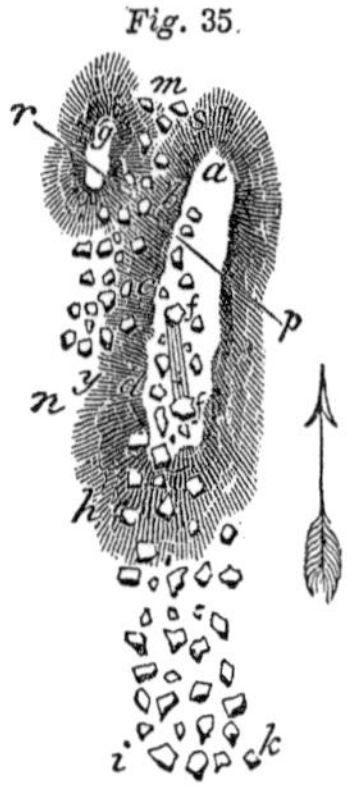

Fig. 35 represents the prominent local features.

A hill, *a h*, rises 150 feet above the valley, *m n*. It consists of fine talcose slate, and is about one fourth of a mile long, with a width of sixty to eighty rods.

On the west is a small hill of the same kind of talcose slate, *g*. The west side, *b c d h*, of the principal hill is very precipitous. A long narrow ledge of very coarse binary syenite, *s y*, lies in the talcose slate, on the south side.

Furrows and scratches on the slate, *f f*, have a direction south 15° east.

The stream of angular stones is prolonged nearly a mile beyond the southern termination of the hill toward *i k*, in the direction of the drift current. The stones were evidently derived from the ledge, *s y*, for they consist wholly of the same peculiar variety of syenite.

The most characteristic example of streams of stones occurs in the western part of Massachusetts, and is represented in *Fig.* 36. There are two distinct streams, which have their sources, A C, between Canaan and Lebanon, in New York, at the summit of the Taconic range of mountains.

The train A B has been traced 20 miles, and the other, C, about 10 miles to the southeast. This is the direction of the drift striæ in the vicinity. They are 300 to 400 feet wide, are nearly parallel, and are from one half to one third of a mile apart. It will be seen in the figure that they pursue a southeast course obliquely across mountains and valleys, with little or no regard to these inequalities of surface. The two ranges of mountains which they cross are each about 100 feet higher than the source of the stones, and are about 800 feet higher than the first valley which they cross.

The stones in these trains consist of a peculiar metamorphic

Why are distinct examples rare? Describe the example in Vermont, and the one in Massachusetts.

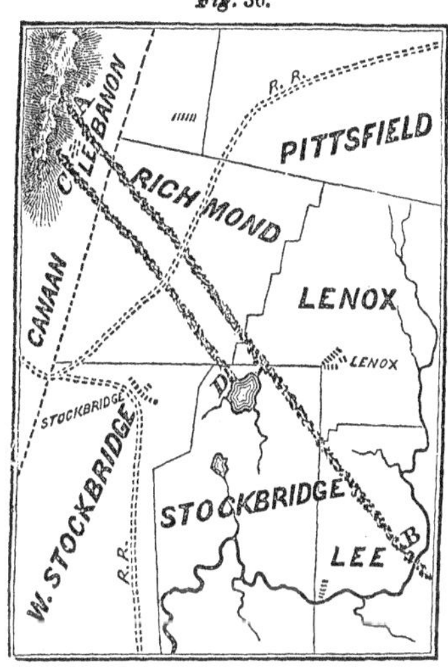

Fig. 36.

feldspathic rock, of a greenish color, which is very easily distinguished from the slates and limestones over which they lie. It is found in ledges only along the crest of the Taconic ridge. The very spot whence the stones were derived is conspicuous. The fragments are very irregular, with their angles but slightly rounded. Most of them are very large, the smaller ones being several feet in diameter. One of the largest is four miles from the source, and weighs 2000 tons.

VI. *Age of the Drift.*—1. The absence in this country of any deposits of the periods immediately anterior to the drift agency deprives us of the means of comparison with such deposits. But the subsequent deposits of the older pleistocene enable us to place it at the commencement of the grand series of events which they commemorate. The older pleistocene is, indeed, so intimately connected with the drift that many writers have failed to distinguish the two periods.

2. We have before remarked that the dispersion of the drift of Northern Europe consisted in a series of radiating movements from the Scandinavian center. The parallel movements of the North American drift must, therefore, have had a different origin. The cotemporaneousness of the two agencies is not, therefore, to be inferred from their similarity alone; proof of their synchronism is found in the fact that each immediately preceded the pleistocene period. It has, indeed, been stated by some writers that the pleistocene deposits of Northern Europe underlie the transported blocks of the drift. But all the unaltered drift of those countries is found by those geologists who have made it a special study to be older than the pleistocene.

What is said of the age of the drift in this country?

Probably the drift of the northern parts of Europe and of North America was cotemporaneous with some deposits which have been referred to the later tertiary of Southern Europe and of our Southern States.

VII. *Theories of Drift.*—The great problem—what was the agency which dispersed the drift, and wore down, smoothed, and furrowed the rocky floor over which the materials moved—has elicited many theories. We shall here notice the outlines of these theories, without attempting to describe all their modifications.

1. The *iceberg theory* supposes that the drift country was submerged below the tops of the mountains not long before the drift agency, and that a polar current floated down icebergs which were loaded with the materials of the drift, and which, melting during their progress into a warmer latitude, strewed the drift along their course, and striated the rocks at the bottom of the sea by the fragments which were frozen into them.

This theory has the great advantage of introducing no more violent agencies than are now in operation. Such a polar current now exists, bearing icebergs, some of which are loaded with gravel, into warmer regions. If this country were submerged, the polar current, which now has a strong westerly tendency, would flow over a large part of the drift region. The hills of gravel are such as would be produced by the stranding of icebergs, either dropping their freight of earth and stones, or crowding up the materials on the bed of the sea, and as they were rocked and urged on by the waves.

It will be seen that this theory involves both submergence and iceberg agency.

It is, however, objected to the theory of submergence that the drift appears to be entirely destitute of fossils. It may also be objected that icebergs could not have taken up masses of rock from submarine valleys, and then carried them over the mountains; that on the tops of the mountains, rather than in the val-

---

What is said of the cotemporaneousness of the American and European drift? of that of the drift and later tertiary? What is the iceberg theory? What is said in favor of this theory? What is said against it?

leys, the scratches should more frequently occur, but this is not the fact; that the source of the materials should be found only in higher northern latitudes, whereas, in fact, with the exception of some scattered bowlders, the great mass of the materials has been removed only a few miles; and that the rocky bed of the ocean, especially in its valleys, would have been more or less protected by a covering of mud from the furrowing agency of stones frozen into the icebergs.

2. The *theory of elevations* supposes that the drift countries were submerged, and that their central regions were subject to violent earthquakes and elevations, oft repeated through a succession of ages; that these convulsions propelled over the northern portions of the globe enormous waves, which bore along the immense icebergs of the polar regions, and strewed the pre-existing loose materials of the surface far to the south of their former position; that immense masses of such materials received a portion of the impulse, and acted on the rocks beneath in the same manner as glaciers.

If a region like the eastern part of Iceland, with 3000 square miles of ice mountains, were exposed to such earthquake action, immense numbers of icebergs would be borne along by the waves.

In support of this theory, it is argued that in earthquake waves there is an actual locomotion of the water, and consequently their momentum must have been inconceivably greater than that of ordinary waves, since it has been proved that a current of twenty miles per hour would transport stones weighing 300 tons.

It is also said that we may thus account for the remarkable similarity between the effects of the drift agency and those of the Alpine glaciers of the present period.

To this theory it is objected that repeated elevations could not go on long enough to produce the drift phenomena before the great central earthquake region would have been elevated above the water.

What is the theory of elevations? What is said in favor of this theory? What is the first objection?

Another objection to this theory is derived from the fact that an immense downward pressure and almost perfect inflexibility of the striating agents are demanded by the phenomena. While a mass of mud, sand, pebbles, and bowlders is moving in a liquid, the upward pressure of the liquid must be nearly or quite equal to the downward pressure of the mass, which therefore would exert a force almost wholly in a horizontal direction. Nor would the pebbles and bowlders be held in a fixed position with that inflexibility which is essential to the production of drift striæ. Such phenomena as we have described on p. 131 seem to indicate a solid agent.

3. *The glacier theory* supposes that by some causes, which it does not profess to demonstrate, a refrigeration of the climate covered the drift region with glaciers, and at length with a vast glacial sheet several thousand feet thick; that in Europe the center of origin was in the Scandinavian Mountains, whence the glaciers proceeded outward in all directions, increasing until they reached the limits of the drift agency; that in North America the glaciers originated in or near the Arctic regions, proceeding in a southerly direction, because in this direction only were they free to move, and increasing until they formed a glacial sheet 5000 feet thick; that vicissitudes of climate during the long periods of drift agency caused retreats and advances of the glacial sheet in directions not exactly coincident.

It has been objected that this supposed glacial agency of the drift period differed from that of the Alps, inasmuch as the latter is limited to the inclined valleys of lofty mountains, and that the theory does not account for the origin of such a glacial sheet. If the latter objection could be removed, the former would be irrelevant, since a glacial sheet of such vast extent would be free to move only outward, as the Alpine glaciers are now free to move only downward.

It has been attempted to account for the origin of such a gla-

---

What other objections are mentioned? Describe the glacier theory. What objection is mentioned? How could the objection be removed? What is said of astronomical causes of the glacial sheet?

cial sheet by astronomical changes, which, although they can not be disproved, are not rendered probable by any known facts of the same kind. Possibly the origin may be found in geographical changes, as a great elevation of the land. Such elevation is rendered probable by the descent of the striæ beneath the ocean, and by the transport of pebbles across regions now covered with unfathomable waters, as Massachusetts Bay, for glaciers never descend into the sea. If now these entire regions were elevated above the general level of land and sea as much as some mountain ranges are, they would be covered with a glacial sheet. Thus the glacier theory furnishes what is at least a possible cause of the origin of the ice.

Some irrelevant arguments have been adduced in reference to the glacier theory. It has been inferred, from the discovery of an elephant and a rhinoceros in the ice of Siberia, that there must have been a great and sudden change from a temperate to a frigid climate. The inference may be correct, but evidence is yet wanting that the event occurred during any part of the drift period.

The argument for the glacial theory is founded on the exact resemblance between the effects of drift agency and those of the Alpine glaciers. It is impossible to distinguish these effects as exhibited in the rounded furrows and striated surfaces of the rocks. In the Alpine regions above the glacial agency, both the ledges and fragments of rocks are angular, in marked contrast to the rounded rocks below. (See *Fig.* 10.) So in this country, the summit of Mount Washington has only angular rocks. Examples similar to that before referred to on page 128 (*Fig.* 32) are found in the Alps: the rolling of angular stones under the glaciers produces similar effects.

Rocks and hills which have been in the path of Alpine glaciers have their stoss and lee sides. The loose materials are

---

What is said of geographical causes? What irrelevant arguments have been adduced? What is the argument in favor of this theory derived from the surfaces of the rocks? from the forms of rocks and hills? from the distribution of the loose materials?

crowded forward, while there is a tendency in the glacier to go over them, so that they are not moved as far as the blocks which are imbedded in the ice or which lie on the top of it. (See p. 31, *Fig.* 7.)

A glacial sheet, at no time exceeding 5000 feet thick in North America, would account for all the phenomena. It would have almost inflexibly ascended the minor inequalities of surface, but have been deflected from its course by lofty mountain ranges and deep valleys. A small portion of a glacier has no perceptible flexibility, while that of the whole mass is so great that they have even been compared to viscid bodies.

VIII. *Fracture of Slate Hills.*—These remarkable phenomena are generally supposed to be the effects of drift agency; but neither the time nor manner of their origin have yet been conclusively established.

Most of the examples yet known in this country occur in the southeast part of Vermont. The layers of slate stand nearly perpendicular, but on the hills they are broken and inclined from the lines of fracture. In some of the fractured masses the laminæ retain their former relative position, without any intervening spaces. Other portions, however, have been broken into loose fragments. The fractures are not confined to the tops of the hills, but in at least one instance occur at the base of a small hill.

Fig. 37.

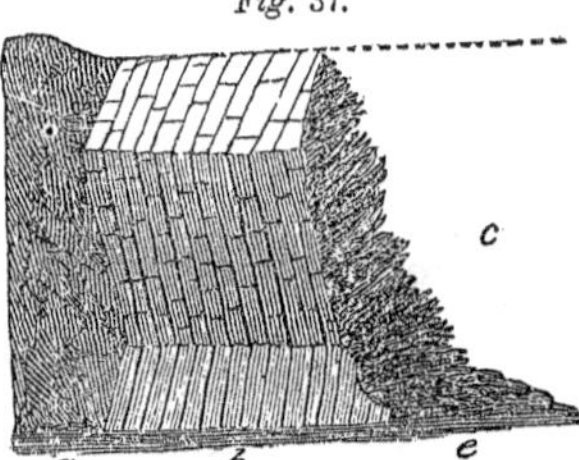

One of the most instructive examples was found in Willard's quarry, Guilford, Vermont, where the top of a hill is quarried for roofing slate. The most conspicuous part of the dislocation has been exposed by quarrying, and is represented in the accompanying figure, of a vertical surface facing south, about ten feet high.

a is the west part of the hill, covered with drift; *b*, slate, the lowest part not fractured, the rest fractured; *e*, loose fragments of slate thrown over to the east by frost; *c*, an opening made by quarrying.

What must have been the character of the glacial sheet? What is said of the origin of the fracture of slate hills? Where do most of the examples occur? Describe the one in Guilford.

Fig 38.

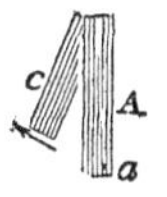

The strata, not being free to yield to either side, were crushed into a zigzag position, and are quite loose, but parallel. Those which have been left, by quarrying, unsupported on the east of this mass, have been thrown over by frost. The zigzag condition of the fractured yet parallel layers, which were not free to move on either side, incontestably points to some force acting almost directly *downward.*

The accompanying figure exhibits the relations of the fractured to the unfractured strata in respect of direction, in a distance of seven or eight miles in the southeast part of Vermont.

a a a a, direction of the strata in each locality, A, B, C, D; *c*, direction of the crushed strata. The arrows indicate the direction in which the force was exerted.

SECTION II.—PLEISTOCENE.

The name of this formation is derived from two Greek words, and expresses the fact that most of the fossils belong to living species.

I. *Lithological Characters.*—The deposits which are referred to this period comprise all which have resulted from aqueous action between the glacial and the historical periods. Many geologists include them in the drift; and no small part of the differences of opinion on theories of drift has been caused by the confounding of the glacial and aqueous deposits of two successive but distinct periods.

The pleistocene deposits in this country are altered drift, blue clay, brown clay, fine sand, and beds of marl.

1. *Origin of the Materials.*—The general character of these deposits will be better understood with a knowledge of their origin, which, so far as they occur in drift regions, is as follows:

The general configuration of the surface of the country having been the same for a long time before the drift period as at present, the streams must have run through the same valleys, and had

What is said of the name *pleistocene?* What is included in this period? What are the deposits? How did the drift agency prepare the materials?

reduced their channels to a level not very different from the present. The drift agency then drifted the enormous amount of loose materials which had been accumulating during former periods, spread them over the surface, and more or less filled the valleys and blocked up their outlets. At the end of the drift period, and during the earlier parts of the pleistocene period, a large part of North America was depressed more than 1500 feet, and in emerging remained for a long period at 400 or 500 feet below its present level. All those parts, therefore, which have now an elevation less than this amount, were beneath the waters of the ocean.

At the close of the drift period, the surface of the drift must have contained a much larger proportion than at present of fine materials. These were washed down into the valleys, from which the streams removed considerable portions to the ocean. The particles of clay being much finer than the sands, the first deposits were mostly of the blue clay. The long continuance of atmospheric agency converted the blue into brown clay in the drift, and the latter deposits are accordingly of brown clay and brown sand. These regular deposits of clay and sand are most abundant in the river valleys, whose outlets were blocked up; and in ponds, and bays, and in sounds like that which formerly extended from the Gulf of St. Lawrence through the valley of Lake Champlain to New York.

While the characters of the drift, which was above the level of the pleistocene seas, were modified by rains and streams, and some new deposits in the adjacent waters were in the process of formation, the action of the waves, tides, and currents essentially modified the submerged drift. Hence resulted what is now called *altered drift.*

In order to understand the action of marine agencies on the drift, it must be remembered that the process of emergence of the land has been very gradual. Each part has been successive-

---

What occurred at the end of the drift period? What was the effect of streams on the drift? of atmospheric agency? of the pleistocene seas? How was the action of marine agencies prolonged and extended?

ly brought nearer the surface of the water, and subjected to marine agencies during long periods of time. The existence, in the valley of Lake Champlain and elsewhere, of extensive deposits of clay and fine sand, which have now a well-defined limit of 400 feet above the sea level, indicates a long stationary period, which interrupted the process of emergence.

By marine agencies the submerged drift was rearranged. The outlines of its surface were remodeled, and small hills and valleys were formed. Most of the pebbles and small bowlders were rolled about until they lost their drift striæ. But the larger bowlders more frequently retain their striæ, and some of extraordinary size probably remain in their original position. Small hills of altered drift and œsars (long, narrow ridges of loose sand and gravel) were formed along the shores by the action of waves and tides.

The knolls, or small hills of altered drift, are common near mountain ranges. In Vermont, in the valley between the Taconic and Green Mountains, they are accumulated in groups, at intervals of five or six miles' distance. Some of them are 150 feet high. They are often very steep, although composed of extremely loose materials. In Hinesburg, in the valley of Lake Champlain, one of these hills is 3000 feet long, 2000 feet wide, and 300 feet high.

Fig. 39.

HILLS OF PLEISTOCENE, AMHERST, MASS.

What were the particular effects? What is said of hills of altered drift?

In Amherst, Massachusetts, may be seen a group of hills of altered drift. *Fig.* 39 represents hills of altered drift in the east part of Amherst Some of the cavities between the hills are occupied by ponds without outlets. They are evidently the effects of marine agency. In the southeast of Massachusetts, similar examples occur at a distance from the mountains. Also in Berkshire county, on the east side of Monument Mountain.

In Andover, Massachusetts, there are œsars of unusual length, *Fig.* 40. These ridges are situated west of the Shawsheen River,

Fig. 40.

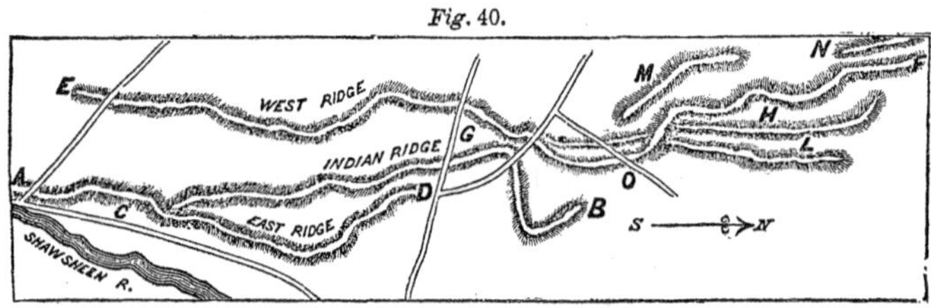

ŒSARS, ANDOVER, MASS.

and about 60 feet above it. They vary from 15 to 30 feet in height, and are nearly semi-cylindrical in form. The length of east ridge is one mile and one third; of west ridge, one mile and three quarters. They are composed of sand, smoothed pebbles, and bowlders. Similar ridges, bowl-shaped cavities, and rounded hills are found on each side of the Shawsheen for several miles.

In Sweden œsars are common, and are of much greater length than in this country.

Fig. 41.

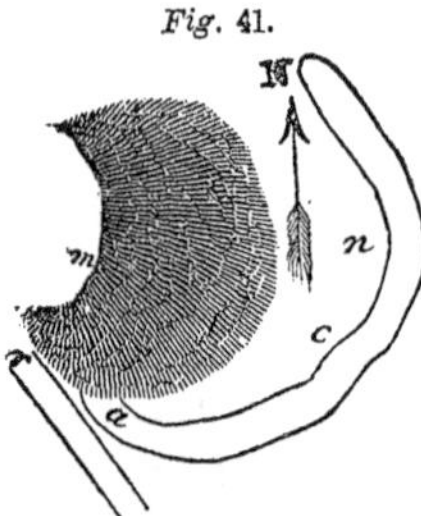

The following example occurs in the town of Peru, Vermont: In *Fig.* 41, *m* is on the east side of a north and south range of hills; the hill at *m* is about 150 feet above its base, *n*, but rises higher to the west; *a* is the southwest extremity of the moraine, which here joins the hill without any intervening depression; at N it terminates abruptly, having a semicircular form, *a c* N. The top varies but little from a level; the total length is 1000 feet; the width of the base is about

What is said of œsars in Andover? in Sweden? Describe the example in *Fig* 41.

six rods, and of the top from three to six yards. It consists of loose yellowish brown gravel with some small bowlders; *r* is the turnpike.

2. *Interstratification of Materials.*—The sands, altered drift, and clays of this formation are variously interstratified, and thus indicate local changes of condition, chiefly in respect of a more or less violent action of the waters.

The following section, *Fig.* 42, through Fort Greene, at Brooklyn, L. I., shows the interstratification of fine sand and altered drift. Fine blue clay often takes the place of the sand.

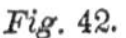

Fig. 42.

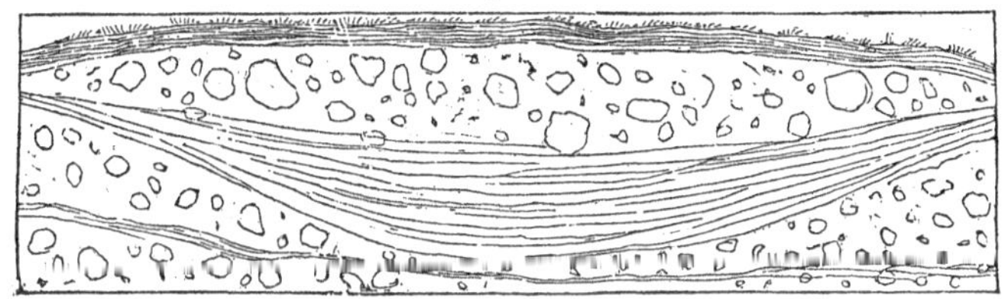

The occurrence of numerous bowlders of large size, in irregularly stratified masses of gravel and sand, is thought by some geologists to indicate the agency of icebergs.

3. *River Terraces.*—The materials which had accumulated during the earlier part of the pleistocene period were more or less rearranged during the latter part. As the land emerged from the ocean, the level deposits in the valleys were exposed. Through them the rivers then excavated channels, and, by the gradual lateral shiftings of the channels, a large part of these older pleistocene deposits were removed to the lower levels and into the sea, while the margins which remain constitute the existing terraces of river valleys.

In general, the origin of river terraces may be described as commencing with the deposition of nearly or quite horizontal plains of sand and altered drift. Flowing through these level plains, the rivers must have formed serpentine channels. Consequently, by increasing the convexity of the bends, and then cut-

What is said of the interstratification of materials? of the origin of river terraces? of the action of rivers on the pleistocene deposits?

ting them off or wearing away their headlands, and shifting their beds, they removed the greater portion of the original plain. It is not necessary to suppose that the distance between opposite terraces is any indication of greater magnitude of the river at any former time, but only of its shifting its channel.

The process having advanced thus far, we have an intervale through which the river flows; and if the channel has entirely cut through the drift, the process is either completed or so much protracted by the difficulty of wearing down the solid rocks, that the progress, during even a geological period, would be scarcely perceptible, and only one terrace will have been formed. But if a terrace has been formed before the complete removal of the obstructions in the channel, the same process must have been repeated within the new and narrower level of intervale. We should thus have a second terrace. Repetitions of the process, in cases where the obstructions were not entirely removed, would occasion a great number of terraces. On some streams four successive terraces may be seen. In some instances, these repetitions have occurred in a valley in which several streams unite, and the changes of their channels have left not only many terraces with irregular margins, but detached portions, like islands, in a horizontal surface in the middle of the valleys.

In the mountainous parts of the New England States, river terraces occur on all the streams. They are characterized by a nearly horizontal surface jutting out from a hill side, or extending from the base of another terrace, with an irregular line of margin, but a uniform slope. Irregularities are common, and are mostly due to existing agencies.

In Scotland, also, there are terraces whose materials accumulated in the pleistocene ocean, and which have been formed by subsequent river agency. The following example (*Fig.* 43) occurs at Dunkeld.

a, present sea level; *b b*, hills forming the valley; *c c*, their ideal continuation; *d d*, the terraces; *e f f*, the course of the Bran; *g*, the Tay; *h*, Dunkeld.

How may one terrace have been formed? how several? What is said of the frequency of terraces? of terraces in Scotland?

Fig. 43.

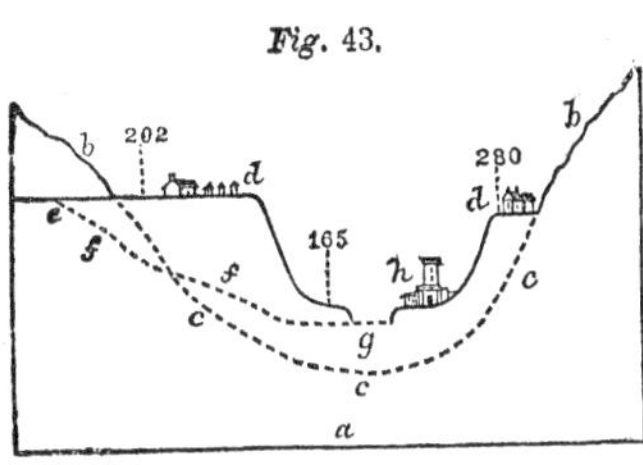

4. *Marine Terraces.* — The prairie terraces of Missouri territory, as the Coteau des Prairies, have no communication with rivers, and are probably the lines of ancient sea or lake coasts. Like the present sea and lake coasts, they have the ground at their basis strewn with blocks of stone. In Chili, in Barbadoes, and in Jamaica, there are terraces and bluffs near and often parallel to the present sea-coast, which were once the lines of coast, but have since been elevated by earthquakes.

In Scotland and in Sweden, marine terraces are common on hill sides, where they are supposed to indicate ancient sea levels In the following example in Glen Roy, Scotland, the three lines 1, 2, 3 (in *Fig.* 44), represent terraces which form a level pathway along both sides of the valleys, with little interruption for five or six miles.

Fig. 44.

GLEN ROY, ACHNAVADDY, SCOTLAND.

II. *Topography of the Pleistocene.*—This formation occurs in the drift-regions of North America and Europe. In South America, the vast plains of the Pampas consists of a pleistocene clay,

What is said of marine terraces? Describe the example in Glen Roy What is said of the topography of the pleistocene?

which has been explored through an extent of 200,000 square miles, but which is probably two or three times larger. A part of the lowlands of the Southern States of this country is pleistocene. Probably the vast level tundras of Northern Siberia are also pleistocene.

III. *Organic Remains of the Pleistocene.*—1. The organic remains of this period *in North America* consist chiefly of the shells of Molluscs and the bones of Mammalia.

Of the marine shells, those which are found in the older blue clays are pelagic. Some of the species are extinct. One of the most common of these is the *Nucula Portlandica.* This species occurs in the valley of Lake Champlain. It is abundant in the vicinity of Portland, Maine, in company with several other shells, such as inhabit rather deep water. *Nucula Jacksonii* (*Fig.* 45) is also extinct. It occurs at Augusta and in other parts of Maine.

Fig. 45.

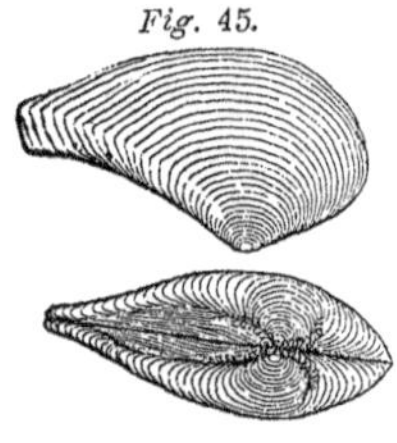

In the sands which overlie the blue clays several species of littoral shells occur abundantly, all of which are now living on the sea-shore. The most common is *Sanguinolaria fusca* (*Fig.* 46), and *Mya arenaria*, the long clam. Both of these species occur in the valley of Lake Champlain, in the same position in which they died. Entire beds of the long clam occur in the same perpendicular position in which this species is now found living in the sandy mud of the sea-shore.

Fig. 46.

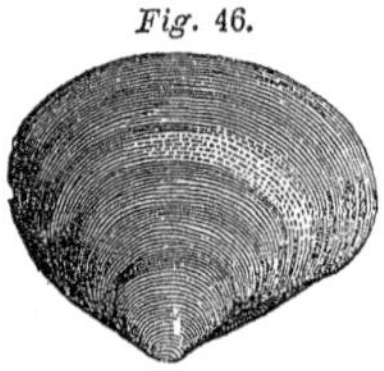

The common oyster, *Ostrea borealis,* is found in Maine, far inland. It occurs also at Brooklyn, New York, beneath thick beds of gravel and bowlders, with numerous other marine species, all of which are found living in the vicinity.

Most of the superficial sands and gravel in the Northern States of this period are too porous for the preservation of fossils, and hence the localities are few.

In the lower parts of the Southern States, marine shells occur

What organic remains occur in North America? What is said of the pelagic (deep water) shells? of the littoral shells? of the pleistocene shells of the Southern States?

in some superficial strata of loose materials, which were deposited during this period. A very large majority of the species inhabit the neighboring seas, and most of the others are now found in the West Indies and the Gulf of Mexico.

The marl beds occupy basin-shaped depressions, which were once occupied by ponds, or they occur in existing ponds. Many are covered with beds of muck, and some with a heavy growth of timber. They consist of fresh-water shells in every stage of decay, of pulverulent carbonate of lime, which has probably resulted from a complete decay of the shells, and of a variable portion of clay. The history of them extends into the historical period, for some of them are yet in the process of accumulation.

Fig. 47.

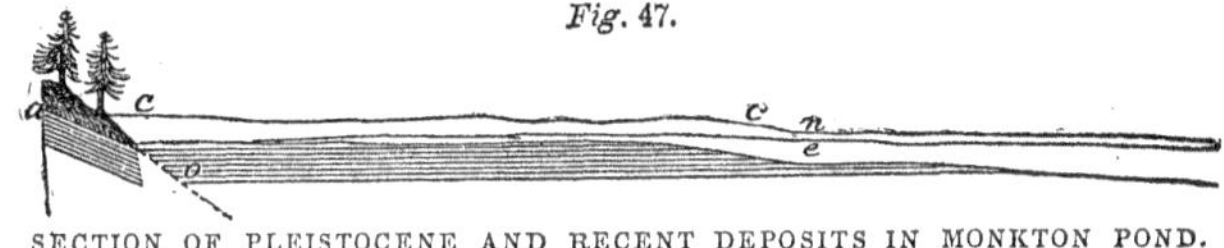

SECTION OF PLEISTOCENE AND RECENT DEPOSITS IN MONKTON POND.

In the accompanying section of a pond in Monkton, Vermont, *c c* is a muck bed, encroaching on the pond *n n*, which is not yet filled up; *e e* is the bed of shell marl, more than ten feet thick; *o o* is a bed of blue clay. The shells which constitute the marl belong to species which are now living in the pond.

It is obvious that we have here a typo of tho usual process. First, the blue clay of the older pleistocene was deposited over drift; then commenced the growth of the mollusca, which, although for the most part less than one quarter of an inch in diameter, and occupying much less space after comminution, have accumulated to the amount probably of 300,000 cords, or more than 6,000,000,000,000 of shells.

Meanwhile the vegetable deposit commenced not far from the margin of the pond, and is now advancing into it over the marl, which, however, is still in progress; thus showing us how, of two deposits superimposed the one on the other, a part of the oldest portions of the upper one may be more ancient than the newest part of the lower bed.

The length of the pleistocene period is strikingly illustrated in

What is said of the origin of marl beds? Describe the example in Monkton.

the Monkton marl bed. A long series of years is required to furnish shells sufficient for a single layer, and yet they have accumulated to more than ten feet in depth. 20,000 years is a very moderate estimate for the time required at the present rate of accumulation, and it is more likely to have exceeded this many fold. Yet it all belongs to the latter part of the pleistocene period, of which it is probably but a small fraction.

Those parts of the Southern States which were not submerged were inhabited by an extinct species of horse, of bison, hippopotamus, elephant, and the great mastodon, the mylodon, and those huge quadrupeds, the megatherium and megalonyx.

It was probably in the later portion of the pleistocene period that the mastodon flourished in great numbers in the Western States, and wandered as far to the northeast as the Hudson River. Their skeletons have been preserved in bogs of shell marl and in the salt licks of the West. From the great salt lick of Kentucky the bones of 100 mastodons have been removed, with those of the extinct elephant and other animals. In August, 1845, was found, in Newburg, New York, an entire* skeleton of the great American mastodon (*Fig.* 48), with the head raised and turned to one

Fig. 48.

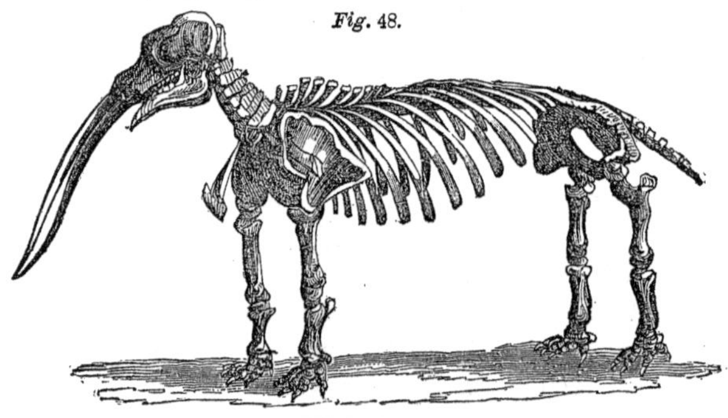

NEWBURG MASTODON.

* The bones of one foot are wanting, and are supposed to have been carried away in the marl, which was removed for agricultural purposes.

What is said of the length of the process? What quadrupeds inhabited the Southern States? Where did the mastodons live?

side, and the tusks thrown upward—the posture natural to a quadruped when sinking in the mire. The stomach was also found, containing leaves and bruised twigs, as had been seen less distinctly in some previous discoveries. The structure of the teeth would have led us to suppose that this species fed on the boughs of trees and on young saplings, since, differing remarkably from those of the elephant, the grinders are covered with large conical elevations, which must have enabled them to grind such food with great facility (see *Fig.* 54, page 156). This peculiar form of the teeth once led to the erroneous idea that the mastodon was a carnivorous animal.

The remains of an elephant have been found in Vermont, where the Rutland rail-road crosses the Green Mountains, at the bottom of a deep bed of muck.

As this was the last of the geological periods anterior to our own, it becomes an object of great interest to form some conclusion respecting the *time* when these gigantic mammalia flourished. In the more northern portion of their range they are found to have been mired in *shell marl*, which consists of the same species of fresh-water shells which now inhabit our waters. That these quadrupeds did not, however, belong to the present period, is obvious from the condition in which the remains are found under beds of muck. Although this is the most favorable situation for preserving the animal, nothing remains but the bones. But the most conclusive fact is the association of the mastodon with many other extinct species. The Indian traditions of living mastodons are doubtless crude geological speculations, founded on the occurrence of the bones.

In the valley of Lake Champlain, in Charlotte, Vermont, the skeleton of a small whale has recently been found. It was brought to light by a rail-road excavation in the blue clay. This animal resembled the Beluga, or white whale (*Delphinapterus leucas*), which now inhabits the Northern Sea. It was 13 feet

---

What is said of the Newburg mastodon? What did they eat? What is said of their teeth? of the Vermont elephant? What is said of the time when these mammalia flourished? What is said of the white whale in Vermont?

long. *Fig.* 49 represents the remains of the head, which was

Fig. 49.

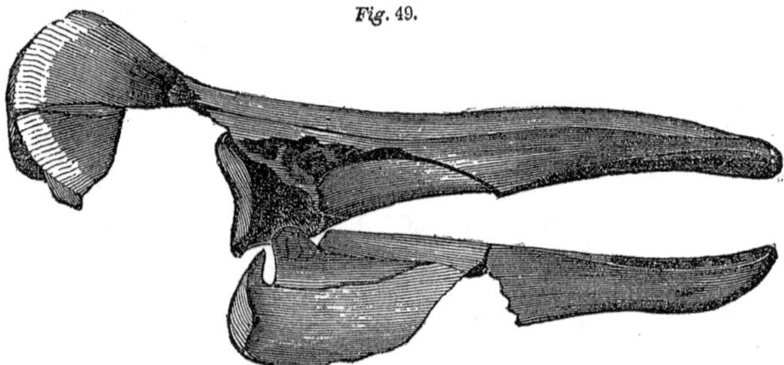

about 22 inches long, the figure being one sixth of the natural size. The upper jaw had 16 teeth, and the lower jaw had 14,

Fig. 50.

Fig. 51.

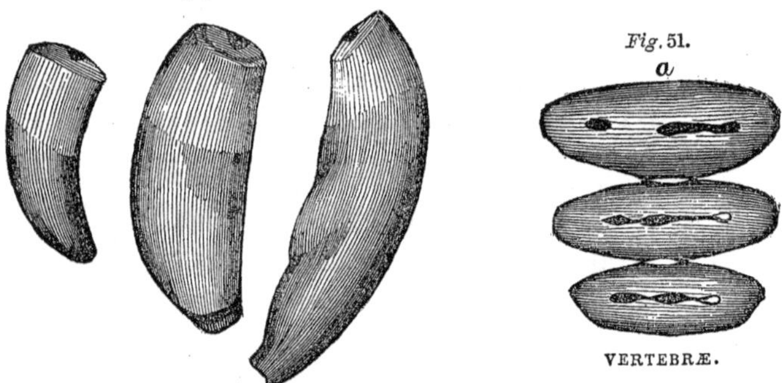

TEETH, NATURAL SIZE.

VERTEBRÆ.

Fig. 50. That the tail had a horizontal fin for vertical motion is inferred from the character of the caudal vertebræ. It is obvious in *Fig.* 51, which represents them as seen from above, that these vertebræ were much more free to move on each other in a vertical direction than laterally. This figure is one half of the natural size.

From the shells and vegetable remains which were found with the skeleton, it may be inferred that the water was at least a few

---

What is said of the motion of the tail fin ?

fathoms deep when the animal was imbedded, and that it subsequently became a salt marsh.

2. *South America.*—The great Pampean formation of South America contains numerous fossil shells, all of which are now living in the neighboring waters.

Several colossal species of quadrupeds, which are now extinct, were very numerous in South America. Among the most remarkable were the megatherium, mylodon, megalonyx, and scelidotherium, some of which had bones much larger than the elephant, although the entire bulk of the animals was not greater. The mylodon and megatherium closely resembled in their structure

Fig. 52.

MYLODON ROBUSTUS.

What is said of the place in which the skeleton was imbedded? of the shells of the Pampean formation? of the quadrupeds of South America? of the habits of the mylodon and megatherium?

the sloths, that feed on the tender twigs of trees, which they climb for this purpose. But while the immense size of these extinct quadrupeds prevented them from imitating their modern representatives in the mode of obtaining their food, their size and peculiar structure enabled them, by tearing down the trees, to bring their food within reach.

The feet, legs, pelvis, and tail of the mylodons (*Fig.* 52, p. 153) were of enormous dimensions, enabling them, as they rested on the firm basis of their massive hind legs and short thick tail, to tear down large trees, whose roots they may have loosened by their enormous claws and fore feet. The teeth were adapted for chewing only the tender buds and leaves of trees. They had only a few broad, smooth grinders, the front of the mouth being destitute of teeth. For chewing twigs, as did the mastodons of North America, more uneven surfaces would have been required. Had their food been on the ground, incisors in the front of the mouth would have been necessary.

The South American megatherium (*Fig.* 53) was larger than he mylodon. Its length was eighteen feet; the breadth across

Fig. 53.

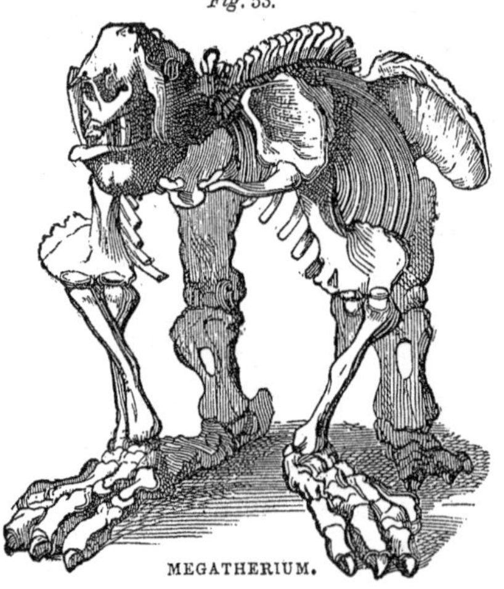

MEGATHERIUM.

What is said of the structure of the mylodon? Describe the megatherium

the pelvis was six feet, and the opening in the several vertebræ for the passage of the spinal marrow was four inches in diameter. The tail was two feet in diameter in the upper part. The bones of the legs were nearly three times as heavy as in the largest elephant, and the foot was enormously expanded, so as to afford a firm basis. The heel bone projects backward eighteen inches, and the whole foot was one yard long and two thirds of a yard wide. Its teeth were adapted for chewing not only the tender buds and leaves, but also the twigs of trees.

The existing animals of this tribe are liable to heavy falls, and their dense covering of soft matted hair prevents any serious injury. For the protection of the brain, the outer and inner plates of the skull are separated by an unusual thickness of air cells, so that the outer one may even be broken without fatal injury. The heads of the mylodon and megatherium had a similar structure, for they were liable to similar injuries when prostrating trees.

A remarkable confirmation of this reasoning occurred in the top of the skull of a mylodon. The outer plate of the skull had been broken in two places, one of which was entirely and the other was partially healed. The animal, therefore, must have survived the blows which caused the fractures. It must, however, have been stunned, and of course temporarily disabled. The carnivorous animals which have inhabited South America overcome their prey, not by the force of blows, but by the pertinacity of their grasp, and could not inflict such wounds. Nor is it likely that they were inflicted by any of the huge animals of the same tribe, since the habits of all the living representatives are remarkably peaceful. Nor is there any evidence that human beings coexisted with the mylodon. Besides, it is extremely improbable that any living adversary inflicted these stunning wounds, for the advantage would have been followed up, and the animal would not have escaped. Nor could he have fallen from a precipice in that region of plains; nor, if he had, would he, with

---

To what accidents are the living animals of this tribe liable? To what were the mylodon and megatherium exposed? What is said of a fractured and healed skull of a mylodon?

a proportionately very small head, have fallen upon it. It only remains, therefore, to ascribe the wounds to falling trees, and accordingly the fissures of the fractured skull proceed from a longitudinal wound, not from a central depression.

3. During the same period, also, there lived in Great Britain and on the continent of Europe, many quadrupeds, most of which, including all the larger species, are extinct; as a species of elephant, of bear, and of hyena, larger than any now living; of horse, lion, tiger, &c. Great numbers of the remains of the elephant are in the shallow parts of the German Ocean, and in the gravel beds of England. These occur with some animals which may have become extinct since the existence of man, as the great Irish elk; and with others, which are yet in existence, as the fox and wolf. Thus it appears that the extinction of species of that epoch was a gradual, and not a sudden and violent process.

4. Probably, also, the gigantic birds of New Zealand existed during the same period, although continued into the historical epoch.

5. In the pleistocene of Northern Asia, the bones of mammoths occur in immense quantities. Sometimes they are found mixed with bones of extinct species of rhinoceros, ox, horse, antelope,

Fig. 54.

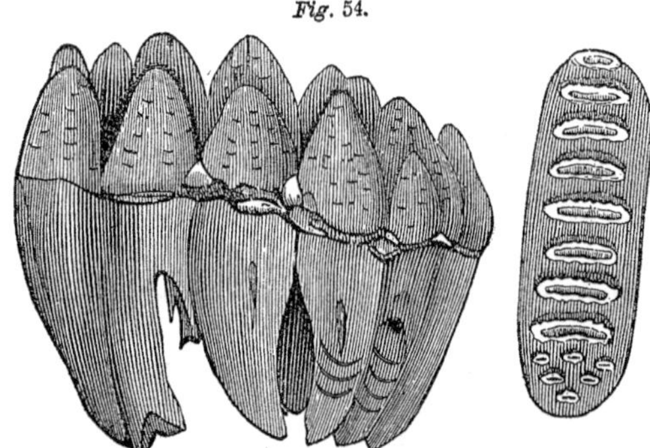

TEETH OF THE MASTODON AND ELEPHANT.

What is said of the European quadrupeds? of the birds of New **Zealand?** of the animals of Northern Asia?

&c., and even with the remains of marine animals. Rarely they occur with shells and corals attached to them.

The mammoth belonged to the same genus with the two existing species, the Indian and the African elephant. But the mastodons constituted a separate genus, which is wholly extinct.

The most characteristic peculiarities of these two genera may be seen in the preceding figures of their teeth, *Fig.* 54.

In less obvious but important peculiarities of the teeth and jaws, the mammoth differed from the existing elephants. The tusks were very large and curved backward, forming almost a circle, *Fig.* 55. It was well clothed with thick-set curly hair.

Fig. 55.

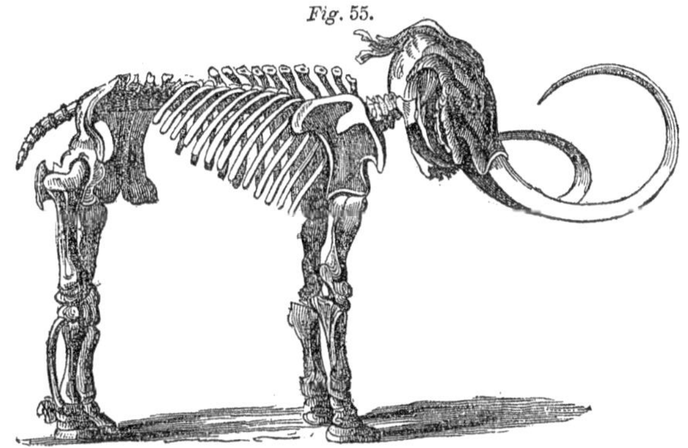

ELEPHAS PRIMOGENIUS.

The abundance of mammoth bones in Siberia is remarkable. In as high latitudes as 65° north, in Siberia, wherever a river undermines its banks, bones are dislodged. The tusks furnish a large portion of the ivory which is used in the arts.

Even as far north as latitude 75°, there is in the Polar Sea, longitude 140° east, an island, Kotelno, north of Siberia, where the hills of the interior contain the bones of horses, buffaloes, oxen, and sheep, in such abundance as to show that these animals formerly lived there in numerous herds.

What is said of the difference between the mammoths and the mastodons? What is said of the form of mammoth tusks? of their abundance? What is said of quadrupeds north of Siberia?

The mammoth, of the skeleton of which the preceding figure is a representation, was found entire, preserved in ice. It was first seen in 1799, by a fisherman who was in the habit of searching along the shores of the Arctic Ocean for tusks. It was then enveloped in blocks of ice, but in 1801 a part of it was fairly exposed to view, and in 1803 the ice melted, so that the enormous carcass fell upon a bank of sand. In 1804 the discoverer cut off the tusks, which weighed 360 pounds, and sold them. Meanwhile the people in the vicinity used the flesh to feed their dogs during a time of scarcity. White bears, wolves, and foxes also fed upon it. In 1806 it was found by Mr. Adams, who was in the service of the Emperor of Russia, and the remains were removed to St. Petersburg.

When found by Adams, the skeleton was entire with the exception of one fore leg. The skin of the head and even the eyes remained, and one of the ears with a tuft of hair. About three fourths of the skin was found, covered with reddish wool and blackish hair, about one and a half inches long. The carcass was nine feet four inches high, and sixteen feet four inches long, not including the tusks. These were nine and a half feet around the curve, and only three feet seven inches from base to point. Much of the hair had been trodden into the ground by the white bears, yet thirty-six pounds of it were collected.

This skeleton is now in the museum of the Academy of St. Petersburg.

IV. *Age of the Pleistocene Period.*—In the drift regions of North America and Northern Europe, the duration of the pleistocene period is well defined geologically, as comprising the interval between the drift and the historical period. The commencement of the formation dates at the dissolution of the glaciers and the submergence of the land. In other regions, the commencement is indicated, with sufficient accuracy, by the in-

---

In what situation was the mammoth found entire? What use was made of its flesh? What was its condition when found by Mr. Adams? How is the duration of the pleistocene defined? What is said of the time of its commencement?

troduction of the marine shells, which were associated with the extinct quadrupeds, but which, with few or no exceptions, have survived to the present time. The end is marked in all countries where deposits are found by the extinction of the numerous species of huge quadrupeds, which most distinctly characterize the latter part of the period.

The synchronism of the deposits in different countries is established with sufficient accuracy by the fact of the association of the extinct quadrupeds with shells, of which few or none are extinct, or even by such shells alone.

In the Southern States, and probably also in Southern Europe, the limits of the pleistocene period are not so well defined as in the drift regions. There appears to have been an uninterrupted series of similar events from the tertiary to the historical period. These events consisted chiefly in the very slow changes of sea levels, in the gradual extinction of a very small minority of the molluscs, and in the entire extinction of the vertebrated animals.

It follows that the pleistocene of the drift regions should be assumed as the fixed point in investigating the age of other pleistocene deposits.

The absolute duration of this period is not easily estimated In the Falls of Niagara we have a rude but grand natural chronometer for the latter part. There is good reason to believe that before the drift period the Niagara River followed a direct course to St. David's, and that its channel was filled during the drift and older pleistocene period with bowlders, clay, &c. Since its emergence from the older pleistocene seas, it has excavated the channel from Queenstown to the present falls. The rate of retrocession at the present time is pretty well known; and although it must have varied, some positions being more and others less easily cut through, the time since its emergence is inferred to be not less than 30,000 to 40,000 years. This is probably but a small portion of the entire pleistocene period.

What is said of the end of the period? of the synchronism of deposits in different countries? of the limits of the period in the Southern States? of the Falls of Niagara?

It is a remarkable fact, that the species of the mollusca have outlived the quadrupeds of the pleistocene period. Yet the duration of quadruped races comprises a longer series of events than has elapsed since the creation of man. We have seen, too, that a small part of the life of the pleistocene mollusca measures 30,000 or 40,000 years. Yet there has been but a very slight change in the species from the commencement of that period to the present day. But the older formations, of which there are several hundred, are distinguished, for the most part, each by very many peculiar species of shells. Consequently, the duration of each of the periods of the older formations must have far exceeded the whole time which has elapsed since the beginning of the pleistocene.

V. *Geography of the Pleistocene Period.*—The geography of this period is susceptible of more complete description than that of any previous period, for each geological formation has been made at the expense of the pre-existing formations. Hence the present limits of the older formations afford comparatively meager information on the distribution of land and water, and dependence is placed more exclusively on the characters of their organic remains and of the materials deposited. But here we have, in addition, the scarcely-altered deposits of ancient seas, and the marks of ancient sea levels over regions of which the inequalities of surface were nearly the same as at the present time

Fig. 56.

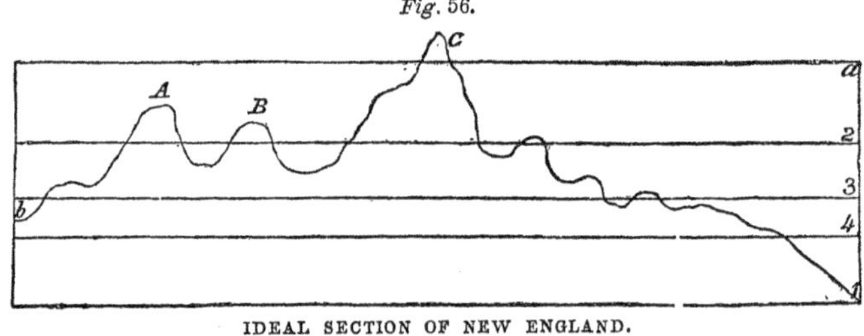

IDEAL SECTION OF NEW ENGLAND.

What is said of the comparative longevity of the species of mollusca and of quadrupeds? What is the inference respecting the older formations? What is said of the geography of the period?

The preceding figure exhibits the relations of land and water during this period in the New England States.

A B. Green Mountains; *C*. Mt. Washington; *a*. Upper limit of the drift agency; *b*. Lake Champlain. 1. Sea level during the drift period; 2. Sea level at the beginning of the pleistocene period; 3. Sea level stationary during a part of the pleistocene period; 4. Sea level at the present time.

Of the different sea levels, No. 2 was more than 1500 feet above the present level of the sea; consequently, most of the continent was beneath the ocean, and the parts which are now mountainous constituted groups of islands. No. 3 has been ascertained to have been, in the valley of Lake Champlain, 400 feet above No. 4, the present sea level. In the valley of the St. Lawrence it is said to have been 100 feet higher. Probably much of the eastern part of North America was submerged to this depth for a long time. A necessary consequence was, that New England and New Brunswick constituted a large island. This was separated from the main land of New York by a strait, which extended from the valley of the St. Lawrence through the valley of Lake Champlain, of the Champlain Canal, and of the Hudson River. The summit level of the canal indicates the most shallow part of this strait, which had a depth of about 125 feet. The western part of Vermont was thickly studded with small islands in a tranquil sound. The exterior portions of the New England States, and extensive districts in the Middle States, constituted a beautiful archipelago of small and picturesque islands.

By a depression of the low land of the Southern States, the Gulf Stream may have flowed over them, and the Gulf of Mexico have covered a much larger area. Subsequently, however, more or less of the Southern States has been more elevated than at present.

The vast plains in South America, which we have mentioned as a pleistocene deposit, were submerged during this period. The fine mud of the great Pampean formation indicates the per-

Describe *Fig*. 56. What was the condition of this continent? What large island in North America at a later part of this period? What is said of the Southern States? of South America?

fect repose of the waters. The subsequent process of emergence, protracted through long periods of time to the present era, is exhibited in the terraced plains which are seen at successively lower levels in approaching the sea.

In Northern Europe the general series of events was similar to that of North America. The accompanying section (*Fig.* 57) represents the different sea levels in Scandinavia. A considerable portion of Northern Europe must have been submerged. There is evidence, also, that during another epoch in this period Great Britain was elevated to a greater height than at present, and that much of the adjacent seas was dry land.

Fig. 57.

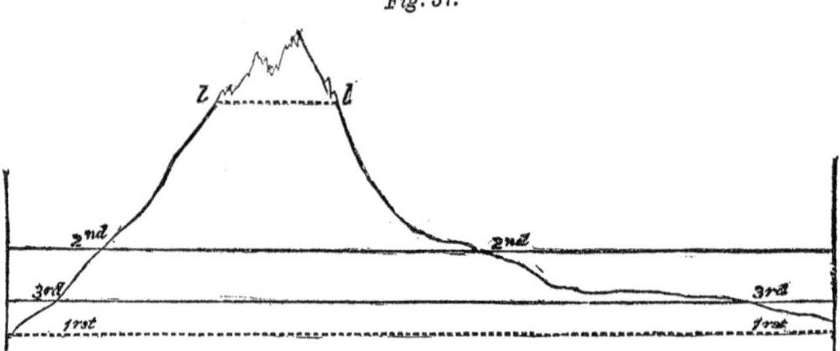

l l. Limit of glaciers.
1*st.* 1*st.* Sea level at the commencement of the drift period.
2*d.* 2*d.* Sea level in the pleistocene period.
3*d.* 3*d.* Present sea level.

The basin of the Caspian Sea was much larger, and communicated with the Black Sea. A large part of North Siberia was, during a part of the period, covered by the Arctic Ocean. This submergence may have occurred in the latter part of the period, subsequently to that of the drift regions in Europe and North America.

Since the numerous islands of Polynesia have been gradually sinking beneath the waters of the Pacific during the present period, and perhaps for a longer time, it follows that during the pleistocene period there was a greater extent of land in those

What is said of Northern Europe? of Great Britain? of the Caspian Sea? of Siberia? of Polynesia?

regions than at the present time. Either during that, or in some more remote period, the immense area of Polynesia may have been occupied with a continent, the mountains of which constitute the present groups of islands.

It is obvious that the extent of submergence appears on a subsequent emergence of the land. It is far more difficult to ascertain what regions which are now beneath the sea may have been dry land during the pleistocene period.

We have seen now that extensive regions on both continents were above and below the level of the sea in different parts of this period. There is no reason to suppose that these events occurred any more rapidly than similar events now take place. Thus the history of one period, which is probably of much shorter duration than any of more ancient date, comprises an immensity of time, which the mind fails to comprehend.

VI. *Climate of the Pleistocene Period.*—The pleistocene climate of *North America* does not appear to have differed in its mean temperature from its present climate. The marine shells which occur in our older clays have been supposed by some European writers to belong to species which are exclusively Arctic. But they are, for the most part, identical with species which now exist abundantly on the shores of New England, as well as further north. They therefore indicate merely the influence of a polar current, similar to that which now chills the waters of our coast.

It is probable, therefore, that the Labrador current, which even now has a westwardly tendency into the Gulf of St. Lawrence, chilled the waters which covered the lower parts of Canada and New England. On the other hand, it is probable that the Gulf Stream flowed over the lower parts of the Southern States. By these agencies the shells now inhabiting Massachusetts Bay and the Gulf of Mexico were mingled in the basin of the Potomac. The waters, at least, of the Southern States were warmer than at present, so as to be inhabited by species of shells which now occur only in the Gulf of Mexico and the West Indies.

The existence of elephants in Vermont does not necessarily indicate a warmer climate, since the species may have had a

---

What is said of the time required for the changes of level? What was the climate of North America during the pleistocene period?

clothing like that of the Siberian elephant. A structure and habits adapted to the climate of the Northern, and Middle, and Western States would have required only *specific* peculiarities in which it should have differed from the existing species of the same genus.

So also the elephants, which inhabited Great Britain and the continent of *Europe*, prove no essential difference of climate. The uniform and mild climate which now characterizes Western Europe may have been extended further eastward by the submergence of large portions of land.

In *Siberia*, the immense numbers of the remains of elephants do not prove the climate to have been tropical, for the species was furnished with a woolly covering. But the climate of Northern Siberia must have been much milder than it now is, in order to the production of a vegetation required by the numerous herds of mammoths.

In *South America*, the pleistocene climate appears, from the fossil shells, to have been the same as at the present time.

CHAPTER V.

TERTIARY PERIOD.

The next chapter in these ancient records, to the investigation of which we now proceed, contains the history of the *Tertiary Period.* The rocks of this period underlie the drift, and must therefore have been deposited previous to any hitherto described.

1. *Geographical Distribution.*— The tertiary of the United States is found upon the Atlantic sea-board and the Gulf of Mexico. Its western limit is at the first or lowest falls of the principal rivers, and is generally marked by the long-leaved pine (*Pinus palustris*), whose distance from the shore is limited by this formation. In South America, according to M. D'Orbigny, tertiary rocks extend from the great plains of the Amazon to the Straits of Magellan, a distance of 2500 miles, with a width in many places of 800 miles.

What is said of the climate of Europe? of Siberia? of South America? How is the tertiary distributed?

In Europe and Asia, the rocks of this period are found principally in basins, apparently deposited in lakes and estuaries of limited extent.

2. *Composition and Structure.*—The tertiary rocks consist of numerous layers of marls, clays, sandstones, and limestones, which generally lie in nearly a horizontal position, and are often consolidated into a hard, compact rock. Most of the rocks are filled with the remains of marine or fresh-water shell-fish, land animals, and vegetables, which more or less resemble existing species; while scoria and streams of lava have in many cases become mingled with the sedimentary deposits.

3. *Classification of the Tertiary.*—This system has been divided by Sir Charles Lyell into four groups, founded upon the proportion of living to extinct species of shells which they contain.

(1.) The oldest or lowest group is called *Eocene, the dawn* or commencement of the existing types of animals and plants, and contains about four per cent. of shell-fish now living.

(2.) The second group is called *Miocene* (*less recent*), and contains about 20 species in 100 of shell-fish, which are identical with those now living in the adjacent lakes and seas.

(3.) The third and fourth groups are called *Older* and *Newer Pliocene* (*more recent*), and contain from 50 to 90 species in 100 of existing shells. These may be described as one group. We shall then have *the Pliocene* or *Newer Tertiary, the Miocene* or *Middle Tertiary, the Eocene* or *Older Tertiary.*

This classification applies to the European better than to the American tertiaries. The eocene of Virginia, South Carolina, and Alabama does not contain a single recent shell, and yet it is probably cotemporaneous with the European eocene.

It is, moreover, doubtful whether the American tertiaries can be classified so as to make three groups. There seem to be but two periods, and those in the northern part of the United States appear to correspond with the eocene and miocene, and in the

---

What is its composition and structure? Mention the principle of classification and the divisions of the tertiary. How does this classification apply to the American continent?

southern portions to eocene and pliocene. It is probable that the miocene and pliocene periods, so distinctly marked in Europe, constitute but one period in the United States, so that it is difficult to determine the exact synchronism of the deposits on the two continents.

SECTION I.—PLIOCENE OR NEWER TERTIARY.

It would be impracticable in an elementary treatise to describe all the deposits included in this division of the tertiary. A few examples, selected from American and European pliocene, will suffice to illustrate its character, and the nature of the mineral and organic changes which occurred during this part of the tertiary period.

I. *Pliocene of the United States.* 1. *Distribution.*—The pliocene of the United States, according to Professor Tuomey, is confined to the Atlantic sea-board. The deposits are found in patches from Maryland to Georgia, but slightly elevated above the sea level.

2. *Composition.*—The pliocene is composed mostly of yellowish sands and marls, which vary in thickness, but rarely exceed 40 feet. The strata are arranged in nearly horizontal layers, and repose mostly on the older eocene, or on cretaceous rocks. The marl contains a large proportion of carbonate of lime, and great numbers of fossil shells.

3. *Geological Position.*—Although these rocks, in South Carolina and Georgia, contain a very large proportion of recent shells, similar deposits in Maryland and Virginia, from the larger proportion of extinct species, have been referred to the miocene. The number of fossil shells identical with existing species constantly increases in going from the northern to the southern portions of strata, which appear to belong to the same period, and the proportion of carbonate of lime also increases; but the transition is not sufficiently marked to enable us to separate those deposits which were made during the liocene from those which

---

How is the pliocene of the United States distributed? What is its composition? its geological position?

belong to the miocene period. Professor Tuomey has referred the South Carolina beds, which rest upon the eocene, to the pliocene. With this exception, and a few other deposits, as one at the mouth of the Potomac, these rocks, from the proportion of recent to extinct shells, would be referred to the older pliocene of Mr. Lyell; but as they repose upon the eocene, they may, with more propriety, perhaps, be described with the miocene.

II. *European Pliocene.*—1. *The Norwich or Mammaliferous Crag* of England belongs to this period. It is composed of shelly beds of sand and loam, deposited at the mouth of a river, and filled with the remains of several species of mammalia; hence its name. Other strata, which overlie the crag, appear to have been deposited in a fresh-water lake, and contain also the remains of mammalia.

But the best examples of the pliocene strata occur in the south of Europe, on the shores of Italy, Sicily, and on many of the islands of the Eastern Archipelago.

2. *The Val di Noto*, lying between Ætna and the southern promontory of Sicily, consists of pliocene hills which are raised from 1000 to 3000 feet above the level of the sea. The rocks are entirely composed of limestones, marls, sandstones, and volcanic products. These hills are separated from Ætna by the low plain of Catanea.

The strata consists of three groups, and are represented in *Fig.* 58.

Fig. 58.

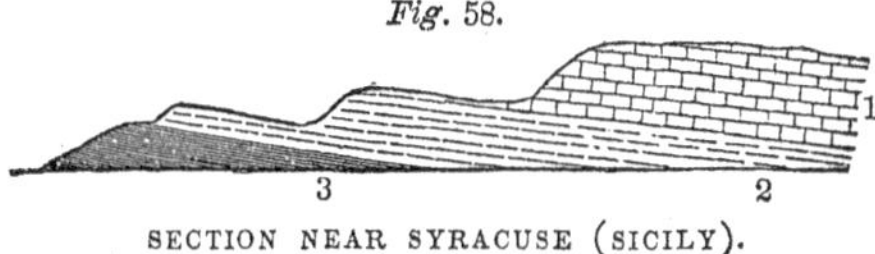

SECTION NEAR SYRACUSE (SICILY).

(1.) *The Great Limestone Formation.*—This is of a yellowish white or pure white color, and has the appearance of a rock precipitated from the waters of mineral springs, being arranged in thick beds, often without distinct lines of stratification. It attains a thickness of 800 feet, abounds in caverns, and contains shells,

What example of the pliocene in England? Where are the best examples found? Of what does the Val di Noto consist?

most of which are identical with those now inhabiting the Mediterranean Sea. From this fact it is inferred that it is the last or most recent of the pliocene deposits.

(2.) *Calcareous Sandstones, Conglomerates, and Schistose Limestones*, several hundred feet in thickness, constitute the next group in the descending order, and these repose upon,

(3.) *Blue and White Marls*, under which are *blue* and *white clays* without fossils.

The limestones and marls are filled with the remains of shell fish, of which a large proportion are recent species.

At the southern base of Ætna, near La Motte, there are alternate beds of clay, sand, and volcanic matter, together with conglomerates and columnar basalt.

The most important deposits in Sicily consist of marls (3), which are of great thickness, and differ from each other but slightly in color and mineral composition. The lower beds are much contorted, and contain sulphur, salt, and gypsum in great abundance; but only a few organic remains, as the skeletons of fishes.

Upon the marls limestones were deposited (2), and upon these blue marls; then sands, sandy limestones, and, finally, the great limestone beds (1), which are found to cap hills in some places 3000 feet in height.

It is obvious that this entire series must have been deposited beneath salt water. The island, therefore, must have formed the bed of the sea, into which the materials were brought from the neighboring continent by fluviatile and volcanic agencies, and very gradually deposited, for we find the strata throughout filled with the remains of marine mollusca.

Several long periods of time are requisite to account for the deposition of sedimentary and volcanic rocks to the depth of 3000 feet, for it should be observed that not a single shell-fish could be imbedded after the island emerged above the surface of the waters. It was not, then, till after the last sediment was thrown down that the island was raised to its present position.

Describe the several deposits in Sicily. What is the evidence that these deposits require long periods of time?

When we consider the magnitude of the changes wrought during this, the most recent portion of the pliocene period, the whole formation of Ætna is truly insignificant, for nearly the whole mountain has been built up to its present height of 11,000 feet, and 90 miles in circumference, since the deposition and elevation of these, the last of the pliocene strata.

3. *Sub-Apennine Tertiary Beds.*—During the same period similar deposits were in progress on the western side of the Apennines in Italy. The rocks here consist of *brown* and *blue marls*, overlaid and interstratified with thick beds of *sandstone* and *volcanic products*. The strata are filled throughout with marine shells, which are sometimes associated with beds of lignite and fresh-water shells. The marls attain a vertical thickness of 2000 feet. The sandy beds pass into conglomerates, and give clear indications of river action during their deposition.

Similar strata are also found on the eastern side of the Apennines, and extend to the Morea, and to the islands still further east. Deposits of this period occur also in Spain, near Olot, so that a large portion of Southern Europe was submerged beneath the Mediterranean Sea during the pliocene period.

4. *Brown Coal Formation.*—Perhaps the most remarkable of the pliocene strata occur in the valley of the Rhine, between Coblentz and Cologne. They consist mostly of lignite or brown coal, and contain the largest quantity of vegetable matter which has been formed since the carboniferous period. The coal rests upon loose silicious sand, which sometimes hardens into sandstone, and sometimes into conglomerate, through which are distributed thin leaves of lignite called *paper coal*, and silicified wood often passing into chalcedony or *semi-opal*. Beds of clay often succeed these silicious deposits, and form the stratum on which the lignites or principal mass of the coal rests. Some portions of the lignite consist of solid wood, of a black color, and so perfectly preserved as to be used for timber, while other portions are mineralized by carbonate of iron, or replaced by coarse quartzose sand. Associated with the lignites, and especially with the

Describe the sub-Apennine beds; the brown-coal formation.

paper coal, are some very remarkable fossils, rather imperfectly preserved; they consist mostly of insects, mollusca, fishes, batrachians, and quadrupeds.

5. *Lacustrine Deposit of the Rhine.*—There is another interesting lacustrine deposit, where the Rhine issues from Lake Constance, consisting of highly fossiliferous marls and limestones.

The lower beds are cream-colored marls, filled with the remains of plants, fishes, and fresh-water molluscs.

Highly fossiliferous beds of fetid marlstone and limestone succeed, in which have been found the skeleton of a fox, allied to existing species; also fishes of large size, a tortoise three feet in length, crustaceans, and insects.

The strata are horizontal, and are raised several hundred feet above the Rhine. It is evident, therefore, that changes in the relative level of sea and land, similar to those in the south, must have taken place in this part of Europe, although the deposits were made mostly in lakes.

III. *Fossils of the Pliocene.* 1. *Vegetables.*—The remains of vegetables are found in most of the pliocene deposits: the most remarkable of which is the brown coal formation already described. Trees in a very perfect state of preservation abound, while the paper coal is mostly made up of the leaves of the poplar, birch, and the same species of plants generally that are found in the existing forests.

2. *Animals.*—(1.) The remains of the lower orders of animals are frequently met with, but they do not differ from existing species sufficiently to require any description.

(2.) *Mollusca.*—The stratum of marl in South Carolina, which is referred to this period by Professor Tuomey, is very rich in fossils, consisting mostly of the remains of *mollusca.*

The following table exhibits the proportion of recent to extinct species in South Carolina:

Describe the lacustrine deposits of the Rhine. What is said of the fossil vegetables? of the animals? of the mollusca?

	No. of Species.	Species Recent.	Per Cent.
Brachiopoda	1	0	0
Gasteropoda	78	39	50
Lamellibranchiata	109	47	43
Cirripedia	2	1	50
Total	190	87	46

The proportion of recent species is probably somewhat larger, and this fact would place these deposits with the Crag of England, which belongs to the older pliocene of Mr. Lyell.

In Europe the remains of shell-fish are the most abundant fossils, of which more than 800 species have been already described. They are very similar to existing species.

(3.) *Articulata.*—The remains of lobsters, insects, &c., are not very abundant. The most noted locality of fossil insects is at Œningen, near Lake Constance, where many species, in the different states of larva, pupa, and imago, have been very perfectly preserved in a stratum of marl. They appear to have fallen into the lake from the flowers upon its bank, and to have become enveloped by the fine marly sediment.

(4.) *Fishes.*—The tertiary deposits have yielded more than 350 species of fossil fishes, but the most interesting of them are found in the miocene and eocene.

The most celebrated locality belonging to the pliocene is at Œningen. The genera are chiefly allied to the modern carp and tench, with an extinct species of pike. Although the number of individuals is large, not more than 20 species have been described. The tertiary marls in South Carolina contain several species of fish, particularly of those belonging to the shark family.

(5.) *Reptiles.*—The remains of batrachians, frogs, toads, and newts are found in the marl beds of the Rhine, at Œningen, associated with a gigantic salamander three feet in length, which was at first described as a "*fossil man!*"

Frogs and other reptiles are also abundant in the brown coal. The following, *Fig.* 59, represents one species of frog:

What is said of the articulata? of fishes? Where is the most celebrated locality? Mention the reptiles of the pliocene.

Fig. 59.

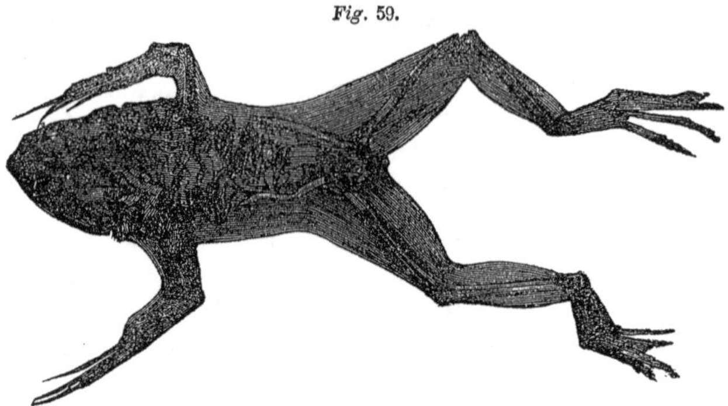

RANA DILUVIANA. (BROWN COAL.)

(6.) *Mammalia.*—The molars and tusks of the mastodon maximus, and the horns of an extinct species of deer, have been found in South Carolina.

The remains of species of elephant, rhinoceros, hippopotamus, horse, ox, deer, and mastodon occur in several European localities, most of them in the English Crag.

The teeth of a porcupine occur in the pliocene of Tuscany. The remains of the bear, hyena, fox, and wolf at Œningen. A water mole, as large as a hedge-hog, on the coast of Norfolk, and a monkey and bat occur in the pliocene of Suffolk, England. The types of organization in all these extinct animals are very similar to those of their existing representatives, but some of the species are of a much larger size.

SECTION II.—MIOCENE OR MIDDLE TERTIARY.

Immediately beneath the deposits described in the last section we find a group of strata of great extent, somewhat similar in composition and structure, but characterized by different fossils, called the *Miocene* or *Middle Tertiary.*

I. *Miocene of the United States.*—The miocene strata occur at different points skirting the Atlantic sea-board, from Massachusetts to the Gulf of Mexico. Mr. Conrad, from the small number of extinct species of shells, has referred the deposits of Yorktown,

Mention the mammalia. Describe the miocene of the United States.

Smithfield, and Suffolk, in Virginia; those of Easton and St. Mary's, in Maryland, and of Cumberland City, New Jersey, to the older pliocene of Mr. Lyell; but as we have already noticed the fact of there being but two principal divisions of the tertiary in the United States, the lowest group undoubtedly belongs to the eocene; and as the other strata were deposited immediately upon it, we have concluded to class them with the miocene, although from the character of the fossils there would seem to have been a long period intervening between the close of the eocene and commencement of the miocene period.

1. *Gayhead.*—A good example of miocene clays occurs at Gayhead, Martha's Vineyard, a section of which is represented in the following diagram (*Fig.* 60), A B:

Fig. 60.

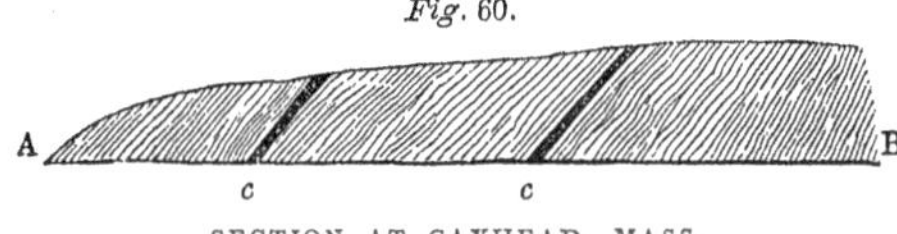

SECTION AT GAYHEAD, MASS.

The strata consist of variously-colored clays, inclined at a high angle, and dipping toward the northeast. They contain beds of lignite (*c c*), sharks' teeth, and other fossils.

The strata are of great vertical thickness, amounting to several hundred feet, and their great inclination indicates considerable disturbance in this part of New England since their deposition.

2. But the principal deposits of this period are found in Mary land, Virginia, and the more southern states.

The miocene of Virginia comprehends all the area extending from the sea-board to Northbury, on the Pamunkey River, and to Coggin's Point, on the James River. The western limit is a wavy line passing nearly north and south through these places.

In the eastern portions the surface is level, and elevated not more than 6 or 8 feet above the tide. The western part attains an elevation of from 30 to 80 feet. These higher regions are separated from the lower by a terrace, which gives the highest evidence of having been the shore of the sea. The surface is

Give examples. Where are the principal deposits found?

generally level, but is cut through by innumerable ravines, and intersected by creeks, which open into the principal rivers.

The cliffs of York River present the following kinds of rocks:

(1.) The lower stratum is a *bluish sandy clay*, containing great numbers of *Turritella alticosta* and *Sytherea Sayana*.

(2.) To this succeeds a layer of *brownish yellow sand*, with but few shells.

(3.) The next stratum consists mostly of *Crepidula costata*, so losely packed together that very little sand or clay can gain admission.

(4.) The whole is overlaid with strata of coarse *highly ferruginous sand*.

In approaching Yorktown the character of the cliff changes, and attains a higher elevation. The lower stratum exchanges its turritella for crepidula, and disappears. Different shells also appear in the other groups.

Higher up the river the shelly bed becomes changed, being composed almost entirely of comminuted shells, so cemented as to form a porous rock 40 feet in height, which thins out toward the west. The fragmentary character of this rock, and the inclination of the shelly fragments, although the beds are horizontal, clearly indicate the action of the surf and of tidal currents. In some places a band of iron ore overlies the fossiliferous strata.

The most important fact connected with these and other miocene strata is the existence of marl beds, composed of carbonate of lime, and containing particles of green sand, a silicate of iron, which render them highly fertilizing manures.

Similar miocene deposits exist on the James River, and the region south of it; between the Potomac and Rappahannock; in Maryland, New Jersey, and North and South Carolina.

A cliff on the north bank of the James River presents the following kinds of rock, which will illustrate the fossiliferous character of these strata (*Fig.* 61).

1. Six feet of sand and clay overlying the shelly strata.
2. One foot of reddish clay.

What kind of rocks do the cliffs of York River present? What is the most important fact connected with the miocene strata? Describe the miocene of James River.

3. A band of small pebbles only a few inches in thickness.

Fig. 61.

4. A layer of sand 3 feet thick, containing *Chama* and *Venus deformis.*

5. A stratum 4 feet thick, consisting mostly of a compact mass of *Chama* and *Arca centenaria.*

6. A stratum 2 feet thick, mostly of large *Pectens.*

7. Another stratum of *Chama*, with *Arca centenaria*, *Panopea reflexa*, about 6 feet in depth.

8. A second layer of *Pectens*, with *Ostrea compressirostra*, 1 foot thick.

9. Another layer of *Chama* 3 feet thick.

10. A layer of *Pectens* and *Ostrea* 5 feet in thickness.

The marl stratum is very extensive; it is found on both shores of the Chesapeake Bay, 100 miles from north to south, and 50 miles wide. In Virginia, and North and South Carolina, the marl strata are of equal and perhaps still greater extent.

3. *General Description of the Miocene of the United States.*—Without attempting to describe the strata in the several localities, it may serve, perhaps, to give a better view of our miocene beds to arrange them in a section. It is proper to observe, however, that the whole series is not found at any one point, and that the subdivisions, as in the preceding section on James River, are often more numerous than are represented in the following figure. Those who would obtain a more minute and extensive knowledge of our tertiaries, will find ample materials in the Geological Reports of the several states.

It will be observed that the miocene properly consists of three groups: the blue marls, the yellow and gray marls, and sands and clays.

1. *The Blue Marls.*—This group is composed chiefly of a fine blue clay of a dark color, and of a soapy feel when wet, but of a bluish gray when dry.

Give a general description of the miocene of the United States. Into how many and what kinds of rock may the miocene of the United States be divided? Describe the blue marls.

Fig. 62.

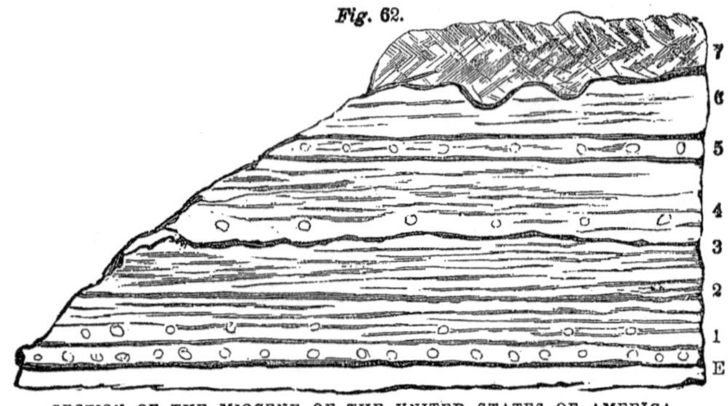

SECTION OF THE MIOCENE OF THE UNITED STATES OF AMERICA.

1. The first and lowest bed, *Fig.* 62 (1), is a thin stratum of pebbles, separating the miocene from the eocene, E.

2. Lower blue marl. 3. Upper blue marl.

4. Lower portion of the yellow marl. 5. Upper portion of the yellow marl.

6. Horizontal beds of sand and clay partially, and in some cases wholly removed by aqueous agency.

7. Pleistocene sand and clay.

The lower beds are more sandy, and contain shells in a very perfect state of preservation, associated with fragments of zoophytes, resembling corals.

The upper portions are characterized by a beautiful little bivalve shell, the *Mactra Modicella,* in such numbers as to constitute in most cases one half the weight of the rock. These shells are also found in the lower beds, and in the marls above, but not to the exclusion of other species.

The perfection of the shells, and their association in groups, prove that they lived and died on the spot, and hence these strata were deposited in quiet water and in a very gradual manner.

2. *The Yellow and Gray Marls* are a series of beds reposing upon the preceding, of a yellowish or grayish yellow color, and composed in part of a very gritty sand. The lower and more eastern beds are clayey in their texture, and of a dark brownish yellow, a color produced by the presence of oxide of iron. In the western portion of these beds they are of a lighter color, and attain a thickness of from 15 to 25 feet.

Describe the yellow and gray marls.

In the upper beds the fossils are not so perfectly preserved, they appear to have been worn by the action of water to a kind of sand, and then firmly cemented by the infiltration of carbonate of lime, forming a tertiary limestone. This limestone, when first taken out of the bed, is soft, but, on exposure to the air, rapidly hardens into a durable compact rock. The layers or fragments are laid upon each other at every inclination, although the strata are as a whole horizontal, plainly showing an agitated state of the waters during their deposition.

3. *Horizontal strata of Sand and Clay* overlie the marls; the lower beds appear to have been quietly and slowly deposited, and as the whole group is destitute of marine shells, and contains pieces of wood, it was probably deposited from the overflowings of the streams, after the manner of the deltas of existing rivers. The above description of the miocene applies particularly to Virginia and Maryland.

These marls contain on an average 40 per cent. of carbonate of lime. In South Carolina the proportion amounts to 70 per cent. The lime, and the grains of *green sand* which are distributed through most of the marls, render them an excellent dressing for the land.

When we consider that in Virginia alone the marl stratum covers an area of from 200 to 300 square miles, furnishing inexhaustible sources of fertility to the soil, we can form some estimate of its great value to agriculture.

II. *European Miocene.*—1. The lower portion of the *Crag* formation is the only representative of this period in the British Isles.

It consists of the *coralline* and the *red crag*. The former extends over an area of 80 square miles, and is about 20 feet thick; the latter exceeds 40 feet in vertical depth. This group is distinguished for the great number of corals which are associated with its mineral contents.

In many parts of Europe the miocene strata are of great thickness, and cover extensive areas. Deposits of this period are

What is said of the marl stratum? What example of miocene in England?

found in the west of France, the great valley of Switzerland, the basins of Vienna and Styria, extending to the plains of Hungary; in portions of Poland and Southern Russia, and in the northern parts of Italy.

2. *Basin of the Garonne.*—From the mouth of the Garonne toward the southeast there exists a series of miocene strata, consisting of quartzose sand mixed with marls. The marls contain great numbers of fluviatile shells, associated with others of marine origin. These fossiliferous beds are separated from the chalk by strata of fresh-water limestone.

In this part of France the miocene rocks may be divided into four groups: 1. The first and highest member of the series consists of silicious sands destitute of shells. 2. The second group is mostly gravel. 3. The third is composed of sand and marl, with shells. 4. And the fourth of blue marl, which is also filled with shells.

The fresh-water limestone near Bordeaux shows that the surface must have been alternately occupied by fresh and salt water, and hence there must have been elevations and depressions of the land during the formation of these rocks.

Fig. 63.

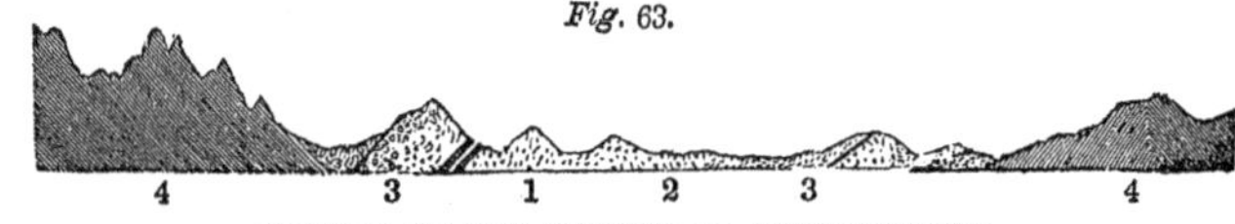

SECTION OF THE VALLEY OF SWITZERLAND.

3. *Valley of Switzerland.*—Between the Alps and the Jura, in the valley of Switzerland, of which *Fig.* 63 is a section, are found extensive deposits of miocene strata. Between the Lakes Geneva and Lucerne, the strata consist of three kinds of rock.

(1) The first is a *coarse conglomerate*, which gradually passes into

(2) A *fine sandstone* called *Molasse.* This sandstone is very thick and rather soft, although some of it is suitable for building materials.

(3) Various bands of *marl* and *lignite*, evidently of fresh-water

---

Describe the miocene deposits in the basin of the Garonne; in the valley of Switzerland.

origin, as they contain the remains of fresh-water shells, are distributed through the molasse. The whole reposes on secondary limestone (4).

The conglomerate and sandstone are at this point destitute of fossils; but further to the north, marine shells and other organic relics occur, from which it has been inferred that the deposits of this basin were made in an estuary opening into the sea toward the northeast.

The molasse forms hills 300 feet in height above the Lake of Geneva; the whole must, therefore, have been elevated with a large portion of the Alps since these strata were deposited.

In the valleys of the Jura and the Rhine similar beds are found containing fresh-water shells. In some parts of Rhenish Bavaria, the strata of sand and marl contain fossils which show a transition from the older or eocene deposits.

The most important examples of the miocene strata, however, are to be found in the central and eastern portions of the continent.

4. *Miocene Basins of Vienna and Styria.*—The strata of these basins consist of conglomerates, sandstones, and marls, overlaid by coralline limestone of great thickness, cotemporaneous with the red and coralline crag of Suffolk, England.

Fig. 64.

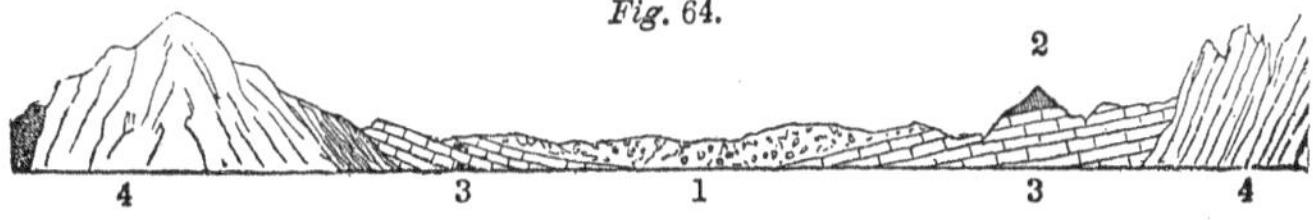

SECTION OF THE BASIN OF VIENNA.

In the basin of Vienna, of which *Fig.* 64 is a section, we find,

First (1), *gravel;* secondly (2), *fresh-water limestone;* and thirdly (3), a kind of marl called *Leitha kalk.* The whole reposes on (4) Alpine limestone, which is a secondary rock.

In the basin of Styria, the upper beds alternate with volcanic ashes ejected from the extinct craters on the borders of Hungary. The coralline limestone, which is mostly covered near Vienna by calcareous marls, has here a vertical thickness of 400

Where are the most important examples found? of what do the basins of Vienna and Styria consist?

feet. Each group is filled with the remains of mammalia and shell-fish.

Numerous other beds of this period are found in other parts of Europe and in Asia; our limits, however, will allow of but one more example.

5. *Miocene Beds of Auvergne.*—In the center of France, in the district of Auvergne, the miocene deposits are noted for the character of their fossils, and for containing beds of alluvium and volcanic rocks.

The following section, *Fig.* 65, exhibits the order of the strata at Mont Perrier. They repose upon the eocene lacustrine strata.

Fig. 65.

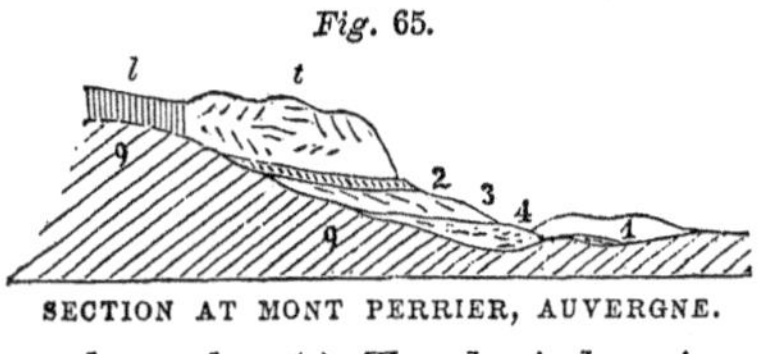

SECTION AT MONT PERRIER, AUVERGNE.

(1) *Newer alluvium;* (2) *Miocene alluvium*, containing bones; (3) *Breccia*, consisting mostly of volcanic products; (4) *Miocene alluvium*, filled also with the bones of many species of quadrupeds; (*t*) *Trachytic breccia* and (*l*) *compact basalt.*

In the two beds of alluvium not less than 40 species of extinct mammalia have been discovered, many of which are peculiar to this locality.

By inspecting this section, and studying the character of the materials and of the imbedded fossils, it is easy to trace the order and succession of changes during the formation of these strata.

1. First, then, after the eocene strata (*e*) had been deposited and covered with lava (*l*), they were elevated and subjected to the action of water, which wore off their edges and formed a basin.

2. Into this basin the materials of the alluvium (4), with the quadrupeds then inhabiting the land, were conveyed.

3. Then succeeded eruptions from the neighboring volcanoes throwing out pumice and ashes, and melting the snows upon their flanks, by means of which great floods of water accumulated in the gorges, bearing along the ejected materials with angular fragments of trachyte, and depositing the mass upon the alluvium, forming the trachytic breccia (3).

Describe the miocene beds of Auvergne.

4. Upon this another bed of alluvium (2), during a period of repose, was deposited, with the remains of similar animals.

5. Finally, another series of volcanic eruptions took place, by which the breccia (*t*) were spread over the whole group.

That volcanic eruptions were frequent during the eocene and miocene periods in this part of France, we have the most con vincing proofs, in the existence of more than 60 extinct craters, some of which are represented in the accompanying figure, so perfectly preserved that the lava streams can be traced to a great distance, as distinctly as those of Ætna or of Vesuvius.

Fig. 66.

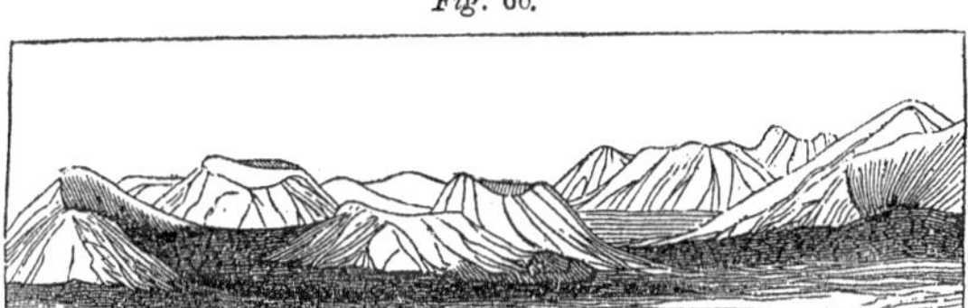

EXTINCT VOLCANOES OF AUVERGNE.

III. *Fossils of the Miocene.*—The miocene deposits contain a larger number of extinct species than the pliocene.

1. *Vegetables.*—The remains of *vegetables* are abundant in many localities. In some cases beds of lignite are interstratified with the miocene clays, as at Gayhead, Massachusetts.

In the miocene strata of Richmond, Virginia, pieces of drift-wood have been discovered, perforated by canals similar to those made by the Teredo. When these fragments of wood are dried, they constitute a true lignite. Some hickory fruits exist in a fos sil state in the same locality.

In Europe, similar vegetable remains are found in most of the miocene beds. The vegetables, however, were very generally allied to existing species.

2. *Animals.*—(1.) The shells of *infusorial* or microscopic animals constitute strata of considerable thickness. A stratum of white clay, from 12 to 20 feet thick, at Richmond, Virginia, con-

What proof of volcanic agency during this period? What is said of the fossil vegetables? What remains of animals are there at Richmond, Virginia?

sists mostly of the silicious skeletons of several species of *Navicula*, *Gaillonella*, *Actinocyclus*, and *Coscinodiscus*, *Figs*. 67 and 68.

Fig. 67. Navicula.

Fig. 68. Gaillonella sulcata. Actinocyclus. Coscinodiscus patina.

The same stratum of marl, containing some new fossil forms, is also found at Petersburg, so that it probably extends over a large area. A similar stratum occurs at Tampa Bay, Florida; the bed appears to be interposed between the eocene and miocene.

The *polishing slate* of Bilin (Germany) forms a series of strata, 14 feet in thickness, entirely made up of the silicious shells of gaillonellæ, so extremely small "that a cubic inch of stone contains *forty-one thousand millions!*"

Professor Bailey, who has examined these minute animals, has observed two species of living gaillonellæ, from Melville Island, identical with the fossils of the Richmond marl; hence the northern seas contain minute animals, whose species extend back over vast periods, to a time when they had a much wider distribution than at present.

The lower classes of animals, the *Radiata*, *Infusoria*, &c., are frequently found associated with innumerable generations of

(2.) *Mollusca*, the remains of which, both fresh-water and marine, are not only distributed through most of the miocene beds, but often constitute the principal portion of their substance. This is the case with the miocene marls of Eastern Virginia and Maryland. Many of the shells are nearly as perfect as their living representatives. The casts of *Chama* are cemented into a hard compact rock, and each bed is characterized by the preponderance of one or more species. The shells are found associated in colonies, which proves that the animals lived and died on the spot where their remains are found. The prevailing genera are *Cha-*

Of what does the polishing slate of Bilin consist? What discovery was made by Prof. Bailey? What fossil shells are found in the United States?

ma, Arca, Pecten, Ostrea, Venus, Crassatella, Crepidula, Pectunculus, and Isocardia.

In the miocene deposits of Europe the same general types are found, although the species are for the most part distinct, and each locality is characterized by different shells. The characteristic shells in the basin of the Garonne belong to the genera *Cardita, Concellaria, Pyrula,* and *Vaginula.* The whole number of fossil species in the various miocene rocks of Europe amount to more than 1000, of which about one fourth are extinct.

(3.) The remains of *Articulata,* such as crabs, lobsters, and insects, are not very abundant, but

(4.) *Fishes,* particularly of the shark family, flourished in great numbers, both in Europe and America, during this and the preceding period.

The tooth of a shark found at Gayhead is much larger than that of any living species: the animal to which it belonged must have been 70 feet in length. A tooth found in North Carolina is still larger, measuring 5 inches in height and 4½ in width: the largest in the collection of Professor Gibbes, of South Carolina, is 6½ inches in height, and 5 inches across the extremities of the root. *Fig.* 69, on the following page, is one of the teeth of the *Carcharodon megalodon,* the largest species of this genus. These animals, if we may judge from the size of their teeth, must have been 100 feet in length. Other teeth are represented among the eocene fossils, page 199. The teeth are rarely altered in composition; the serratures of their edges, and their entire structure, are as perfect in most cases as those of existing sharks. The great abundance of the teeth which have been found does not necessarily prove that these animals were more numerous than some other families of fishes, for the teeth of sharks are arranged in several rows, and are very durable, so that a single individual would furnish a large number. But, from the greater number of species, it is obvious that they attained a much larger development, and prevailed in greater numbers than at the present time.

Describe the fossil shells of Europe. What are the principal remains of fishes? Give examples.

Fig. 69.

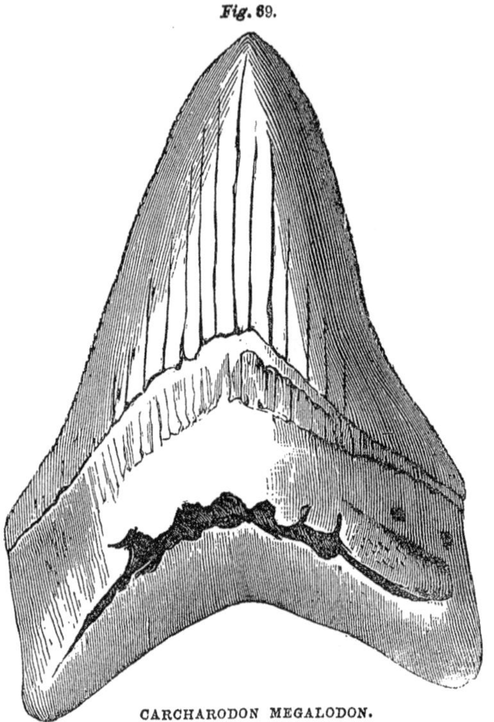

CARCHARODON MEGALODON.

Fig. 70.

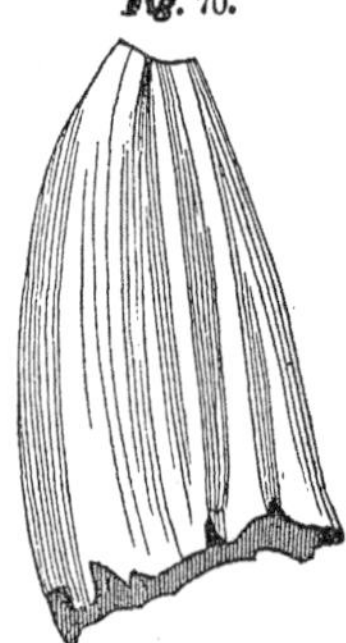

(5.) *Reptiles*, such as batrachians and tortoises, are frequently met with in the lignites, although their remains are not so abundant in any of the tertiary rocks as in those of the secondary periods. *Fig.* 70 represents a tooth of a reptile, evidently a species of crocodile. Some of the teeth of another reptile allied to the crocodile were found in company with the preceding, but the exact locality is not given. These remains probably belong to the miocene of Virginia.—See *Journal of Science, September*, 1850, p. 233.

(6.) *Birds.*—In Auvergne and in Ascension

What is said of the reptiles of the miocene ? of the birds?

Island, a few vertebræ and the eggs of *birds* have been discovered, but their remains are quite rare in the miocene rocks, although this fact does not prove any deficiency of the feathered tribes during this period. See page 102.

(7.) *Mammalia* are abundant in several of the miocene deposits.

The most characteristic mammal of this period was found in the valley of the Rhine, near Eppelsheim, and is called the *Dinotherium*, *Fig.* 71.

Fig. 71.

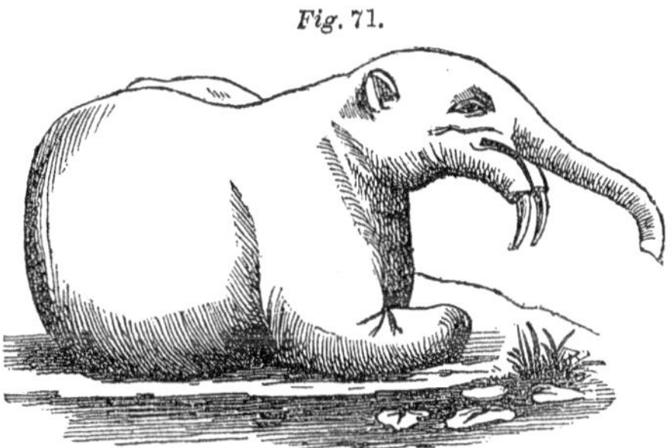

DINOTHERIUM.

From certain fragments of the teeth and head the animal has been restored. It was 18 feet in length, and resembled in its proportions the American tapir. It had two tusks in the lower jaw, similar to those in the upper jaw of the walrus. Fitted by its structure and magnitude for living mostly in water, this animal appears to have formed a connecting link between the aquatic mammalia and the land pachydermata. Associated with these remains were those of thirty extinct species of fossil mammalia.

In the miocene alluvions of Auvergne, more than forty species have been described. The large pachydermata are very abundant, as species of elephant, mastodon, hippopotamus, rhinoceros, tapir, bear, and horse.

There are also the bones of many other families, as of the ox tribe, many species of deer, three species of the cat family (*Felis*), and three species of bears. This locality has also furnished the remains of a species of dog, otter, beaver, hare, water-rat, glut-

---

Describe the most characteristic mammal. What other mammals?

ton, and of a peculiar genus, the agnotherium, which was allied to the dog, but as large as a lion.

The tooth and other remains of a seal were discovered by Dr. Burton in the miocene of Richmond, and have been described by Dr. Wyman in Silliman's Journal for September, 1850. Also the tooth of the *Phocodon* (Agassiz), *Fig.* 72, an animal allied to the basilosaurus of Harlan. See page 203.

In the same deposits Dr. Burton discovered the remains of a *Cetacean*, allied to the whale, a portion of the jaw of which is represented in *Fig.* 73, *a.*

Fig. 72.

Fig. 73.

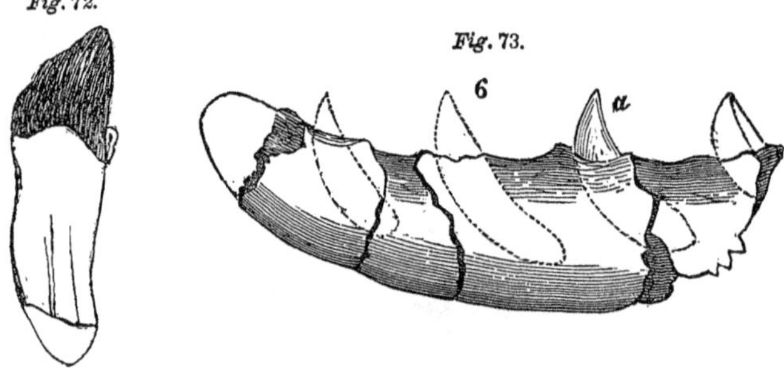

The seas of the pliocene and miocene periods were filled with marine mammalia, such as species of walrus, whale, dolphin, seal, and lamartin; but every species of mammalia, and even some of the genera which flourished at the time, and whose remains are distributed through many beds of rock, have long since ceased to have any living representatives.

They were created and adapted to the conditions under which they were to live; they enjoyed their brief period of existence, and having left in the rocks the imperishable memorials of life and death, have been succeeded by other forms suited in turn to the altered conditions of the earth's surface.

What remains of cetaceans have been discovered? Have they any living representatives? What reason is assigned for this?

SECTION III.—EOCENE OR OLDER TERTIARY.

The strata embraced in the eocene division of the tertiary are better defined, and more extensive and important, than either of the preceding groups.

I. *Geographical Distribution.*—In the United States eocene deposits underlie the miocene, extending from Maryland through Virginia, North and South Carolina, Georgia, and the states bordering the Gulf of Mexico.

In South America there are two extensive deposits, called the Patagonian and Guaranian.

Cotemporaneous strata are spread over the southeast of England, occupying the London and Hampshire basins, the Paris basin, and district of Auvergne in Central France, the basin of the Netherlands, Aix in Provence, the southern flanks of the Alps, the eastern portion of the Grecian Archipelago, and the sub Himalayan range of mountains in Northern India. Of these extensive deposits we can present but a few examples.

II. *Eocene of the United States.*—In South Carolina there are three principal groups.

1. *The Buhr-stone Group.*—This is mostly composed of sand, clay, marl, silicified shells, and bands of iron ore. The following section, *Fig.* 74, from Aiken to Horse Creek, exhibits the order and character of this group :

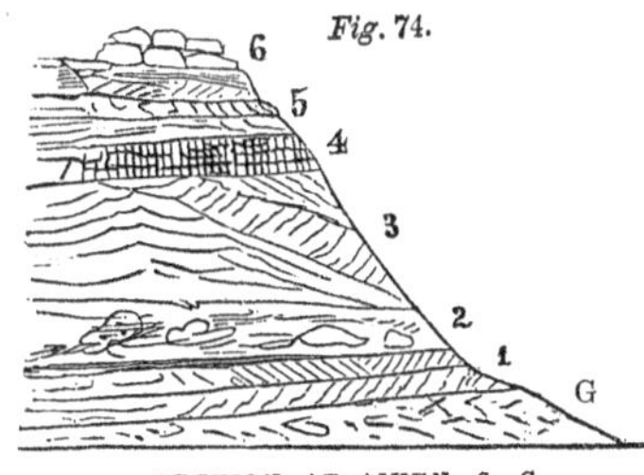

Fig. 74.

SECTION AT AIKEN, S. C.

G. Granite.

1. Beds of sandstone and grit.

2, 3. Beds of sand, gravel, colored clays, &c.

4. Silicious clay bed.

5. Silicified shells.

6. Beds of sand and iron ore.

This section embraces a vertical depth of 200 feet, and the rocks belong to the lower division of the eocene.

How are the eocene deposits distributed? Describe the buhr-stone group. What is said of the section at Aiken, S. C.?

The most important stratum is that containing the silicified shells, marked 5 in the section. It is about 30 feet thick. The shells have left their hollow casts, which are filled with silica, giving a porous structure to the rock, on which account it is admirably fitted for *millstones*. Porcelain clay, of a superior quality, is found in the beds below, the lower part of which exhibits marks of considerable disturbance during its deposition. Associated with the clays are pieces of silicified wood, shells, and bones of extinct animals.

2. *Santee Beds.*—The Charleston basin is about 75 miles long, and 60 miles wide. The eocene beds here consist of white limestone, marls, and green sand. The strata are highly calcareous, and the marls are in some places hard, and in others clayey. This group also occupies the lower portion of the eocene.

3. *Eocene of the Ashley and Cooper Rivers.*—The eocene beds on these rivers occupy a higher position, and, in connection with the Santee beds, are 600 or 700 feet in thickness. They are composed of various marls, and are noted for the remains of sharks, chelonians, cetaceans, and quadrupeds.

The following section shows the relative position of the eocene rocks in South Carolina.

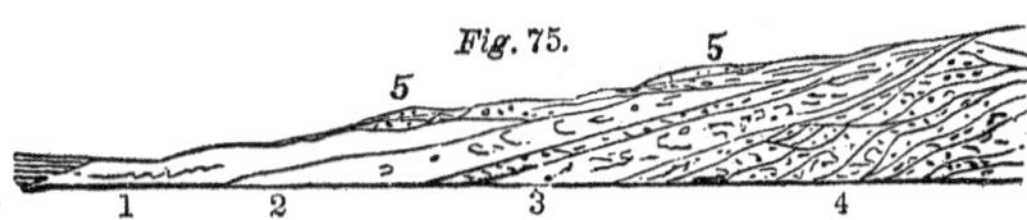

SECTION OF THE EOCENE OF SOUTH CAROLINA.

1. Ashley and Cooper beds.
2. Coralline marl.
3. Santee beds.
4. Buhr-stone formation.
5. Pleistocene.

4. *The Eocene of the James's River, Virginia,* consists also of clays and sands of a greenish tinge. They often contain green sand, gypsum, shells, &c., and repose upon sandstone. The latter is much channeled, showing the long action of aqueous currents previous to the eocene period. Similar beds occur on the Potomac.

Which is the most important stratum? Describe the Santee beds; the eocene of the Ashley and Cooper Rivers; of James River.

5. Eocene deposits are found in all the states bordering the Gulf of Mexico. The most celebrated localities are at Tampa Bay in Florida, Claiborne in Alabama, Vicksburg in Mississippi, and Washita River in Louisiana.

6. The following table will give a chronological view of the eocene of the United States, with the characteristic fossils:

Divisions.	Localities.	Characteristic Fossils.
Upper or Newer Eocene.	Vicksburg, Miss., white limestone of St. Stephen's, and of Claiborne, Al., and part of that of the Charleston basin, S. C.	Scutella Lyelli. " Rogersi. Pecten Poulsoni. Nummulites Mantelli.
	Limestone in the vicinity of Tampa Bay, Florida.	Nummulites Floridana. Crestellaria rotella. Ostrea Georgiana.
Lower or Older Eocene.	Fossiliferous sands of Claiborne and St. Stephen's, Al.; of the Washita River, near Monroe, La.; Pamunkey River at Marlbourne, and eocene green sand on James's River, below City Point, Va.; Fort Washington, Piscataway, and Upper Marlborough, Md.	Cardita planicosta. " Blandingi. Crassatella alta. Ostrea sellæformis. Turritella Mortoni. *Conrad.*

III. *European Eocene.*—The most important tertiary deposits in Europe, as well as in America, were formed during the older eocene period.

In England the eocene is chiefly found in basin-shaped de-

Fig. 76.

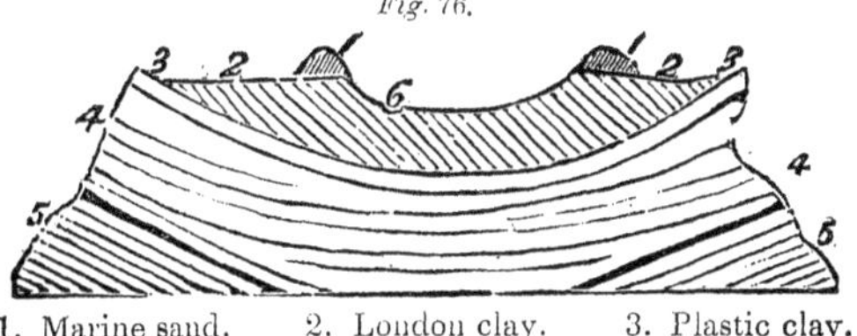

1. Marine sand. 2. London clay. 3. Plastic clay.
4. Chalk. 5. Green sand. 6. River Thames.

1. Marine or Bagshot sand, 400 feet thick.
2. London clay, 350 " "
3. Mottled clays and sand, or plastic clay, 85 feet thick.

What is said of the European eocene? What kinds of rock in the basin of the Thames? Describe them.

pressions of the chalk. There are three principal basins—the basin of the Thames, of Hampshire, and of the Isle of Wight.

1. *Basin of the Thames.*—The deposits here consist mostly of what is called the *London clay.* The strata are divided into three groups, and are represented in the preceding section, *Fig.* 76.

(1.) The beds of the lower division, which form the base of the London clay, rest on the chalk, and often contain rolled fragments of chalk and flint. The deep indentations in the chalk, and the irregularity of the surface, give striking evidence that it was subjected to the action of water for a long time previous to the commencement of the eocene period. In the lower beds, fresh-water shells, drift-wood, and various species of plants have been discovered, but the great mass consists of

(2.) *London clay,* which varies from 300 to 600 feet in thickness. It is of a blackish color, tough, and often mixed with a greenish earth, white sand, and flattened masses of clayey lime stone, called "septaria." The latter contain cracks filled with calcareous spar, and are used for *Parker's cement.*

At the mouth of the Thames, on the Isle of Sheppey, cliffs are exposed more than 200 feet in height, and the total thickness of the clay is not less than 600 feet. This locality has been long celebrated for its fossils. More than 50 species of fishes, a number of crustaceans, and other invertebrate animals, bones of large serpents and of birds of prey, are found; but the most interesting are the fossil fruits, berries, and woody seed-vessels of several hundred species of plants. These fossils are strongly impregnated with iron pyrites, which renders them very brittle and per ishable.

The character of the vegetable remains indicates a tropical climate, and the existence of numerous spice islands in the vicinity during the deposition of these strata.

(3.) *The Bagshot sands* are mostly silicious, but sometimes marly, and are nearly destitute of organic remains. These strata are found capping the Highgate and the Hampstead Hills, near London, and were no doubt formerly continuous over the whole

What is said of the London clay? of the Bagshot sands?

surface, but have been mostly carried away by the subsequent agency of water.

2. *In the Hampshire Basin* the strata are much thicker, and vary somewhat in composition. The following table by Mr. Prestwich exhibits the order and average thickness of the several groups :

(1.) Fluvio-marine and fresh-water series	350	feet.
(2.) Barton clays	300	"
(3.) Bracklesham sands	700	"
(4.) Bangor beds	250	"
(5.) Mottled clay and sand	150	"
Total	1750	

The first three groups lie above the London clay, and the others are cotemporary with it. In the upper groups of this basin and in the Isle of Wight, beds of fresh-water origin alternate with those deposited in salt water. It is evident, therefore, that the deposits were made in an estuary opening into the sea.

3. *Paris Basin.*—The eocene basin, upon which the city of Paris is built, is about 100 miles from east to west, and 180 miles from northeast to southwest. The total thickness is 700 feet.

The following section of the Paris basin, *Fig.* 77, will serve to exhibit the order of the several groups of strata :

Fig. 77.

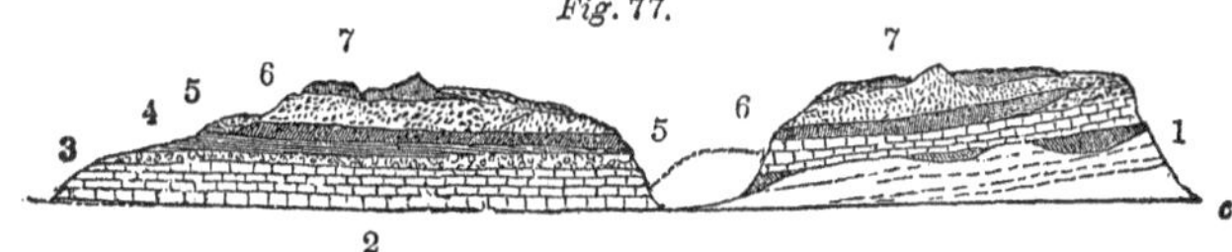

SECTION ON THE BANK OF THE SEINE—PARIS BASIN.

6. Upper fresh-water sands.	7. Upper marine sands.
4. Gypsum.	5. Green marls.
2. Calcaire grossier.	3. Calcaire siliceux.
1. Plastic clay.	*c*. Chalk.

(1.) *Plastic Clay.*—The lowest and oldest stratum is of fresh-water origin, and reposes upon the chalk (*c*). It is composed of

What is the depth and character of the deposits in the Hampshire basin? What is the extent of the Paris basin? Describe the plastic clay.

clay and lignite, and contains rolled pebbles of chalk and of flint, associated with fresh-water shells. The portion of clay called "*plastic clay*" is extensively employed in pottery, and some of it is used for the finest porcelain ware.

(2.) *Calcaire Grossier.*—This stratum is a marine limestone, occupies the northern part of the Paris basin, and is remarkable for the great numbers of fossil shells. More than 400 species have been obtained from a single locality at Grignon, near Paris. The shells are imbedded in a sand composed chiefly of fragments of shells. Nearly 140 species of the genus *Cerithium* have been described. There are also strata of argillaceous and calcareous marls, alternating with limestones, some of which, in the lower beds, are *Nummulitic.* (See page 197.)

(3.) *Calcaire Siliceux.*—The next stratum is a silicious lime stone, a fresh-water formation, apparently deposited from mineral springs. It is almost destitute of organic remains; but its structure is peculiar, being filled with small empty cavities, on which account some of it is employed for millstones (*Buhr-stone*).

(4 & 5.) *Gypseous Series.*—A series of white and green marls, containing gypsum, and also limestone, reposes upon the calcaire grossier, and constitutes a remarkable group of fresh-water origin. The celebrated plaster of Paris quarries are found in this group, and in the center of the basin.

The gypsum is supposed to have been chemically precipitated from the waters of the southern rivers. How the water became impregnated with such large quantities of this mineral is not certainly known; but it was probably due to volcanic agency, which we know was active about this period in Auvergne, near the sources of the rivers. Fluviatile shells, wood, the bones of fresh-water fish, reptiles, birds, and quadrupeds, are abundant. From these strata Cuvier derived the materials, *bones of mammalia,* which formed the basis of his great discoveries, and laid the foundation of the science of *Palæontology.*

(6.) *Upper Marine Sands.*—The next group consists of marls, micaceous and quartzose sands, and beds of sandstone filled with

---

Describe the calcaire grossier; the calcaire siliceux; the gypseous series. What important fact connected with these beds? What other groups in this basin?

marine shells. These beds are separated from the gypsum by a thin bed of oyster shells, and are generally situated upon the summits of the hills.

(7.) *Upper Fresh-water Sands.*—The highest and last group of this basin consists of fresh-water marls, which are interstratified with layers of flint, and contain numerous animal and vegetable remains, especially the seed-vessels of aquatic plants. These deposits appear to have been formed in marshes and fresh-water lakes after the basin had been nearly filled, and portions of it elevated above the water.

The alternations of marine and fresh-water deposits in the Paris basin indicate several elevations and depressions of the land during this period, and yet a few of these groups may have been cotemporaneous, for some of the marine strata were deposited in that part which was open to the sea, while the fresh-water deposits were thrown down near the mouths of rivers.

Fig. 78.

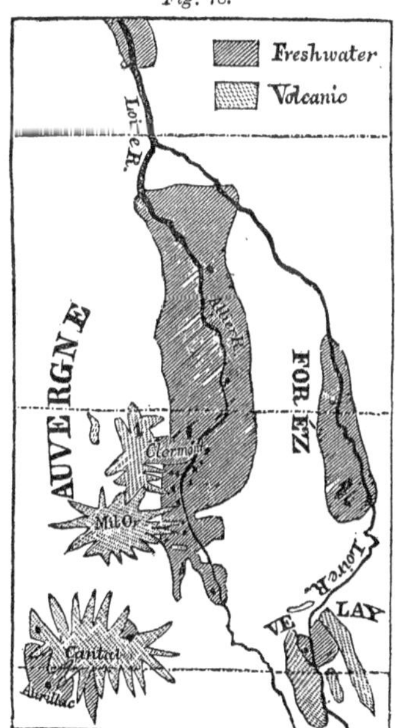

We have then, in this part of France, an ancient gulf of the chalk, opening into the sea toward the northeast. Into this gulf large rivers from the south emptied, bearing along the land animals, plants, and fresh-water shells, which inhabited the land and the water at the time. By a series of elevations and depressions, the sea was alternately admitted and driven out, until the whole basin was filled and elevated to its present position, a monument of the agencies of life and death during inconceivable periods of the past.

4. *Eocene Basins of Auvergne.*—In the center of France there

---

Give a summary of the changes. Describe the eocene basins of Auvergne, Cantal, and Velay.

are three lacustrine deposits of the tertiary, represented in the map on the preceding page, *Fig.* 78, Auvergne, Cantal, and Velay. The strata consist of nearly horizontal beds of sandstone, marls, clays, and limestones, which are all of fresh-water origin, and repose on granitic rocks.

The most northern is in the valley of the Allier, and is developed on each side of that river for a considerable distance, with an average width of 20 miles.

The rocks consist of four groups.

(1.) *Sandstones and Conglomerates* abound upon the borders of the basin. The band, however, is not continuous around the entire margin, but interrupted at different points, apparently by torrents from the neighboring high lands. The beds contain angular and rounded fragments of the primary rocks of the region, granite, gneiss, mica, and clay-slates. Lignite and pieces of wood are found, but no shells.

(2.) *Red Marl and Sandstone,* somewhat resembling the red sandstones of an older period, abound in some parts of the basin.

(3.) *Green and White foliated Marls* next occur near the center of the basin, composed of the same materials, but of very fine texture. The marls are not less than seventy feet in thickness, and thinly foliated. This structure arises from thin scales of the *Cypris,* a genus of small crustaceous animals, species of which now inhabit stagnant pools, and annually cast off their crustaceous envelopes.

These yearly deposits have given rise to divisions in the marl, so that the leaves are as thin as paper, and what most astonishes us is the fact that this structure extends through several hundred feet of vertical depth, indicating a clear and tranquil state of the water, and pointing to the immense periods of time during which the beds were forming.

The three groups were probably in the process of formation at the same time; for while the marls were thrown down in the center of the basin, the coarser materials would be deposited near the margin, and be formed into conglomerates and sandtones.

(4.) *Limestone, Travertine, Oolite.*—The preceding rocks frequently pass into limestones, which in some cases are *oolitic,* that is, are composed of concretions which resemble small eggs. Some

---

Of how many groups do the rocks consist in the valley of the Allier? Describe them

of the limestones are filled with the cases or *indusiæ* of the larvæ of Phryganeæ, and are called *indusial limestones.*

The larvæ of this family are accustomed to attach to their dwellings small shells, grains of coarse sand, &c. The extinct species of the same genus had similar habits. The tubes were thus left, when the animal had passed through its metamorphosis, on the bottom, to be covered with mud, and to give character to the strata. (See page 198.)

(5.) *Gypseous Marls.*—On the right of the Allier there are strata more than 50 feet thick of laminated gypseous marls similar to those in the Paris basin. These beds rest on strata of green cypriferous marls, alternating with grits. The whole constitutes a stratum whose vertical depth is more than 250 feet.

In Velay and Cantal the fresh-water formations are similar to those in Auvergne, although in Cantal the deposits are more silicious, contain flints, and can hardly be distinguished from the chalk.

5. *Nummulitic Formation of the Alps.*—Some of the higher Alps are formed of *nummulitic limestones,* with overlying strata of dark slates. This group was formerly supposed to belong to a much older period, but lately it has been referred to the older eocene, many of the fossils being identical with those of the Paris basin; hence it is inferred that the upheaval of the Alps, with all the contortions and dislocations of the strata, have occurred since the deposition of a considerable portion of the eocene; and that while the London clay was in the process of accumulation, the ocean still rolled its waves over the space now occupied by some of the highest summits of the Alps.—*Lyell.*

IV. *Fossils of the Eocene.*—At the commencement of the eocene period, types of animals and vegetables, closely resembling those which now exist, were brought upon the earth's surface; we should expect, therefore, that these fossils would be similar to those already described. In general this is the fact, but there are some which are characteristic of this period, and some which are common also to the *miocene.*

Of what are some of the higher Alps composed? What reason for placing them in the eocene? What types of animals and vegetables were introduced with the eocene?

1. *Vegetables.*—In the eocene deposits of the United States and Europe, the remains of vegetables often form beds of lignite, in which the leaves and portions of the wood of trees are found in abundance; particularly is this the case in fresh-water basins; but the vegetable remains which have excited most attention are those of the London clay (Isle of Sheppey). Several hundred species have already been determined, all extinct, yet indicating by their structure the tropical character of the climate at the time they flourished.

The remains of the seeds and seed-vessels give to these fossils a peculiar interest. The following figures represent some of the most common of their fruits:

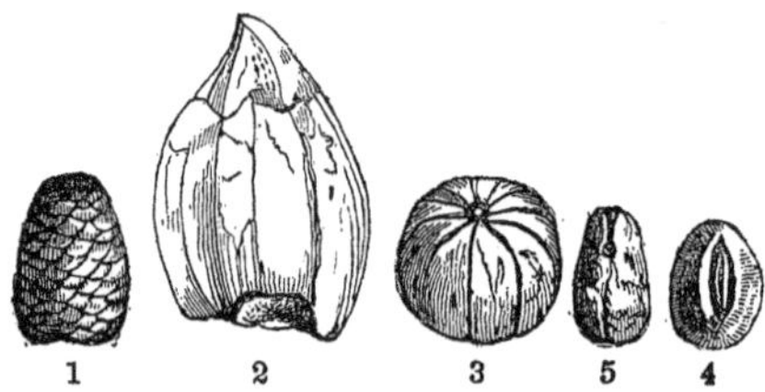

1. Pterophiloides Richardsoni, $\frac{1}{3}$.*
2. Nipadites cordiformis, $\frac{1}{3}$.
3. Cucurmites variabilis, $\frac{1}{2}$.*
4. Wetherellia variabilis, $\frac{1}{3}$.*
5. Faboidea semicurvilinearis, $\frac{1}{4}$.*

The most abundant of these fruits are those called "*fossil figs*" (2). They are allied to the fruits of the *Pandanus* or screw pine, and also to the *Nipa.* The latter is a small tree growing in marshy and clayey places in the Indian Archipelago. From this resemblance the fossil fruits are called *Nipadites.* Thirteen species have been described by Mr. Bowerbank.

Some of these fruits have a striking resemblance to coffee (4). Others are referable to the cucumber and gourd family (3). Others, still, resemble the common scarlet bean (5), with some peculiarities which distinguish them from any living plant. There have been at least 35 species of this latter genus found at Shep-

* Fractional parts of natural size.

The most remarkable locality of fossil vegetables? What is their character? Describe the fossil fruits of the Isle of Sheppey.

pey. There are two bean-like fruits, the one allied to the *Aca-cia*, and the other to the common *Laburnum*.

Mr. Bowerbank has found other fruits allied to the pepper, the cardamom, the cotton-plant, the cypress, and the coniferæ (1).

Most of these families are now represented by the existing tropical plants, although all the species and at least one genus have long since passed from the surface of the earth.

2. *Animals.*—(1.) The lower and microscopic animals were very abundant during the eocene period, although they do not appear to have formed as extensive beds of rock as were deposited during the miocene period; but a family of very minute animals, allied both to shell-fish and polyps, called *Foraminifera*, have contributed to the formation of extensive strata, not only in the tertiary, but in the older rocks.

Some of the shells of these animals are microscopic, and others are of the size of a small coin, and are called *Nummulites*. *Fig.* 79 represents several forms.

Fig. 79.

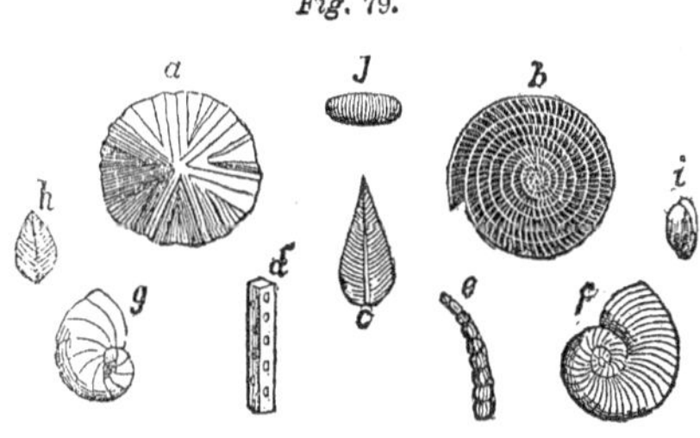

GROUP OF FORAMINIFERA.

a. Lineolaria.	*f.* Operculina.
b. Nummulites.	*g.* Polymorphina.
c. Frondiculina.	*h.* Frondiculina.
d. Vincularia.	*i.* Triloculina.
e. Marginulina.	*j.* Alveolina.

Nummulites are the largest and most widely diffused; their shell is divided into a vast number of chambers, which communicate with each other. Many of the eocene rocks are little else than aggregations of these shells. Several thick beds of the *cal-*

What is said of the lower forms of animals during this period? What family is the largest and most widely diffused?

caire grossier are entirely composed of a small species called the *Miliola*, from its resemblance to millet-seed.

(2.) The London clay contains some corals, but they are generally of small size. Corals and sponges have been observed in all the eocene deposits of the United States. The species indicate a shallow sea in which the animals lived.

(3.) *The Mollusca* of the eocene present us with many species which are not found in the other divisions of the tertiary.

Many of the strata are made up of comminuted shells.

The *buhr-stone* of the United States appears to have been composed at first of shells, but their places have become supplied with silex, although the form of the shell is still retained.

The European eocene is characterized by similar shells, although the species differ, and some 3 or 4 in 100 are identical with those which now inhabit the adjacent seas.

(4.) *The Remains of Crustaceans.*—Lobsters, crabs, &c., abound in the Isle of Sheppey, at which place more than thirty species have been discovered.

A very small crustacean, *Cypris* (*Fig.* 80 is a living species of this genus), contributed largely to the formation of the fresh-water marls in Auvergne (see p. 194). With the marls are associated limestones, formed by an aggregation of the *indusiæ* of the Caddis-worm (*May-fly*). These are little tubes coated with small shells, and bits of wood or straw, and remain after the larva have passed into the perfect state. *Fig.* 81 represents one of these tubes

Fig. 80.

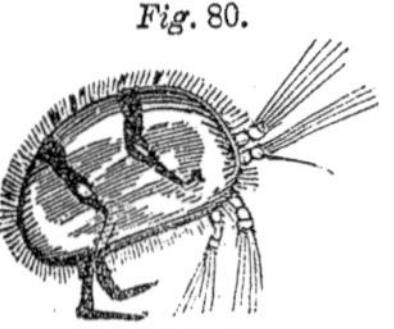

CYPRIS.

Fig. 81.

CADDIS-WORM.

(5.) *Fishes.**—The remains of fishes are abundant throughout

* The number of species of fossil fishes determined by M. Agassiz amounts to 1500.

What is said of the mollusca of the United States? of Europe? Describe the crustacea; the fishes.

the eocene rocks. The most interesting are those of the squalidæ or shark family. In the miocene and eocene of the United States the teeth of several genera, and of a great number of species, have been described by Professor R. W. Gibbes, of South Carolina. From his monograph of this family we select the following examples, *Fig.* 82.

Fig. 82.

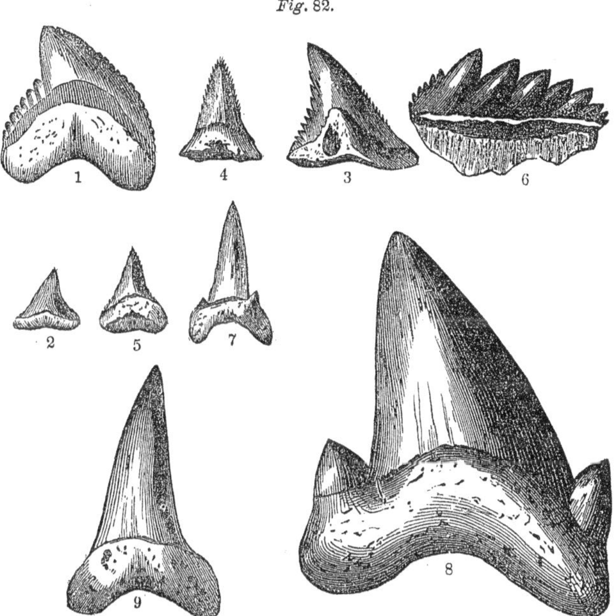

For Carcharodon megalodon, see page 184.

1. Galeocerdo aduncus.
2. " minor.
3. Hemipristis serra.
4. Glyphis subulata.
5. Sphyrna prisca.
6. Notidanus primigenius.
7. Lamna elegans.
8. Otodus obliquus.
9. Oxyrhina hastalis.

Classification of Fishes.—M. Agassiz has divided fishes into four orders, founded upon the peculiar structure of the scales.

I. ***Placoid*** *(broad plate)*.—The skin in this order is covered with enameled plates, as in the shark family. *Fig.* 83, *No.* 1, represents a plate, and *No.* 2 the prickly tubercles of ray fishes.

What fishes most prevailed? What evidence of this? What division has been made of the fishes? What is the basis of the classification?

Fig. 83.

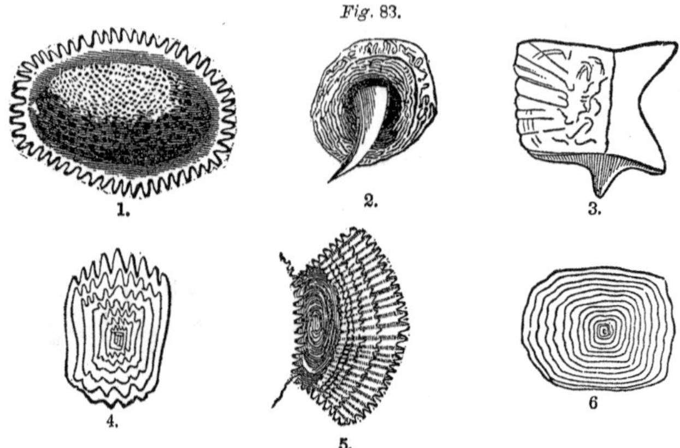

II. *Ganoid* (*splendid*).—The scales are composed of plates of bone or horn, covered with a thick layer of enamel, identical in structure with the teeth, *No.* 3. The sturgeon belongs to this order.

III. *Ctenoid* (*toothed or comb-like*).—The scales are formed of plates, which are toothed like a comb, *Nos.* 4 and 5. The perch is an example of this family

IV. *Cycloid* (*circular*).—In this order the scales are plates of bone or horn, without enamel, *No.* 6. The borders are smooth, but the external surface variously marked. The mullet, salmon, and carp belong to this order.

The fishes belonging to each of the orders mentioned above flourished in very different proportions during the successive geological periods. Those of the Ganoid and Placoid orders were first introduced, and prevailed exclusively till the cretaceous period, when the Ctenoid and Cycloid orders were introduced, and have continued till the present time. They include five sixths of existing species. All the fossil fishes found in the palæozoic rocks are distinguished by having the vertebræ of the tail prolonged into one of the lobes of the tail, so that the lobes are unequal, while most of the fishes since that period have *equally bilobate* tails. The tails with unequal lobes are said to be *heterocercal*, and those with equal lobes *homocercal.*

Existing examples of fishes with heterocercal tails are sharks and sturgeons.

Coprolites.—Associated with the fishes are their coprolites, or

How are the orders distributed through geological times? What are found associated with the remains of fishes?

fœcal remains; some of them are of large size, and indicate by their structure the character of the animals which existed at the time. One of the Coprolites found at Richmond was $6\frac{1}{2}$ inches in length, and 3 in diameter The remains occur in most of the fossiliferous rocks.

Two species of *saw-fish* (*Pristis*) have been found in the Isle of Sheppey, and one species at least in the United States, of which *Fig.* 84 represents a tooth.

Fig. 84.

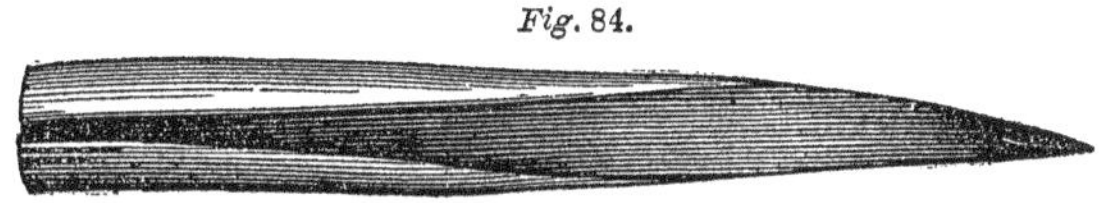

TOOTH OF PRISTIS AGASSIZI.

The genus *Myliobatis* (*Eagle Ray*), of which there are five living and fifteen fossil species, was allied to the preceding. Their teeth occur in the Isle of Sheppey and in the eocene of the United States.

Fig. 85 represents the appearance of the teeth of *Myliobatis Holmesii*, from South Carolina.

Fig. 85.

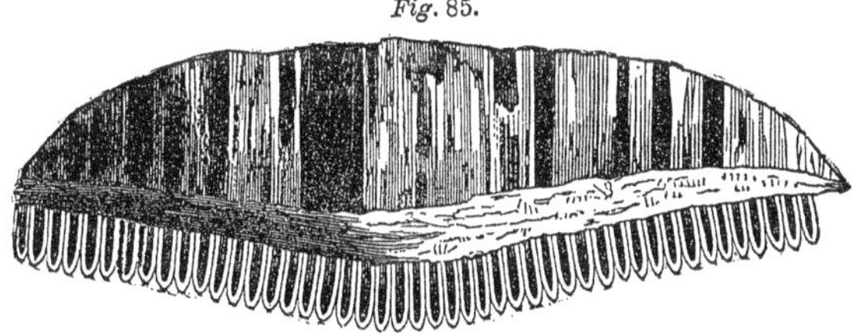

MYLIOBATIS HOLMESII.

"The genus is characterized by broad, transverse teeth, on a flat plate, bounded laterally by three rows of narrow hexagonal teeth of equal length, united by fine sections. The arrangement resembles that of a tesselated pavement."—*Gibbes.*

At the celebrated locality of Monte Bolca, the remains of a gigantic torpedo was discovered, with a large number of fishes belonging to several families.

At Aix, in Provence, a small slab of marl was found which

Describe the fishes found in South Carolina. Mention other localities.

contains multitudes of small fishes as perfect as if recently imbedded in soft mud.

The fishes of the tertiary period resemble those that now exist; but all of the species and many entire genera (about one third of those belonging to the eocene) are extinct.

(6.) *Reptiles.*—The remains of reptiles are not so numerous in the tertiary as in rocks of an earlier date, but the vertebræ of a serpent found in the Isle of Sheppey is the earliest indication of the ophidian tribes on the surface of our planet. It was allied to the *Boa*, and must have been 20 feet in length. The crocodiles of the London clay were more nearly related to those now living in the island of Borneo than to those of the Nile. Turtles are more numerous, and some of them attained a great size. The remains of a turtle found in the sub-Himalayan tertiary measured twenty feet across the curve of the shell. The fossil turtle is distinguished from those that now exist by possessing a combination of the jaws and buckler of the fresh water with the bony helmet of the marine species.

Birds.—The remains of birds are rare in the fossil state. From the gypseous quarries of Montmartre, near Paris, Cuvier discovered nine or ten species, related to the pelican, sea-lark, curlew, woodcock, owl, buzzard, and quail. In a few cases the skeleton was nearly entire. The bones of a bird allied to the vulture occur in the Isle of Sheppey.

Mammalia.—In the United States the remains of a remarkable cetacean mammal were first noticed by Dr. Harlan, of Philadelphia, who, in 1832, described a gigantic vertebræ, from the Washita, weighing 44 pounds, and referred it to a new genus, which he called *Basilosaurus* (*king of Saurians*).

In 1843, nearly an entire skeleton was found in the eocene limestone of Alabama, and is now in the possession of Dr. J. C. Warren, of Boston. The length of the animal, when living, could not have been less than 70 feet. Dr. R. W. Gibbes, of Columbia, South Carolina, has described three species.

What reptiles are found in eocene deposits? Why are birds rare in the fossil state? What remarkable mammal in the United States?

The following figures are representations of the teeth and bones of these remarkable animals :

Fig. 86.

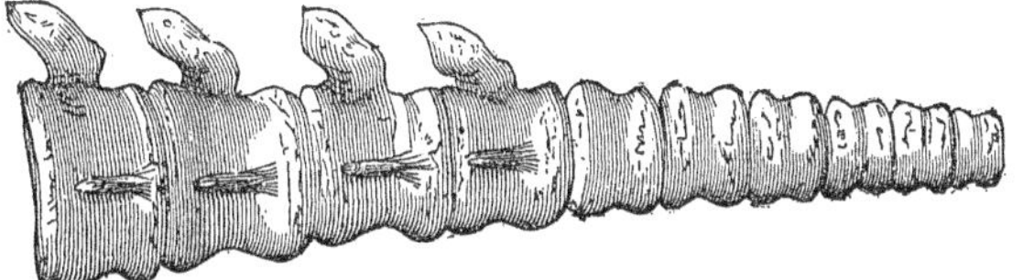

Caudal Vertebræ of Basilosaurus Cetoides.

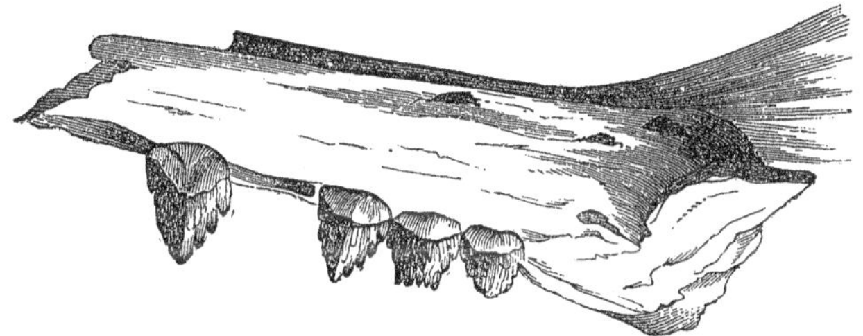

Fragment of Upper Jaw of Basilosaurus Squalodon.

Fig. 87.

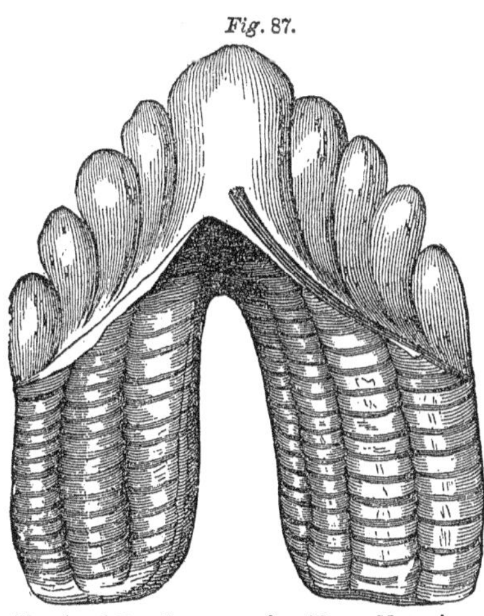

Tooth of Basilosaurus Cetoides. Nat. size.

The reference of these animals to the cetacean mammals, or whale tribe, has been shown satisfactorily, not only from the character of the teeth, and their insertion in the jaw (*Figs* 87 and 88), but from the structure of a cranium discovered by Prof. Lewis R. Gibbes, and described by Prof. Tuomey. The geological position of these remains in South Carolina is in the older, and also in the upper eocene beds.

Mr. Koch, a German collector, a few years since obtained a large number of vertebræ, belonging to many individuals, with the bones

---

What reason for referring these animals to the cetacean family?

Fig. 88.

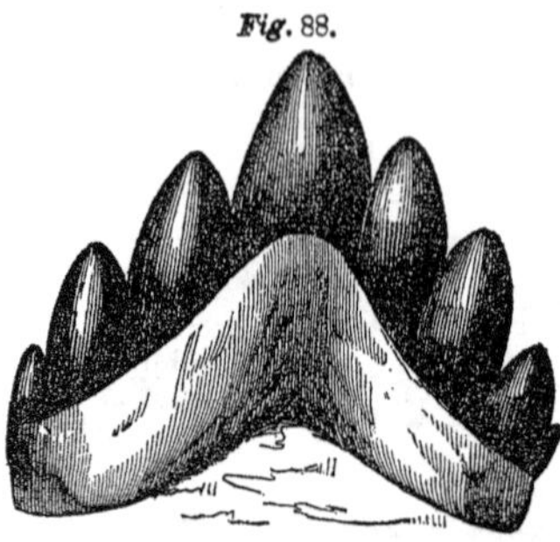

Basilosaurus Serratus, Gibbes.

of other animals. With these and other materials, he constructed a skeleton of enormous length, and exhibited it to the public with the name of *Hydrarchus*. The imposition was at first successful, until seen and exposed by Prof. Wyman, of Boston.

Since these bones extend throughout the eocene strata, these animals must have lived during a period sufficiently long for the deposition of several hundred feet of rock.

The most characteristic mammals of this period are the *Pachydermata*. Four fifths of the fifty species of quadrupeds found in the Paris basin belong to one division of this order, represented at the present day by only four living species, of which three are tapirs.

The two genera, *Palæotherium* and *Anoplotherium*, are only found in this early part of the tertiary period. The *Palæotherium*, *Fig*. 89, was intermediate between the tapir and rhinoceros. Twelve species of this genus have been described, varying in size from that of the rhinoceros to that of the pig. The animal had a short fleshy trunk, and was adapted to live in swampy districts, feeding on coarse vegetables, like the tapir of America and Asia, to which it was very nearly allied.

Fig. 89.

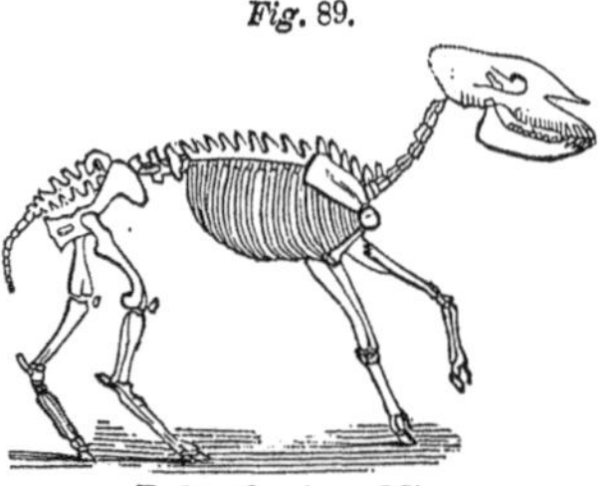

Palæotherium Minor.

The genus *Anoplotherium* contains but six species, varying much in size, from the dimensions of a pony to that of a hare.

Anoplotherium Commune is most abundant in the Paris basin. It was 4 feet high and 8 feet long, of which 3 feet belonged to the tail. It resembled the otter, and lived in marshy places.

Anoplotherium Gracile was a much more graceful and slender animal, resembling in its structure the gazelle. *Fig*. 90.

During what period did these animals live? What is said of the pachydermata? Describe the palæotherium and anoplotherium.

Fig. 90.

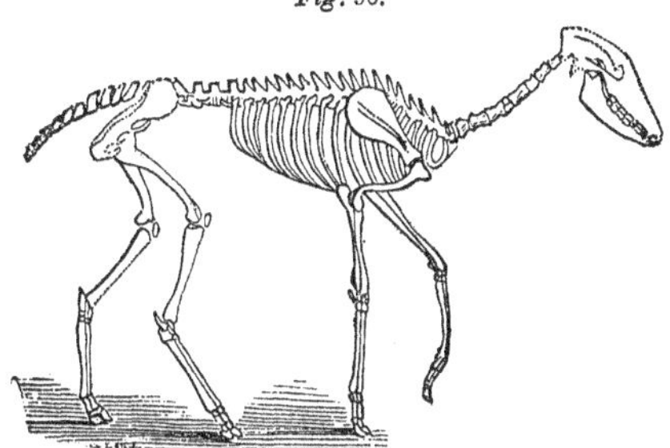

Anoplotherium Gracile.

Anoplotherium Leporinum, a much smaller species, resembled the hare. The bones of another small pachyderm, still more like the *hare*, have been found in the London clay, called *Hyracotherium*, and a very rare genus called *Chæroptamus*, similar to the preceding, in the Paris basin.

Of the remaining mammals of the eocene, two species of dormouse and two of the squirrel, an opossum, and a bat, have also been found in the Paris basin. The remains of monkeys occur in Brazil, India, and in England.

In the Sewalik Hills, in Northern India, many remarkable fossils have been discovered, in which are many species of pachydermata, and of the other orders of mammals. Of the pachydermata there are two species of mastodon, two new species of elephant, several new species of rhinoceros and hippopotamus; several allied to the hog; three new species of horse, one of which had the slender and graceful figure of an antelope.

The same district contains the remains of many *Ruminants*, as two or three species of giraffe, two species of camels, and many species of antelopes, deer, oxen, and other existing genera. But besides these, there is a new genus, the *Sivatherium*, a ruminant larger than the rhinoceros, having many affinities with the elephantoid animals. Its head was large, and provided with two pairs of horns, with a large upper lip, probably extended into a

What other animals belong to the eocene? What animals have been found in Northern India?

short trunk, small eyes, and teeth fitted to masticate leaves and twigs rather than grass. It appears to have been intermediate between the ruminants and pachyderms.

Most of the other orders of mammals are found among the fossils of this remarkable region.

We find in the tertiary rocks all the orders of mammalia, with the exception of the human species; the remains of man have never been found in any but the most recent deposits, and hence the changes which we have described as belonging to the tertiary period were all antecedent to his existence on earth.*

SECTION IV.–CLIMATE AND PHYSICAL GEOGRAPHY OF THE TERTIARY.

1. *Climate.*—The climate of the United States during the tertiary period, as indicated by the character of the organic remains, was considerably warmer than at the present time. This fact is indicated by the character of the fishes, which are generally limited in their distribution by temperature. They are allied to those existing within the tropical seas. The sharks, whose remains are found in Massachusetts, Maryland, and Virginia, belonged to the larger genera, which now live much nearer the equator.

In South America, according to M. D'Orbigny, the climate was the same as at present. In Europe, the character of the fossil fruits of the Isle of Sheppey indicates a tropical climate. This difference of climate was due to two causes: volcanic agency, and the greater preponderance of water in the northern hemisphere; in other words, to

2. *Physical Geography.*—The proofs of changes in physical

* According to Mr. Tennant's catalogue of British eocene fossils, there are of plants 100 species, and the following species of animals:

Zoophytes	4	species.	Conchifera and brachiopoda	235	species.
Echinoderms	5	"	Cephalopoda and gasteropoda	267	"
Foraminifera	8	"	Fishes	97	"
Annelida	11	"	Reptiles	14	"
Cirrhopoda	3	"	Birds	1	"
Crustacea	4	"	Mammals	14	"

Are all the orders of animals found in the tertiary? What is said of the remains of man? What was the climate of the United States during the tertiary period? How is this known? What changes in physical geography?

geography since the commencement of the tertiary, and which may be traced directly to volcanic action, are found in the north of Italy, in Sicily, in Spain, in Greece, and the neighboring islands, while in the center of France and on the banks of the Rhine numerous extinct volcanic craters, with their erupted materials, still exist. (See page 181.)

In the West India Islands, in Central America, and in many parts of the Andes, the changes produced by volcanic agency during this period are no less clearly indicated than in the eastern continent.

It has lately been shown by Sir Charles Lyell that the Alps have been uplifted since the commencement of the eocene period.

According to M. de Elie de Beaumont, there have been five principal directions of elevation during the tertiary period; and Johnson, in his physical atlas, has given the following view of the changes on sea and land:

"During the epoch, the Mediterranean, or another great and corresponding inland sea, covered the deserts of Sahara, Lower Egypt, and part of Arabia; for not till long after the commencement of the eocene were the present contours defined, and the lagunes and ancient shores left dry.

Later still, the Strait of Gibraltar did not exist, and the waters of our inland sea mingled through the channels of the Red and Persian Gulf with those of the Indian Ocean, which seems to explain the analogy of the fossils of the middle and higher tertiary Mediterranean beds, with creatures still living in the Red and Indian Seas, and with fossils of corresponding age in the great basin of the Black Sea and Caspian. At the same epoch, too, the North Sea and the Baltic spread over the plains of Northern Europe, and another ocean stretched from the recesses of Siberia and joined with the Mediterranean by the Black Sea.

Asia Minor contained small isolated basins, though the Black Sea on the south and east was confined by its present banks.

When were the Alps uplifted? How many directions of elevation? What is said of the Mediterranean Sea? of the Straits of Gibraltar? of other portions of Europe and of Asia?

In the south of Asia a broad sheet severed the peninsula of India, then a triangular island, from the chains of the Himalaya and their dependents; and there existed also a great fresh-water basin in the peninsula beyond the Ganges, two other basins at least in China, one on the banks of the Lower Amour, and two in Siberia. As in the case of Europe, the center of this continent was covered by an inland sea, which has now wholly disappeared. Other aqueous masses covered Persia, and probably formed, later even than the tertiary epoch, one basin dependent on the Caspian, and another annexed to the Indian Sea. In another district of the continent, large portions of the Isles of Sunda, the Philippines, Borneo, New Guinea, and Australia were at this period under the waters; and many volcanic peaks, now existing and belonging to great areas of elevation, had not yet risen above the surface of the Indian or Malay Seas.

Turning to the map of America, we discern evidences of changes equally singular and extensive. The Gulf of Mexico then penetrated deep into Mexico, Florida, the lower basin of the Mississippi, and also into the basin of the northern rivers of South America. It washed the southern extremity of the Alleghanies as well as the foot of the Ozark Mountains, and the Mexican and Columbian platform. Further north, a great interior ocean overspread a part of this continent, comprehending the higher Mississippi and the great lakes.

[The ocean washed the eastern margin of the whole Appalachian chain in the United States.]

The Gulf of Mexico already contained a few islands composed of older formations, probably of much larger size than those whose shores it now washes, but its volcanic isles sprung into existence subsequently, during that series of subsidences and elevations (of *écroulements*) of the chains along the ancient shores of South America, which drove the sea from the Ozark Mountains and the Alleghanies, and fixed its limits further south.

The northern part of the continent had three islands, the basin

---

What were the general outlines of the American continent? the Gulf of Mexico?

of the St. Lawrence separating the district of the Alleghanies from dry land on the banks of Hudson Bay, and perhaps bend ing round to the icy sea.

"The platform of Mexico and Guatemala formed an appendix of the long isle of the Rocky Mountains, and the Ozark chain advanced into the waters.

"The volcanoes of Continental America, as we see them now, were cotemporaneous with the formation of Mexican and Mediterranean basins.

"In South America we discern abundant proof that at the tertiary epoch the Atlantic covered the great strait between Brazil, the Andes, and Central Guayana, as well as between the Panama and the chain beyond the Orinoco; whence the mingling of the tributaries of this and the Amazon, as well as the mode of the division of the waters between certain affluents of the La Plata and the Amazon. South America was then composed of three great islands, for the Isthmus of Panama did not exist."

It is obvious that such changes in physical geography as took place during this period must have produced corresponding changes in climate.

CHAPTER VI.

SECONDARY PERIODS.

Having in the preceding chapter described those deposits which were made mostly in estuaries, shallow seas, and fresh-water lakes, we now approach the shores of those great oceans in which were accumulated the materials that constitute a large portion of our present continents and islands.

The groups of strata which lie immediately below the tertiary, and extend to the base of the triassic, or red sandstone series, are

---

Of what was South America then composed? In what respects does the secondary differ from the tertiary?

called *Secondary*, also *Mesozoic*, because deposited during the middle period of geological history. The character of these groups, mineral and organic, indicate conditions under which they were formed, somewhat different from any hitherto considered.

In passing from the oldest of the tertiary to the newest of the secondary rocks, there appears to be a wide gap, which the investigations of geologists have not as yet succeeded in filling up. The chalk strata give abundant evidence of the action of aqueous currents, by which their surface has been eroded and large portions carried away, leaving the basins which contain the oldest tertiaries; but the deposits formed by this denudation of the chalk have not been found—possibly they may be beneath the present oceans. But, according to our present knowledge, the close of the secondary period terminated the ancient order of the earth's surface, with all its inhabitants, and the beginning of the tertiary was the commencement of an entirely new creation of animals and plants, and of essential modifications in the physical geography of the earth.

In pursuing our investigations, therefore, of the secondary rocks, we take leave of almost every thing that would remind us of the present state of the earth. The remains of the large mammalia are no longer found, and the types of animal and vegetable life are altogether new and peculiar. We appear to enter upon a period of the earth's history when all of its ancient types were wearing out, so that not a single species (if we except a few shellfish of doubtful species), and but few genera, had sufficient vitality to stretch across the gulf of time, which separated the secondary from the tertiary epoch.

The rocks also are more uniform in structure and composition, and extend over much larger areas. The tertiary is characterized by its *individuality;* the secondary by its *universality.* The rocks of the secondary epoch are arranged in five divisions or systems, each of which is variously subdivided:

What is said of the period between the tertiary and secondary? What changes are observed in entering upon the secondary periods?

Newer secondary	1. Cretaceous system.
Middle secondary	2. Wealden. 3. Oolitic series.
Older secondary	4. Liassic series. 5. Triassic series.

SECTION I.—CRETACEOUS SYSTEM.

The deposits of the cretaceous period are more fully developed, and their character has been more thoroughly studied in Europe than in this country. It may serve, therefore, to give a clearer view of the American cretaceous system, if we commence with the European.

I. *European Cretaceous Series.*—The several groups of strata which compose the cretaceous system are better developed in England than any other part of Europe. The following table will exhibit the order of the several groups, and indicate their geographical distribution:

	English, Danish, and Belgian series.	French series.	German series.
Upper Series.	1. Chalk of Faxoe, Denmark. 2. Mæstricht beds, Belgium. 3. Upper chalk, with flints, England.	Terrain senonien.	Upper Quader sandstone.
	4. Middle and lower chalk and clunch, do. 5. Chalk marl, do.	Terrain turonien. 1. Craie chloritée. 2. Glauconie crayeuse. 3. Craie tufau.	Upper Pläner limestone. Middle Pläner. Lower Pläner.
	6. Upper green sand, do. 7. Gault, do.	Terrain albien.	
Green sand Series.	8. Argillaceous beds, do. 9. White and green sandy beds and clay, do. 10. Kentish rag, do. 11. Atherfield clay, do.	Terrain aptien. (Argile plicatules.) Terrain néocomien. 1. Argiles bigarrées. 2. " ostréennes. 3. Calcaire a spatangus.	Lower Quader sandstones. Hilsthon. Hils-conglomerat.

The following section at Lulworth Cove represents the order of this group in England, and its connection with the Wealden:

How is the secondary divided? Where has the cretaceous system been most thoroughly examined?

Fig. 91.

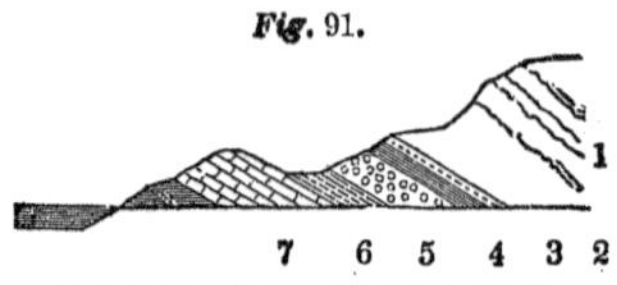

SECTION AT LULWORTH COVE.

1. Upper chalk
2. Lower chalk.
3. Upper green sand.
4. Gault.
5. Lower green sand

Wealden............. { 6. Purbeck beds. 7. Hastings's sand.

1. *Upper Chalk.*—The upper beds of the cretaceous system in England consist of white chalk, which is nearly pure carbonate of lime, 56·5 lime, 43·0 carbonic acid, and 0·5 water. When examined by a microscope, it is found to be almost wholly composed of the disintegrated remains of infusoria, zoophytes, and small shells. Associated with the beds of chalk are bands of black flints. The flints are generally in nodular masses, having an organic nucleus of delicate sponges, and of small animals. They appear to have been formed after the deposition of the chalk, partly by segregation, and partly by the passage of water holding silica in solution, through the chalk beds, analogous to the deposition of bog-iron ore around organic bodies at the present day.

2. *Middle and Lower Chalk.*—The middle and lower beds of the chalk contain but few flints. The grains of silica are distributed through them with argillaceous matter and iron, on which account they are much more compact, and portions are used for the internal work of cathedrals and other public buildings. This is locally called *Clunch.*

3. *Chalk Marl.*—The next group is similar, being covered with green particles of silicate of iron, or dark red particles of oxide of iron.

Beds of the same age in Germany are marly limestones of a gray color, and contain about 75 per cent. of carbonate of lime.

The chalk strata pass through England from Hampshire, in a northeasterly direction, to the Yorkshire coast, separating into

---

Describe the upper chalk. What are associated with it? How were they formed? Describe the middle and lower chalk; chalk marl. Where is the chalk found?

two spurs toward the east, which form the North and South Downs. The thickness is about 1000 feet. Across the Channel, white chalk forms cliffs on the shores of France, opposite Dover and Beechey Head; also in Denmark, Poland, the south of Russia, and the Caucasus Mountains. Cotemporaneous rocks in Germany consist of loose sandstones, with but a few, about 30 species of fossils, while nearly 3000 have been found in the English chalk alone.

4. *Upper Green Sand.*—(3.) Immediately beneath the chalk marl we find a group of argillaceous and silicious limestones, colored green by silicate of iron, and hence called *green sand.* This group is represented in Saxony and France by sandstones; some of those rocks are used for lining furnaces, and hence are called "*fire-stone;*" others are quarried for "*whet-stones.*" In the Isle of Wight the beds attain a thickness of 100 feet.

5. *Gault.*—(4.) The gault is a fine blue clay, much used for bricks. This clay, in its lower beds, is filled with iron pyrites, and in the upper with silicate of iron. The total thickness is 150 feet.

Fig. 92.

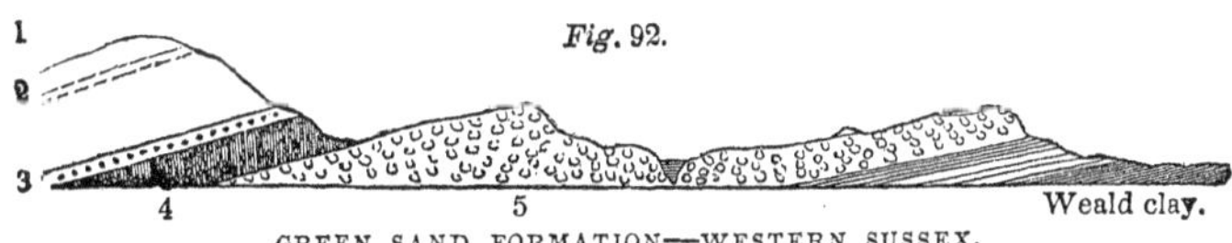

GREEN SAND FORMATION—WESTERN SUSSEX.

6. *Lower Green Sand.*—The lower green sand, marked 5 in section, contains the following kinds of strata:

(1.) The first are *argillaceous beds*, containing considerable quantities of iron and sand.

(2.) The second group consists of *white* and *green* sand, alternating with beds of clay.

(3.) The third stratum is a coarse limestone, called the *Kentish rag*, which is employed as a rough building-stone.

(4.) Beds of dark clay, lamellated in structure, and containing nodules of clay iron-stone, form the fourth and last group, and

---

What is said of the cotemporaneous rocks in Germany? Describe the upper green sand; the gault. Mention the kinds of rock in the lower green sand.

these repose directly upon the Wealden, forming the base of the cretaceous system. The cotemporaneous rocks in France and Germany are mostly sandstones and limestones.

II. *American Cretaceous System.*—Deposits of the cretaceous period extend in the form of a crescent from New Jersey, through North and South Carolina, Alabama, Tennessee, &c., to Texas, a distance of three thousand miles. The strata dip beneath the tertiary, and lie a little west and north of that formation. Cretaceous rocks also occur high up the Missouri River, as far as 50° of north latitude, and spread out toward the Rocky Mountains and New Mexico, embracing an area larger than any other of this formation on the surface of the earth. The strata, however, belong to the newer portion of the cretaceous system, lying between the Mæstricht beds and the gault. In Mexico older beds occur, and also in South America, attaining a thickness in New Granada of 5000 feet.

They extend along the Andes to the extreme part of the continent, in latitude 53°. The strata are met with high up among the Andes, attaining an altitude of 13,000 feet above the sea; but they do not reach the plains either to the east or west.

Character of the Strata.—White chalk is not found in the United States; but instead of it, we have sandy limestones and argillaceous beds. The upper green sand has an equivalent in the green sand of New Jersey, which is a stratum 30 or 40 feet thick, spread over the southern half of the state. The green sand is made up mostly of grains about the size of gunpowder, of a greenish color, and consists of the following substances:

Silica	49·83	Water	9·80
Alumina	6	Loss	0·89
Magnesia	1·83		100
Potash	10·12		
Protoxide of iron	21·53		*Seybert.*

The silicate of iron and of potash render it one of the most fertilizing substances for the land. Further south, in North and

Where are the principal deposits of the American cretaceous system? What is the character of the strata? What is the composition of green sand? For what is it used, and why?

A piece of the Buttonwood tree that stood in front of No 1151 Broadway between 26th & 27th Streets

It was planted in 1620 during the settlement of Manhattan island by the Dutch, and was situated upon the old Varian farm. Was cut down Nov 3rd 1869. Measured 5 feet 10 inches in diameter, nearly 18 feet in circumference and about 150 feet in height.

C. Mathews.

South Carolina, we find thick beds of limestone, the proportion of carbonate of lime increasing, until in Texas the strata consist of compact silicious limestones. Lignite and numerous fossils are found in many of these beds.

Fig. 93 is a section showing the position of the strata in South Carolina.

Fig. 93.

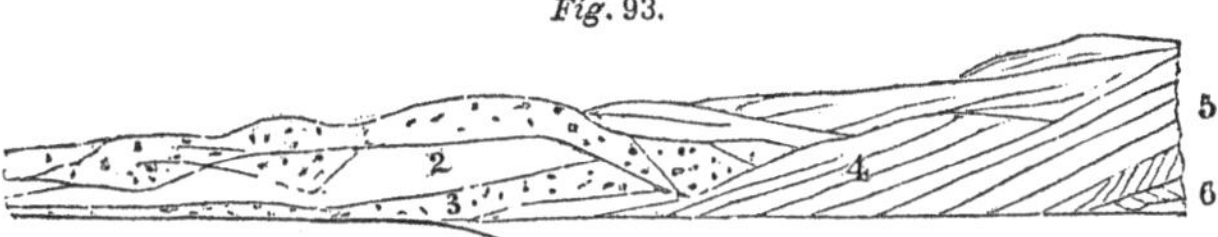

SECTION OF THE CRETACEOUS ROCKS OF SOUTH CAROLINA

We have (1st) the *tertiary beds;* (2d) a bed of *marl;* (3d) *marl stone,* a stratum about three feet thick; (4th) *gray marl;* (5th) *shale;* (6) *sand* and *blackish clay.* The thickness of these strata is very inconsiderable.

There is abundant evidence of the erosion of these rocks by water before the tertiary was deposited. A similar fact was noticed in reference to the upper cretaceous rocks in Europe.

III. *Fossils of the Cretaceous System.* 1. *Vegetables.* — The remains of vegetables are not very abundant in the cretaceous rocks. Beds of lignite, however, occur in the United States, and many species of Xanthidium have been found incased in the flint of the chalk.

Fig. 94.

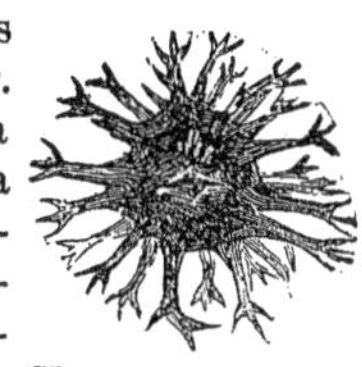

XANTHIDIUM RAMOSUM.

Fig. 94 represents one of these delicate vegetable bodies very highly magnified. These were formerly supposed to be animals. No. 2 is a thin transparent chip of flint, which, on being viewed by transmitted light through a microscope, exhibits five Xanthidia. These bodies vary from $\frac{1}{300}$th to $\frac{1}{500}$th of an inch in diameter. Some of the flint pebbles are filled with them, a group of 20 having been found in an area of a line in diameter. They appear as globular bodies, with minute branches, as in the figure opposite. It is probable that they formed a nucleus around which the silicious particles accumulated, and thus were formed into permanent

What fossil vegetables occur, and in what form?

herbaria, which the blights of time can neither tarnish nor destroy.

2. *Animals.*—We have already observed that the white chalk was mostly composed of the remains of infusoria, small corals, and shell-fish.

(1.) According to M. Ehrenberg, a cubic inch of chalk may contain upward of one million of well-preserved animalculites and shells. In addition to these microscopic animals, great numbers of *Foraminifera*, small shells, spines of sponges, and of echinoderms, the scales and teeth of fishes, pass through the field of the microscope. The most abundant of the microscopic animals belong to two genera, *Rotalia* and *Textularia*.

These are minute chambered shells, resembling the nautilus in form, but very different in structure, as their viscera occupy the separate chambers of the shell.

Fig. 95 (2) represents fragments of a branched sponge in the hollow of a flint nodule, and *Fig.* 96 the convex surface of a small coral. *Fig.* 97 represents one of the corals common in the United States, *Anthophyllum Atlanticum* (Morton).

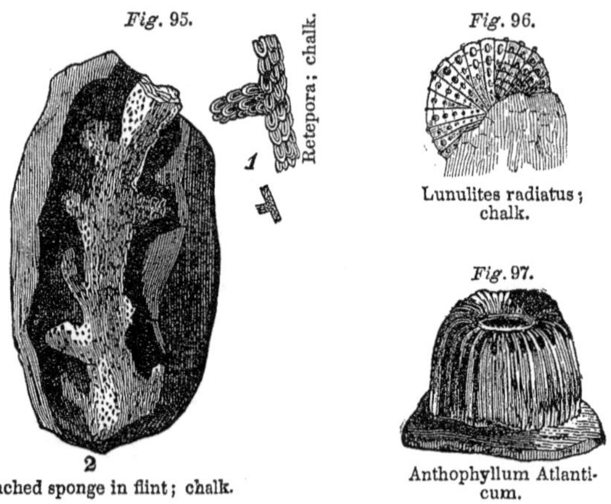

Fig. 95. Branched sponge in flint; chalk.

Fig. 96. Lunulites radiatus; chalk.

Fig. 97. Anthophyllum Atlanticum.

Many of the flints derive their forms from the form of the pol·

What are the principal fossil animals of the chalk? How have they been preserved? From what have the flints derived their forms?

ype, which constitutes the nucleus around which the silicious particles were collected. *Fig.* 98 represents two of these forms, derived from a *ventriculite* called "*petrified mushroom.*" *Fig.* 99 represents the perfect animal in chalk.

Fig. 98.

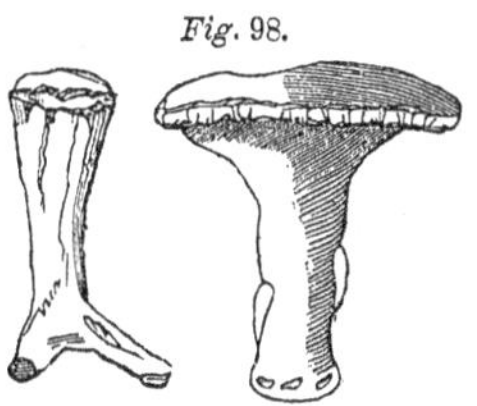

Ventriculites in flint.

Fig. 99.

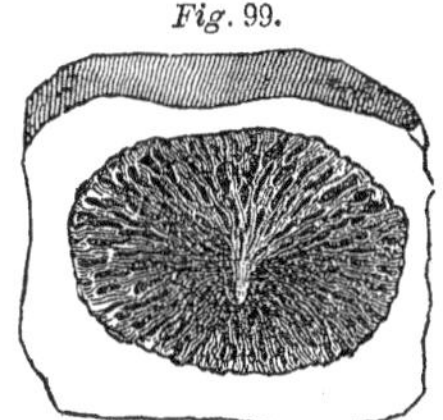

Ventriculites radiatus; chalk.

(2.) *The remains of Echinodermata* are very abundant in the upper part of the cretaceous rocks; stone-lilies, star-fishes, and sea-urchins are the principal families, and are represented in the accompanying figures of English fossils. *Goniaster semilunata* (chalk, 100); *Spatangus cor-anguinum* (chalk sands, 101); *Cidaris diadema* (upper chalk, 102); *Ananchytes ovatus* (upper chalk, 103).

Fig. 100.

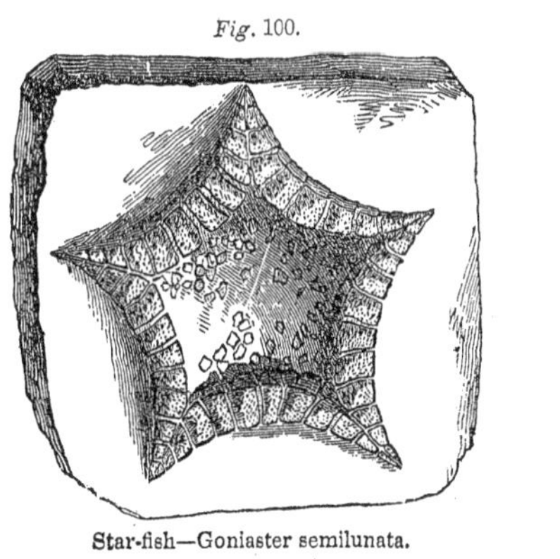

Star-fish—Goniaster semilunata.

Fig. 101.

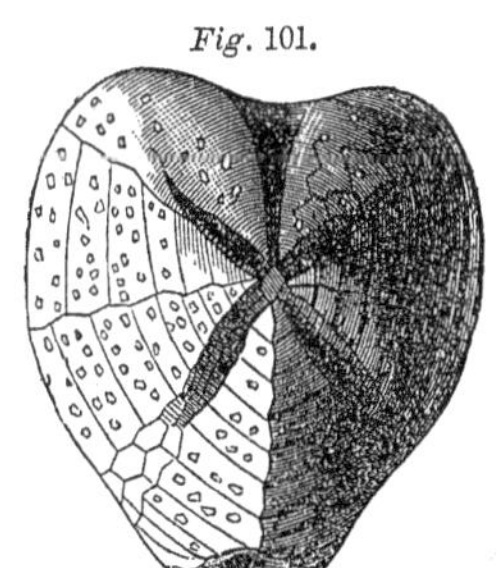

Spatangus cor-anguinum

Fig. 102.

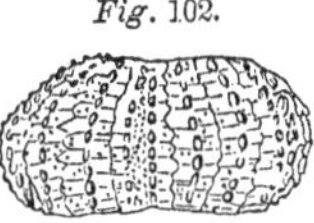

Cidaris diadema.

Where are the remains of echinodermata abundant? Mention the principal families.

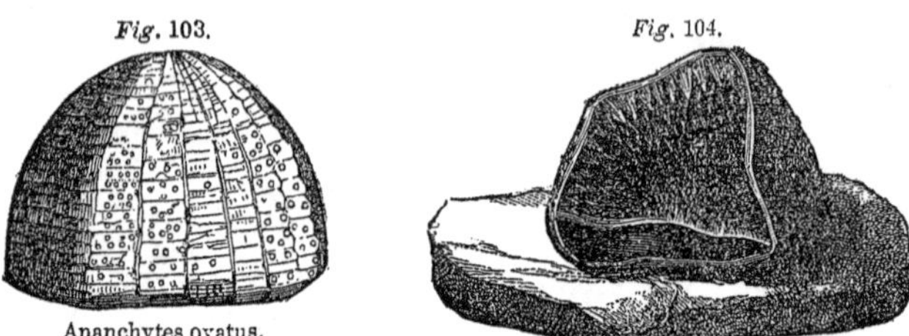

Fig. 103. *Fig.* 104.

Ananchytes ovatus.

Fig. 104 represents the interior of an ananchytes, with flint on the bottom and crystals of calcareous spar around the sides and top.

(3.) *Mollusca.*—The fossil shells of the cretaceous period are very numerous, and are confined mainly to salt-water species. The remains of *Terebratulæ* abound in the white chalk and green sand. These, with several species of ostrea, *Fig.* 105, and plagiostoma, *Fig.* 106, appear to be characteristic of this group.

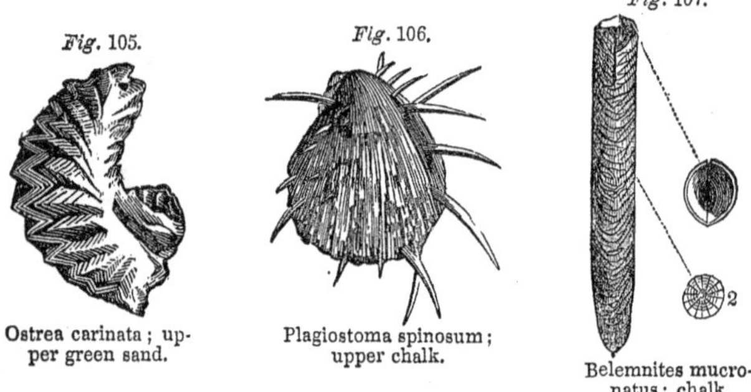

Fig. 105. Ostrea carinata; upper green sand.

Fig. 106. Plagiostoma spinosum; upper chalk.

Fig. 107. Belemnites mucronatus; chalk.

Many of the fossil shells of this period are allied to the nautilus and to the cuttle-fish.

The most remarkable of them are the *Belemnites,* which are long cylindrical stones of calcareous spar, somewhat pointed at one extremity, and containing a hollow cup at the other, *Fig.* 107. Many species are transparent, others nearly opaque. When broken, they present a radiated structure (2). The ani-

---

What was the character of the mollusca of this period? What are the most remarkable shells of this period?

mal resembled the cuttle-fish; the sheath or pen containing an ink-bag was inclosed within the body of the animal.

The bed of marl marked 4 in *Fig.* 93 is so filled with belemnites that they appear to be driven into the bluff as thickly as possible.

Of those related to the nautilus, the two principal families are the *Nautilidæ* and the *Ammonitidæ.* The following figures present several forms, and are good representations of those families.

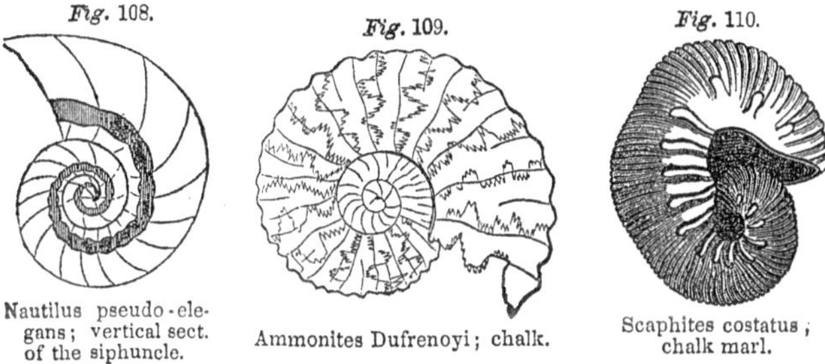

Fig. 108. Nautilus pseudo-elegans; vertical sect. of the siphuncle.

Fig. 109. Ammonites Dufrenoyi; chalk.

Fig. 110. Scaphites costatus; chalk marl.

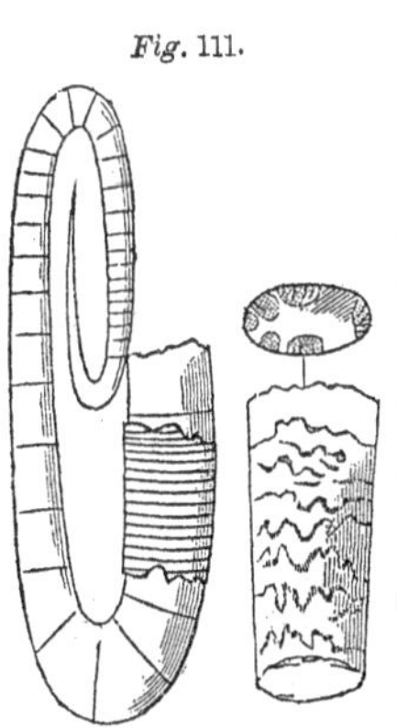

Fig. 111. Hamites cylindraceus.

Part of the stem of same species.

All the shells of these families are divided internally into many compartments, by partitions, through which a tube extends from the outer open chamber to the innermost cell. This tube is called the *siphuncle.* The outer chamber, which is much larger than the others, was occupied by the animal.

The siphuncle of the nautilus is situated either in the center or inner margin; that of the ammonitidæ upon the outer margin or back of the shell. When the shell of the ammonitidæ is regularly coiled up, as in *Fig.* 109, the genus is *Ammonites;* when each end is rolled up, as in *Fig.* 110, *Scaphites;* when coiled, as in *Fig.* 111, *Hamites;* and when straight, *Baculites.* Other genera have different forms.

How are the shells of these families divided? How are they distinguished?

Some of these shells are peculiar to the gault, and others to the chalk. They extend from the earliest palæozoic through the secondary periods, but not into the tertiary period.

(4.) *Articulata.*—A very rare species of *Balanus*, supposed to belong to the chalk, is represented in *Fig.* 112. Crabs occur in

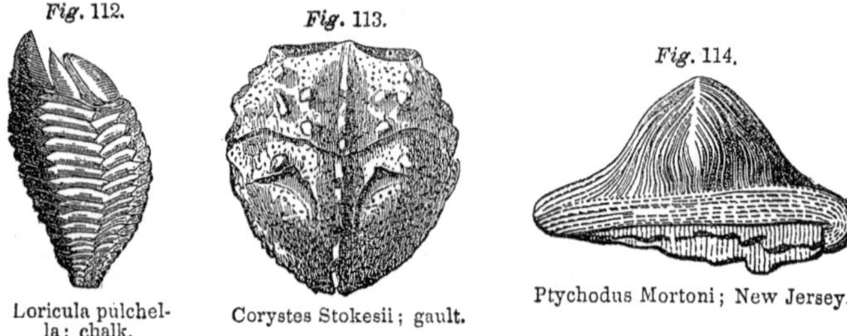

Fig. 112. Loricula pulchella; chalk.

Fig. 113. Corystes Stokesii; gault.

Fig. 114. Ptychodus Mortoni; New Jersey.

the gault, *Fig.* 113, and lobsters in the chalk. *Fig.* 115 represents the claw of *Astacus Sussexiensis;* chalk; England.

Fig. 115.

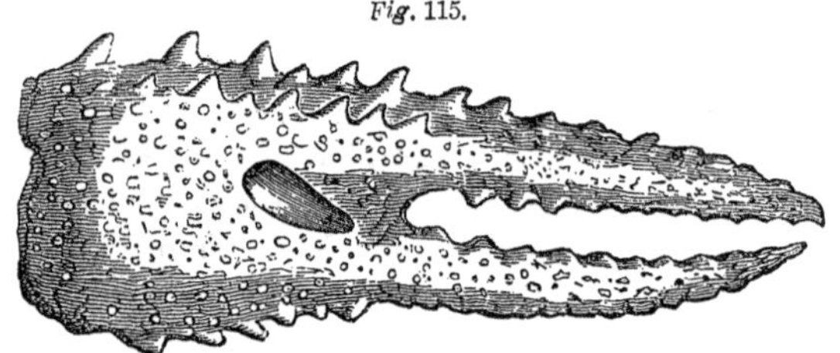

Chelate claw of Astacus Sussexiensis.

(5.) *Fishes.*—The teeth of sharks are very common in the chalk and the cretaceous rocks of the United States. They belong to the same genera with those already described, but the species are believed to be all different; *Fig.* 114 is the tooth of Ptychodus Mortoni, N. J. The vertebræ, jaws, scales, and teeth of many other families are frequently met with. Some very singular forms are found in the United States, and in the chalk of England. The fossil fishes of the chalk are obtained with great difficulty, in consequence of the friable nature of the rock. The

Where do they belong? Mention examples of articulata. What teeth are common in the cretaceous rocks? What is said of the fossil fishes?

most remarkable remains of fishes are their *coprolites* (fecal remains). It has been suggested that a large portion of the chalk has been passed through the bodies of fishes.

Darwin found that certain fishes in the Pacific Ocean fed upon the coral zoophytes, and that, when they were caught and opened, the substance contained within them could hardly be distinguished from chalk.

(6.) *Reptiles.*—The remains of crocodiles and of several other reptiles are common to the cretaceous system and the oolites; but the most characteristic reptile of the chalk is the *Mososaurus*. Teeth of this animal have been found in the United States, and four or five vertebræ in the English chalk; but the head, with the teeth, was discovered in a quarry at Mæstricht, *Fig.* 116. It is now in the museum at Paris, having been carried thither bv the savans under Napoleon.

Fig. 116.

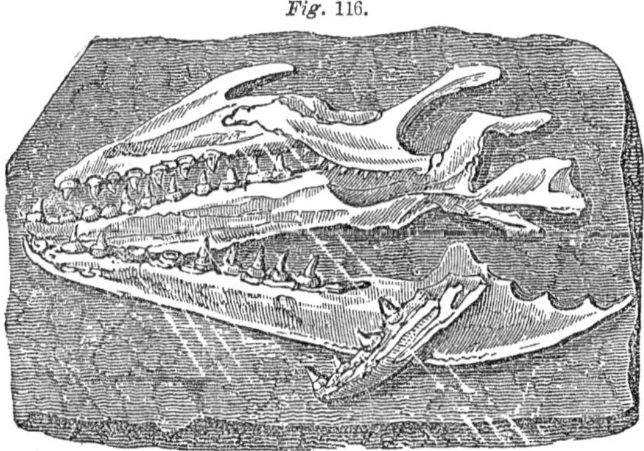

MOSOSAURUS.

This aquatic reptile was about 25 feet in length. It probably had webbed feet, and a tail suited to propel it through the water. Cuvier supposed that it inhabited the ocean.

The *Leiodon* was similar, and several reptiles more nearly allied to lizards than crocodiles occur in the chalk; also the remains of turtles, consisting mostly of the shells of the animals.

What other remains of fishes? What observation of Darwin? What reptile peculiar to the chalk? What other reptiles abound?

Some specimens found in the lower chalk, at Burnham, Kent, are remarkable for their fine state of preservation.

The remains of birds and mammalia are not found in the chalk. As it is a marine formation, we should not expect to find any but those of seals, and whales, and other cetaceans; but these probably did not exist during this period—they commenced with the tertiary, which was the age of mammals, while the deposition of the chalk terminated the reign and the age of reptiles.

SECTION II.—WEALDEN FORMATION.

The rocks of this group are of fresh-water origin, and with a few exceptions are confined to the southeast of England.

The Wealden is interposed between the oolitic and the cretaceous system, but from the character of its fossils it is considered by most geologists as a part of the oolitic group.

I. *Composition.*—It is composed of a series of clays and sands, and rests on beds of imperfect limestone and shells, through which are distributed fresh-water shells, terrestrial plants, and the teeth and bones of reptiles and fishes. Univalve shells are mostly confined to the upper, and bivalve to the lower division, while the remains of reptiles occupy the intermediate beds. It appears to be an estuary deposit. The following are the principal divisions:

1. Weald clay. 2. Hastings sand. 3. Purbeck strata.

1. *Weald Clay.*—This group consists of beds of sandstone and shelly limestones, with layers of argillaceous iron stone. The limestone is called *Sussex marble*, and is found in layers from a few inches to a foot in thickness, separated from each other by thin seams of clay and coarse limestones. It is composed almost entirely of the shells of paludina, cemented by lime into a compact marble. The clay is only remarkable for being favorable to the growth of the oak, and hence called by Dr. Mantell "oak-tree clay."

2. *Hastings Sand.*—This group is of fresh-water origin, and is

---

Are there any birds or mammals found in these deposits? Where is the Wealden group developed? Of what is it composed? Describe the Weald clay; the Hastings sand.

the most important of all the Wealden deposits, not only on account of its greater thickness and extent, but also for the fossils which it contains. It consists of numerous strata of sand, sandstones, grit, and shells. The upper beds form a group the most noted and interesting, because of the extensive quarries in Tilgate Forest, near Horsham, in which have been discovered the most remarkable fossils of this period.

3. *Purbeck Strata.*—The Purbeck strata constitute the base of the Wealden, and in many places seem to pass into the next group, so that it is often difficult to determine to which system many of the beds belong.

The following section of the quarries in the Isle of Portland will exhibit the order of these strata:

Fig. 117.

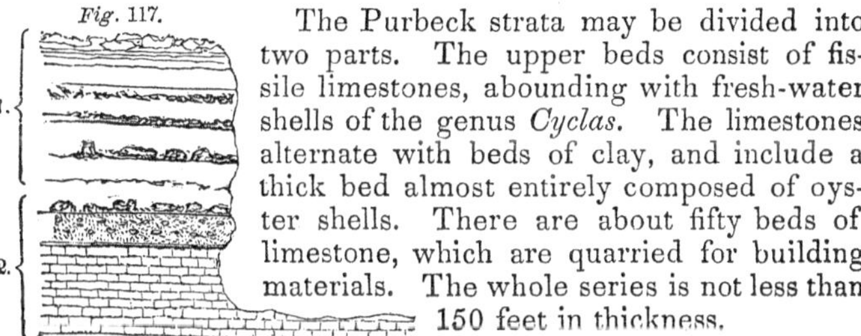

The Purbeck strata may be divided into two parts. The upper beds consist of fissile limestones, abounding with fresh-water shells of the genus *Cyclas.* The limestones alternate with beds of clay, and include a thick bed almost entirely composed of oyster shells. There are about fifty beds of limestone, which are quarried for building materials. The whole series is not less than 150 feet in thickness.

SECTION. ISLE OF PORTLAND.

1. Purbeck series. 2. Portland beds.

The lower beds are coarse, fissile limestones, used for roofing purposes, and associated also with beds of clay.

The section at Lulworth Cove, page 212, gives the relations of the Wealden to the cretaceous and also to the oolitic system. The Wealden is principally found in the counties of Kent and Sussex, and in the Isle of Wight.

In the northwest of Germany, a series of strata 800 feet in thickness has been referred to this period, but the deposit has a very limited geographical distribution, being confined mainly to a narrow strip in England.

Describe the Purbeck strata. How is the Purbeck strata divided? Are there other strata belonging to the Wealden?

II. *Fossils of the Wealden.*—The fossils of this period belong to land animals and vegetables, or to species inhabiting fresh-water rivers, estuaries, and lakes. The number of species is not large, but the individuals are very abundant, although rather imperfectly preserved.

1. *Vegetables.*—The fronds of ferns or of cycadeæ, and the silicified trunks of large coniferous trees, are frequently met with in the Wealden strata; but the most characteristic genera are clathraria and endogenites.

The *Clathraria, Fig.* 118, had a thick epidermis or false bark, formed by the union of the basis of the leaves, and covered by distinct scales.

Fig. 118.

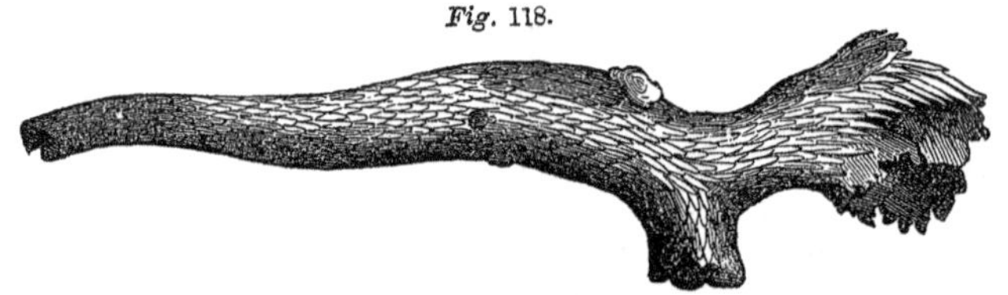

CLATHRARIA LYELLI (MANTELL).

The *Endogenites* are elongated flattened bodies, tapering at both extremities, and sometimes measuring 8 feet across. They consist of two portions, a stony nucleus and a crust of lignite.

2. *Animals.*—The fossils of this period belonging to the invertebrate animals do not possess much interest, as they differ but little from those inhabiting the rivers and estuaries of the present day.

The shells of molluscs are chiefly referable to extinct species of cyclas, melanopsis, and paludina. Entire beds are often composed of single species of each of these genera.

Fig. 119.

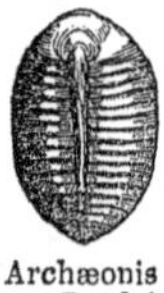

Archæoniscus Brodei; Wealden.

Crustaceans are abundant, but are mostly confined to the genus Cypris, and a new genus of *Isopoda, Fig.* 119, which is allied to the trilobites of the palæozoic periods.

The insects are chiefly beetles, and a few genera belonging to families which live on plants or hover over the surface of water.

What was the character of its animals and plants? What vegetables are characteristic of it? What is said of the invertebrate animals?

The remains of *fish* consist mostly of imperfect fragments, which have been referred to 13 species. They appear to have inhabited shallow and muddy water.

The most interesting and remarkable fossils of the Wealden are

The Remains of Reptiles.—About one sixth of the whole number of extinct reptiles belong to this period, and five genera belonging to three orders are peculiar to it. The *Suchosaurus* and *Goniopholis* are referred to the crocodilian order of Owen. The *Iguanodon* and *Hylæosaurus* to the order Dinosauria. The *Tretosternon* and some other turtles to the order Chelonia. A few other reptiles existed in this period, which also are found in the oolite.

The *Suchosaurus* was a long-snouted crocodile resembling the gavial of the Ganges. The crowns of its teeth were slender, compressed, and acute.

The *Goniopholis* was more nearly allied to the short-snouted crocodile, with teeth whose crowns were round and obtuse. This singular animal was covered with large bony scales, more perfectly and powerfully arranged than those of any known reptile living or extinct.

The *Hylæosaurus* (Wealden lizard) was a land saurian 15 feet in length, and covered with elliptical scales. Dr. Mantell obtained a large portion of the skeleton of this animal from the quarries of Tilgate Forest.

The *Iguanodon*, the largest and most remarkable of the extinct reptiles either of the Wealden or of any other period, was first discovered by Dr. Mantell in the quarries of Tilgate Forest. It was at first described by him as a reptile 100 feet in length, and in consequence of its resemblance to the modern iguana he called it *Iguanodon*. Its structure, however, differs from any living reptile, and later discoveries have shown its length to have been much less than it was at first supposed. But it is still regarded as one of the most gigantic of the reptiles of the ancient world. It was herbivorous, and from 25 to 50 feet in length.

What is said of the reptiles? Mention the several orders of reptiles. Describe them in order Which was the largest?

The length of the head .... 3 feet.
" trunk..............12 "
" tail................13 "
28.—*Owen.*

The bones are very large, the legs and feet* being of gigantic proportions, suited to sustain the immense weight of its body.

The *Tretosternon* had a shell similar to turtles, although very flat, measuring in one species 17 inches by 13½. Turtles are common in the quarries of Tilgate.

Birds are few, and their remains are imperfectly preserved.

The characteristic fossils of this period, it will be seen, are reptiles which are so far removed from existing forms that it is difficult to derive any satisfactory account from them, either of the physical geography or of the climate of the period, during which they appear to have been the undisputed lords of the habitable earth.

SECTION III.—OOLITIC SYSTEM.

The term *oolite*† is applied to a series of strata constituting a formation, on account of the structure of some of the limestones, which are made up of small egg-shaped particles. The oolite is composed essentially of limestones, clay, and sandstones. Some of the lower beds contain valuable seams of coal.

I. *Oolite of the United States.*—The only deposit in this country which has been referred to this period is found in Eastern Virginia. The strata consist of conglomerates, sandstones, and beds of coal. The coal-field, of which *Fig.* 120 is a section, constitutes the most extensive deposit of coal that has been discovered in any of the formations which are newer than those of the carboniferous system. The area has been estimated at 185 square miles. The depth,

Fig. 120.

SECTION OF THE COAL-FIELD, E. VA.

* The *metacarpals* were 2 feet 6 inches in length, and the last joint of the toe 5½ inches.

† ὠὸν, an egg; λιθος, a stone. This structure is not confined to the rocks of this system, but occurs in the tertiary and some other rocks.

From what is the term oolite derived? Describe the oolites of the United States.

excepting at the skirts, is unknown. The central portions are covered by conglomerates, and the coal rests on a series of sandstones. Three seams have been explored, the total thickness of which varies from 11 to 40 feet. The coal is bituminous, and is divided into thin horizontal layers, as in the older coal beds. "Sometimes thin layers consist alternately of highly crystalline and resinous coal, with a bright luster, and of other portions exactly resembling charcoal in appearance."

II. *European Oolite.*—In England we have the most perfect development of this system, consisting of three ridges, running northeast and southwest, with extensive intervening plains. The rocks are divided into three groups, *the upper, the middle,* and *the lower oolite.*

1. *Upper Oolites.*—This group consists of three kinds of strata, Portland stone, Portland sand, and Kimmeridge clay.

(1.) The *Portland Stone,* found in the Isle of Portland, includes several beds of coarse earthy limestone, which repose upon silicious beds containing green sand, and hence called *Portland sand.* The lower limestones are coarse, alternately hard and soft, and 50 or 60 feet in thickness. Upon these repose three beds of fine building stones interstratified with clayey and silicious bands.

The upper part of the series contains a remarkable deposit, called by the workmen the *dirt bed.* It consists of bluish loam, containing the petrified roots and stumps of trees in the position in which they grew, *Fig.* 121, B. The stratum is only about one

Fig. 121.

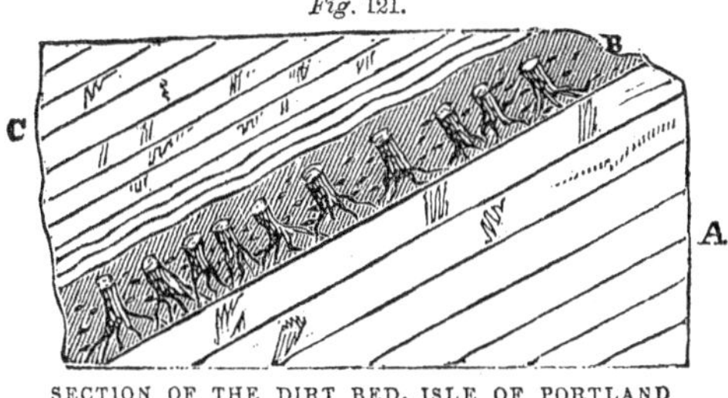

SECTION OF THE DIRT BED, ISLE OF PORTLAND

Mention the oolites of England. How are they divided? Describe the Portland stone. What remarkable deposit in the upper portion?

foot in thickness, but extends over a large area. It appears to have been formed about the close of the oolitic period, the earth having been elevated above the water sufficiently long for the formation of a foot of soil, and for the growth of *large trees;* after which it was converted into an estuary or fresh-water lake, and the rocks covered by the Wealden deposits.

(2.) *Portland Sand.*—This stratum is a silicious sand, mixed with green particles, attaining a thickness, in the western part of the Isle of Portland, of 80 feet, and interposed between the Portland stone and

(3.) The *Kimmeridge Clay*, which is of a bluish or grayish-yellow color, and contains selenite, with vegetable and animal impressions. At Kimmeridge Bay, from which the clay derives its name, beds of highly bituminous shale alternate with the clay. The shale is combustible, and known as *Kimmeridge coal.*

Beds supposed to be cotemporaneous with the upper division of the oolite are found on the coast of Normandy; also in the upper strata of the Jura, in Switzerland; at Solenhofen, in the north of Bavaria; and on the banks of the Donitz, in Southern Russia.

2. *Middle Oolites.*—In this division, *Fig.* 122, we find:

(1.) *Calcareous Grits*, or sand beds more or less abounding in calcareous matter and fine seams of clay.

Fig. 122.

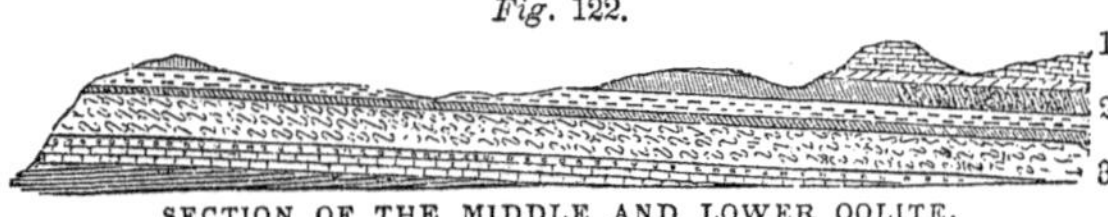

SECTION OF THE MIDDLE AND LOWER OOLITE.

(2.) *The Coral Rag*, which consists mostly of calcareous deposit, derived from the decomposition of animal remains. Corals are distributed through this rock in great abundance, particularly in Wiltshire, where the bed attains a thickness of 40 feet. The whole appears to have been a coral island in an open sea.

(3.) *Oxford Clay.*—The next group is a stiff pale-blue clay,

What is said of the Portland sand? of the Kimmeridge clay? What rocks cotemporaneous with the upper oolites in England? What kinds of strata belong to the middle oolites?

containing calcareous matter, iron pyrites, and numerous organic remains. These remains are sometimes found in the clay, but generally form the nucleus of the pyrites. The thickness of this bed is about 500 feet. It extends over a large portion of England. The Oxford clay usually rests on the *corn-brash*, and forms the basis of the middle oolites, but in a few cases a stratum from 3 to 5 feet thick, consisting of calcareous sandstone, and abounding in fossils, intervenes; the same is true in Normandy, but in Switzerland the middle oolites rest on argillaceous rocks, and consist of strata which contain oolitic iron ore of great economical value, and also of a limestone called *Nerinæan*, because it is composed almost entirely of univalve shells of the genus nerinæa.

3. *Lower Oolites.*—This division of the oolitic series is vari ously subdivided in England, but the subdivisions do not apply to the group in other parts of Europe.

(1.) *Corn-brash* is a series of clays, sandstones, and limestones. The name is derived from the excellent corn land formed by the disintegration of the rock.

(2.) *The Forest marble* is a limestone, sometimes crystalline, and sometimes marly, so filled with organic remains that in some cases they constitute nearly the whole substance of the rock.

(3.) *The great Oolite* consists of a series of shelly limestones, alternating with fine soft freestone, destitute of fossils. In the upper strata the limestones are hard, of a yellowish-white color, and filled with fossils; lower down they are shelly, and the lowest are very fine grained and almost crystalline in structure. This is the most important of the whole series, both from its furnishing building materials and for the beauty of its fossil shells.

(4.) *The Bradford Clay* is nearly of the same age with the preceding. It is a pale grayish clay, inclosing thin slabs of a tough brownish limestone. It is about 60 feet in thickness, and often entirely absent, or interstratified with the underlying fuller's

How are the lower oolites subdivided? Describe the several kinds of rock in order.

earth. It is remarkable for containing a peculiar fossil, the ***Api-ocrinite.***

(5.) *Fuller's Earth.*—This is a peculiar kind of clay, used in the manufacture of cloth; associated with this is a flagging-stone, known as *Stonesfield slate.* The latter occurs in two beds, separated by a loose calcareous sandstone. This strata has long been celebrated for the only remains of mammalia anterior to the tertiary.

(6.) *The Inferior Oolite* consists of about 40 or 50 feet of calcareous freestone, reposing on friable sands, which constitute the base of the oolitic system in England.

The lower oolites are represented in several parts of Europe, and in Eastern Virginia, United States, already described as containing coal. The Brora coal is referred to the older oolitic period. This coal has been mined for 250 years. It has two workable seams, the main seam being about three and a half feet thick. There are several coal-fields in India, which probably belong to this period.

4. *Oolites of the Jura.*—The Jura Mountains are separated from the higher Alps by the great valley of Switzerland. In the northern part of this range we have a beautiful development of the *Oolitic* or *Jurassic system.* The rocks here are inclined at a high angle, *Fig.* 123, and are divided into three groups.

Fig. 123.

SECTION ACROSS THE JURA CHAIN.

1. Coralline limestone. 2. Argile d'Oxford, or Oxford clay. 3. Inferior oolite.

(1.) The upper beds resemble the Portland rocks and the Kimmeridge clay, but contain great quantities of iron ore, which abounds through a vertical depth of about 40 feet of the strata.

(2.) The middle beds are more than 300 feet thick, and differ

---

Where are the inferior oolites found, and what do they contain? Where are the lower oolites represented? Describe the oolites of the Jura.

much from the corresponding beds in England. These most important strata consist of an oolitic iron ore in beds of marl one hundred feet thick, and the *Nerinæan* limestone, which corresponds to the coral rag.

(3.) The lower oolites of the Jura are about 300 feet thick, and are subdivided into four groups, consisting mostly of limestones, alternating with marls and yellowish clay, and also with shaly bands and red oxide of iron.

In the north and center of Germany, instead of the Kimmeridge clay there occurs in Bavaria, near Solenhofen, a very fine-grained limestone of a rich cream color, which is extensively used for the purposes of lithography, and exported to most parts of Europe. This lithographic rock rests on a stratum which is celebrated for its caverns, and for the beauty of the fossil insects that have been preserved in the stone.

III. *Fossils of the Oolite.*—We have noticed, page 226, the extensive beds of coal derived from the plants which flourished during this period. The vegetables differ from those found in the older coal-beds, but are very similar to those of the succeeding periods. The *Cycadeæ*, a tribe of plants now flourishing in the tropics, were most abundant. They resemble the palms.

Of the Animal relics, corals and salt-water shell-fish are very abundant. Most of the families of fishes and reptiles which are found in the chalk existed during this period.

The echinodermata were represented by many singular animals, some of which, the crinoideans, were star-fishes that were attached each by a long stem to the rocks beneath.

Sea-urchins, as echini, and the genera diadema, *Fig.* 124, and spatangus, which are very nearly allied, abound in various parts of the oolitic system.

Fig. 124.

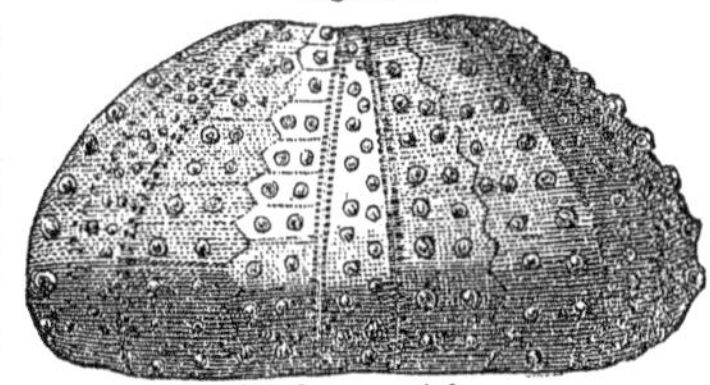

Diadema seriale.

The higher orders of mollusca were very abundant. One spe-

What important deposits in Germany? What is said of the fossil plants of this period? What families of mollusca are abundant?

cies of ammonite, from the Solenhofen slate, *Fig.* 125, is so perfectly preserved as to give the most distinct outline of the termination of the aperture. *Fig.* 126 is a less perfect specimen.

Fig. 125.

Ammonites semicanaliculatus.

Fig. 126.

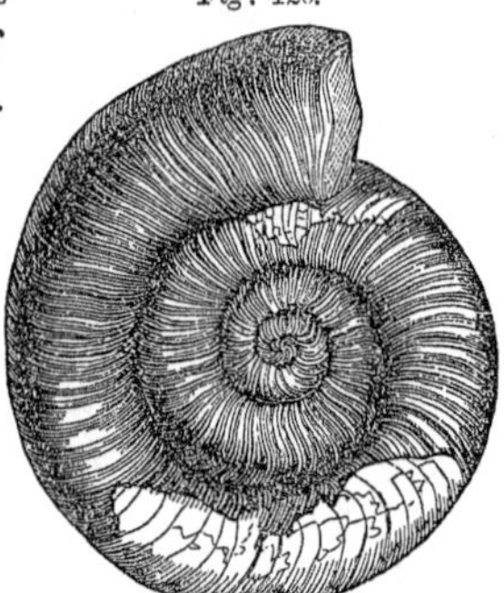

Ammonites striatulus; great oolite.

In the Oxford clay, specimens have been found both of the ammonite and belemnite, which have removed all doubt in respect to the character of these animals.

A specimen of belemnite was found in the Oxford clay, *Fig.* 127, "in which not only the ink-bag, but the muscular mantle, the head and its crown of arms, are all preserved in connection with the belemnitic shell." Another specimen presents the contour of large sessile eyes, the tentaculæ armed with a double alternate series of horny hooks, and the remains of two lateral fins. "The belemnite," says Professor Owen, "having the advantage of its dense but well-balanced internal shell, must have exercised its power of swimming backward and forward with great vigor and precision. Its position was probably more commonly vertical than in its recent congeners. It would rise swiftly and stealthily to infix its claws in the belly of a supernatent fish, and then perhaps as swiftly dart down and drag its prey to the bottom and devour it."

Fig. 127.

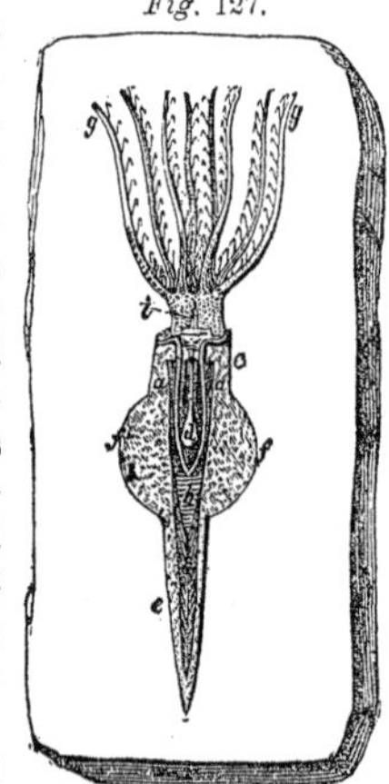

Animal of the Belemno-sepia; Oxford clay.

The ink-bag was doubtless used like that of the cuttle-fish, to discolor the water, and thus enable it to escape from its enemies.

What was the character of the belemnites?

Crustaceans and insects are admirably preserved in the lithographic limestones of Solenhofen.

The fishes and reptiles, however, of this period are objects of the greatest interest. More than fifty species of fishes, belonging mostly to the sauroid family, have been found in the various oolitic strata of England. This family includes those genera which resemble the saurian reptiles. The gar-fish of Lake Champlain, *Lepidosteus oxyurus*, is one of the few living species which represent this ancient family.

Marine reptiles were still more abundant during this period. The largest of these tribes was named by Professor Owen *Pliosaurus*. This animal must have rivaled in size the largest whales. The length of that part of the jaw in which the teeth were inserted was three feet, and the conical teeth were seven inches in length. The head must therefore have been of great size. The neck was of massive proportions, some of its vertebræ being eighteen inches in circumference, although only one inch in length, a structure which unites great strength and flexibility. This animal had paddles like the whale. Of several other genera which more nearly resembled the crocodile of the present epoch, the *Cetiosaurus* was one remarkable for its great size, the largest species being not less than 60 feet in length. This genus lived in the water, and was provided with webbed feet and a broad vertical tail, as organs of motion. Some of the *terrestrial saurians* were allied to the iguanodon. Of these,

The *Megalosaurus* was the largest, attaining a length of 30 feet. It was provided with strong double-edged teeth, curved backward, and serrated on both edges like a saw, which admirably fitted them for holding, tearing, and cutting up their prey. The largest relic of this animal is the jaw, *Fig.* 128, but this is sufficient to determine the character and form of the reptile; like the iguanodon, it was probably elevated several feet from the ground on massive legs, which sustained a body broader and deeper than that of any modern saurian.

What is said of the crustaceans? of the fishes and reptiles? Describe the marine reptiles. Which was the largest?

Fig. 128.

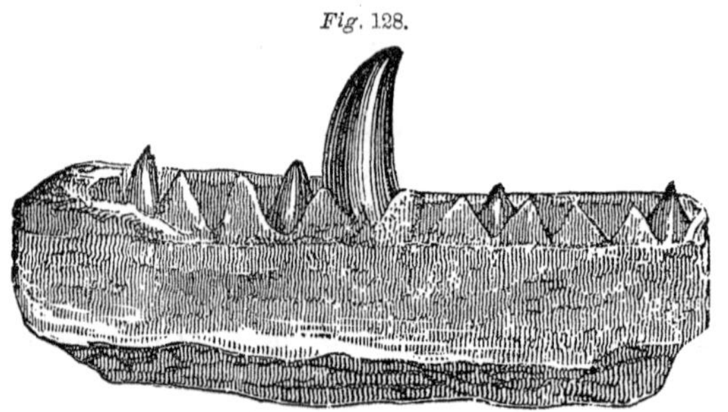

JAW OF THE MEGALOSAURUS—OOLITE.

Altogether the most extraordinary reptile of this period was the *Pterodactyl*, *Fig.* 129, an animal with the form and general

Fig. 129.

appearance of a bat, except its head, which was very long, and resembled that of the crocodile, to which also the internal anatomy of the body corresponded. Although the neck had only seven vertebræ, it was so flexible that the head could be thrown over upon the lower part of the back. Instead of four toes extended and enveloped in skin, as in the bat, the fifth toes only were prolonged, and the skin extended along the side of the body and legs. By this structure the animal could fold its wings and walk, or swim, or rise into the air, as its necessities might require

---

Which was the most extraordinary reptile of this period?

The *most characteristic* fossils are the remains of the marsupial quadrupeds, found in the Stonesfield slate, and referred by Owen to two genera, *Amphitherium* and the *Phascolotherium*. The jaw of the latter is represented in the accompanying figure. These

Fig. 130.

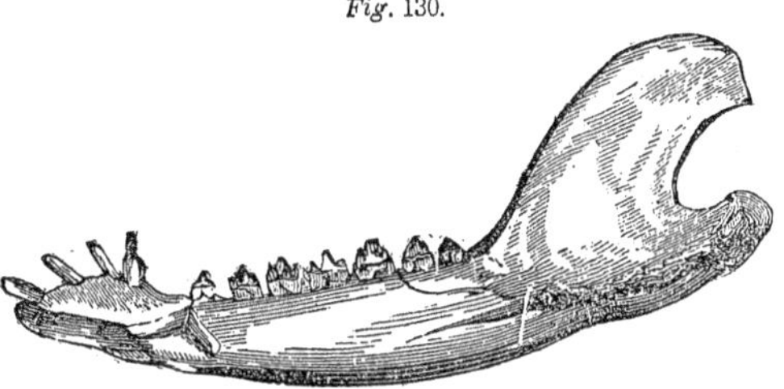

JAW OF PHASCOLOTHERIUM (OOLITE)—STONESFIELD SLATE.

remains are the earliest proofs of the existence of mammals on the surface of our planet.

SECTION IV.—LIASSIC SYSTEM.

The lias formation is found in various parts of Europe and in South America. The strata are argillaceous throughout, although a considerable portion of calcareous matter is mingled with the clay, forming in some places bands of argillaceous limestone.

I. The subdivisions of this formation in England vary at different points, but generally we find three principal groups, which are represented in section, *Fig.* 131, the upper, middle, and lower lias.

Fig. 131.

SECTION FROM THE SEVERN TO THE COTTESWOLD HILLS.

1. The *Upper Lias*, also called *alum shales*, consist of three portions. The first and highest beds are shales of a slaty structure, and have long been celebrated for the remarkable fossils found at Lyme Regis. The middle portion is a hard stratum of

What fossils are most characteristic of this period? Where is the liassic system found? How many groups of strata are there? Describe the upper lias.

shale about thirty feet thick, and contains a large quantity of carbon in the form of *jet*, with large fragments of the bituminized wood of cone-bearing trees. The lower beds are soft shales, filled also with numerous fossils.

2. The *Middle Lias* consists of sandy shales about 130 feet in thickness, through which are distributed bands of argillaceous iron nodules.

3, 4. The *Lower Lias Shales* consist of thick layers, finely laminated with numerous calcareous bands and concretions, the whole either reposing on a whitish sandstone of the triassic system, as in the middle of England, or upon bluish marls, which compose the upper layers of the new red sandstone at Lyme Regis, in the south of England.

In the lower division (4) there is a thin stratum almost entirely made up of fossils, mostly the remains of fishes. This bed passes occasionally into sandstone, which is destitute of fossils. The fossiliferous bed is not more than three inches in thickness, but extends over a distance of 100 miles, affording convincing proofs of a long period of time for their accumulation.

II. *Fossils of the Lias.*—The liassic system is characterized by the most remarkable remains of animals of any in the whole series of fossiliferous strata.

The structure and composition of these rocks are well suited both to preserve and to exhibit the remains of organized beings.

1. The plants do not differ essentially from those that flourished in the periods immediately preceding or succeeding, although some new and extraordinary forms have been discovered.

2. *Animals.*—In the absence of corals the seas appear to have been filled with echinodermata, among which the crinoideans were most abundant.

Fragments of *Pentacrinites*, *Fig.* 132, constitute extensive beds some inches in thickness, so perfectly preserved that their stony skeletons can be accurately traced, and the character of the an-

---

Describe the middle and lower lias. What stratum in the lower division? What is said of the fossils of the lias? of the plants? of the animals? of pentacrinites?

Fig. 132.

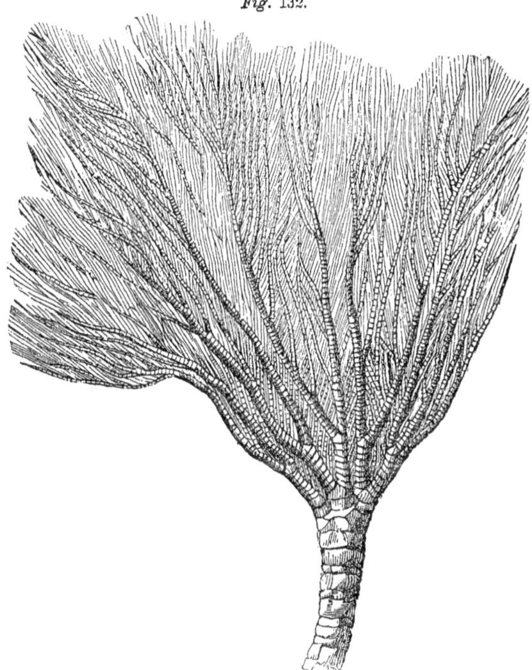

PENTACRINITES SUBANGULARIS—LYME REGIS.

imal easily deduced. This animal had a five-sided stem, from the summit of which arms were given off in innumerable ramifications, giving it the appearance of a plant. The number of parts of which the skeleton was composed was not less than 150,000. "This animal is supposed, by Dr. Buckland, to have been capable of withdrawing itself readily from any substance to which it was attached, and, after floating about in search of a new and more convenient resting-place, fixing itself again upon the lower surface of some floating piece of timber."

The stomach formed a funnel-shaped pouch, composed of contractile membrane, and terminated by a small aperture, which

---

Describe the animal.

could be elongated and formed into a proboscis for taking hold of the food brought within its reach by the tentaculæ, or arms and fingers.

Many species of shell-fish occur in this formation, and three species of the genus *Spirifer* are characteristic of it.

Ammonites, belemnites, and nautili abound. Not less than 65 species of ammonites, 12 of belemnites, and 5 of nautilus, have been discovered in the lias strata.

From the *ink-bag* of one species of belemnites found at Lyme Regis, a beautiful pigment was prepared, similar to the India ink which is now obtained from the cuttle-fish, and this fact has completed the chain of evidence necessary to establish the character of these ancient races.

Fishes and reptiles are the only classes of vertebrated animals which have been found in the lias. The number of species of *fishes* from Lyme Regis described by Prof. Agassiz is not less than 60; all of them are now extinct, and most of them referable to the ganoid and placoid orders. Two genera, lepidotus and dapedius, *Fig.* 133, are the most common. The remains of fish

Fig. 133.

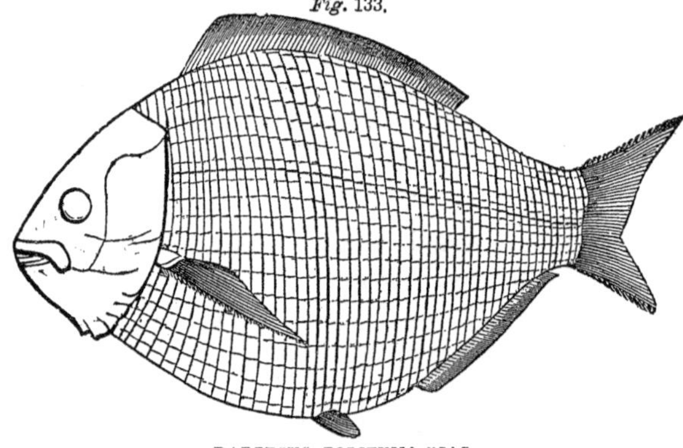

DAPEDIUS POLITUM; LIAS.

belonging to the placoid order, called ichthyodorulites, are the bony rays of the fins, *Fig.* 134.

What evidence of the true character of the belemnites? What vertebrated animals existed? What remains of fishes?

Fig. 134.

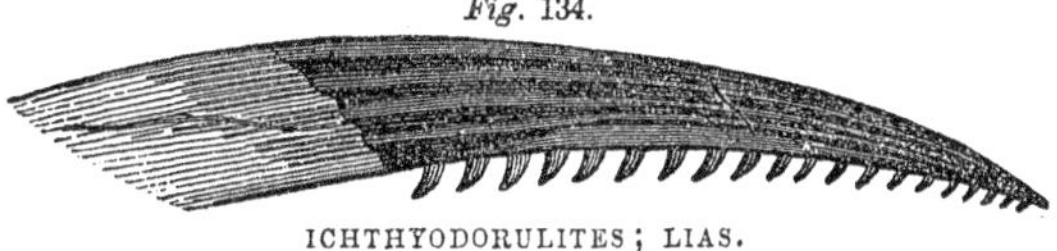

ICHTHYODORULITES; LIAS.

The Reptilian remains of this period belong mostly to two genera.

The Ichthyosaurus (*fish-Lizard*), *Fig.* 135, was similar to

Fig. 135.

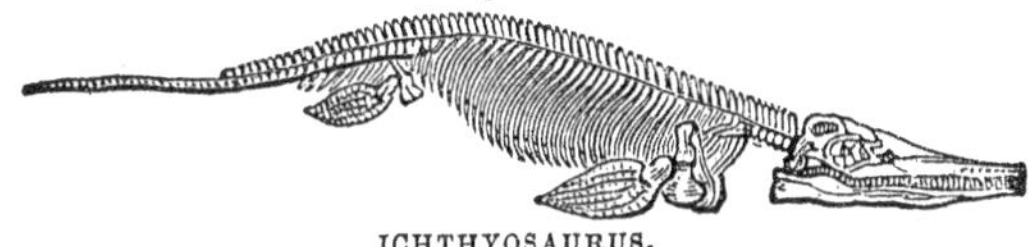

ICHTHYOSAURUS.

some of the large predatory fishes, but had a much larger head and more powerful tail, a skin without scales, covered with smooth, fine wrinkles. The head was very peculiar. The jaws were so long that the gape must have exceeded seven feet, and, to give them the requisite strength, they were made up of a number of parallel plates of unequal thickness, to enable them to resist any sudden shock produced by snapping at their prey. Rows of powerful teeth extended along both sides of the jaw, and a provision made, as in the crocodile, for renewing them when broken or worn out. The number of teeth found in one individual was 180.

The eye was placed far back, and its orbit was 18 inches in diameter, having circular bony plates around the pupil to defend it, and to adapt it to long and short-sightedness, so that the animal could see equally well in the water and on the land. The breathing-holes were placed on the top of the head, in order that most of the body might remain under water during respiration.

The vertebræ were like those of fishes, and the ribs were small and numerous. The extremities were paddles, like those of the whale, though so constructed that the animal could move upon the land.

The food of the ichthyosauri was fish and reptiles. This is

What remains of reptiles? Describe the ichthyosaurus. What was their food?

proved by the fact that their half-digested remains have been found in the stomach of the animal, and not only so, but coprolites abound in this formation similar to the remains found in the abdomen; in fact, these *coprolites*, loaded with the scales, teeth, and bones of fishes and reptiles, constitute strata many miles in extent, presenting the appearance of a conglomerate rock composed of the exuvia of these and other animals which flourished during this period.

About twenty species of this genus have been described. They commenced with the lias, and flourished until near the close of the cretaceous period.

The Plesiosaurus, *Fig.* 136, was another genus of reptiles still

Fig. 136.

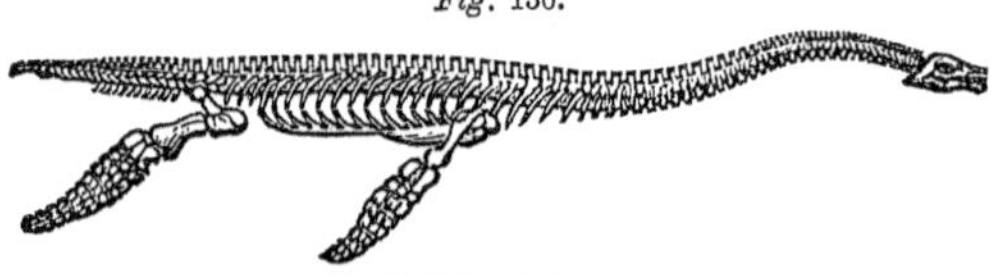

PLESIOSAURUS.

more peculiar in structure. Its head resembled that of a lizard, and was quite small. Its neck was three or four times the length of the head, and contained upward of 30 vertebræ. This gave it great flexibility and rapidity of motion. The body was small and resembled a fish, and its paddles more nearly resembled those of the cetaceans than those of the ichthyosaurus. Its length was about 17 feet. Eighteen species have been described, of which six belong to the oolitic and cretaceous groups.

With the power also of creeping on land, it possessed extraordinary powers of swimming, which enabled it with ease to overtake its prey, while the powerful and rapid motions of its wedge-shaped head, wielded by its long and flexible neck, rendered it a match for most of its enemies. That it was, however, sometimes overpowered by the ichthyosaurus, is proved by the half-digested fragments of its bones in the coprolites of the ichthyosaurus.

What is said of the coprolites and other remains of the animals of this period? Give the characters of the plesiosaurus. How many species have been discovered?

SECTION V.—TRIASSIC OR NEW RED SANDSTONE SYSTEM.

The new red sandstone system has received the name of *Triassic*, from the three well marked divisions of it found on the Continent of Europe.

I. *Geographical Distribution.*—In the United States this group extends from Vermont, in the valley of the Connecticut River, through Massachusetts and Connecticut; and from New Jersey to North and South Carolina. In England it is principally developed in the valleys of the Dee, Mersey, and Weaver. It is also found in France and Germany. The strata appear, for the most part, to have been deposited in broad *estuaries*.

II. *Composition and Structure.*—The rocks of this system are sandstones, conglomerates, and shales, and are generally characterized by their color, which is of various shades of red and brown. The strata are very regular, and the rocks are often composed of fine sand, and are admirably suited to building purposes.

III. *Trias of North America.*—The strata in the United States are separated, broken through, and overlaid in some places by a ridge of greenstone, which forms mountains several hundred feet in height, as those of Mount Holyoke and Tom in Massachusetts, East and West Rock in Connecticut, and the Palisades in New York. The structure of this greenstone is often columnar, like that of the Giant's Causeway and Fingal's Cave, which belongs to the same period.

Fig. 137.

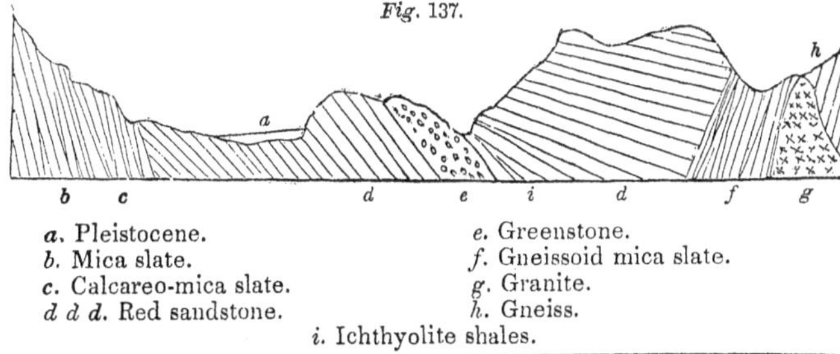

a. Pleistocene.
b. Mica slate.
c. Calcareo-mica slate.
d d d. Red sandstone.
e. Greenstone.
f. Gneissoid mica slate.
g. Granite.
h. Gneiss.
i. Ichthyolite shales.

From what is the name *triassic* derived? What is the geographical distri bution of this group in the United States? in Europe? What is said of the color and structure of the rocks? of intruded greenstone?

In Massachusetts, the trias lies in a basin of metamorphic rocks, as represented in the preceding section, *Fig.* 137, p. 241 (from Dr. Hitchcock's Report), across the valleys of the Connecticut and Deerfield Rivers. The intruded greenstone (*e*) is seen in the middle of the section.

In New York, the red sandstone is similarly associated with greenstone and with metamorphic rocks, as may be seen in the following section, *Fig.* 138, across the Palisades.

Fig. 138.

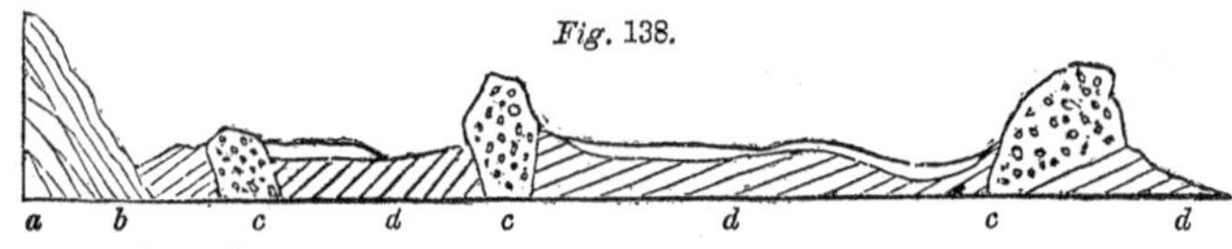

a. Gneiss and granite. *c c c*. Greenstone.
b. Red sandstone and conglomerates. *d d d*. Red sandstone.

In New Jersey, the red sandstone is associated in the same manner with metamorphic strata and with trap ridges.

It is obvious that the age of these sandstones of the Northern and Middle States can not be inferred from their position. From their lithological characters, and from the fossils, they have rightly been regarded as belonging to the triassic system. They are deficient in salt and gypsum, which in Nova Scotia and in Europe occur more or less abundantly in the rocks of this period.

Some geologists have thought it probable that these rocks belong to the lias rather than to the triassic system. The fossil fishes are, according to Mr. Redfield, more nearly allied to liassic types than to any types older than the trias.

IV. *European Trias.*—The following section, *Fig.* 139, across a new red sandstone basin, from Oswestry to the north of Staffordshire, in England, clearly exhibits the position and age of these rocks.

Fig. 139.

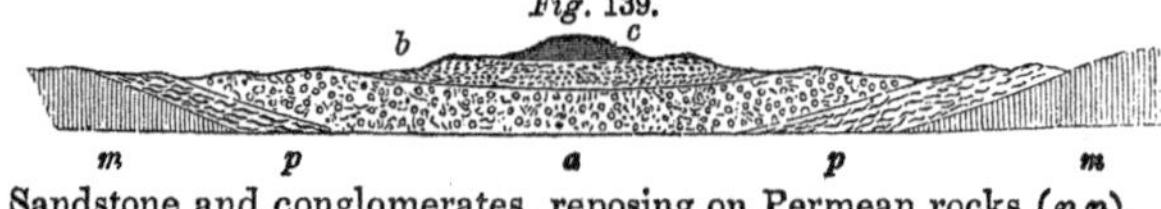

a. Sandstone and conglomerates, reposing on Permean rocks (*p p*).

Describe the position of the trias in Massachusetts; in New York, and New Jersey. What is said of the age of this system in the United States?

b Sandstone and marls, which contain salt.

c. Lower oolites, reposing, in the absence of the lias, on the trias. Cheshire is celebrated for its beds of rock salt, the total thickness of which is sixty feet. They alternate with beds of gypsum, and of blue, red, and brown indurated clays, and with red sandstones.

m m. Coal measures.

V. *Fossils of the Trias.*—1. *Plants.* In the new red sandstone of Connecticut, fossil trunks of trees have been found. In Massachusetts, a mass of greenstone of the trias, near Mount Hol yoke, appears to have enveloped, while in a melted state, a vegetable stem two inches in diameter and several feet long. The cavity left by the destruction of the stem was subsequently filled with mineral matter, which is accordingly destitute of organic structure. This extraordinary specimen is now in the museum of Amherst College. Probably the process by which the cavity was formed was similar to that described on page 56, as having been witnessed during the eruption of Kilauea.

In the European trias, the remains of ferns have been found with the marks of fructification. Plants of the Zamia tribe also flourished in this period.

In the triassic rocks, some minute seams of carbonaceous matter have been found, but they are not regarded as indications of workable coal, for no beds of coal have yet been discovered in this system.

Fig. 140.

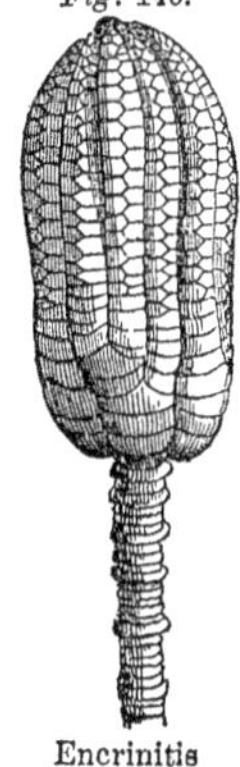

Encrinitis moniliformis.

2. *Radiated Animals.*—Radiated and other animal remains are found chiefly in the muddy deposits of the trias. Coarse sands are unfavorable both to the growth and to the fossilization of organic bodies, and their remains are not common in sandstones. Corals, however, do not flourish in muddy waters, and hence they are rare in all the rocks of this system.

One of the most beautiful of the Crinoideans of this period was the Lily Encrinite, *Fig*. 140, remarkable for the elegance and symmetry of its form and for its

---

Describe the European trias. What is said of the fossil plants in the United States? in Europe? of coal? In what kind of strata are the animal remains abundant? in what are they rare? Describe the Lily Encrinite.

complicated skeleton, which consisted of not less than twenty-six thousand pieces. The body was supported on a slender column, which was attached at the base to some hard substance at the bottom of the sea.

3. *Mollusca.*—A large portion of the shells are such as must have inhabited shallow water. The following occur in the muschelkalk (a triassic formation) of Europe.

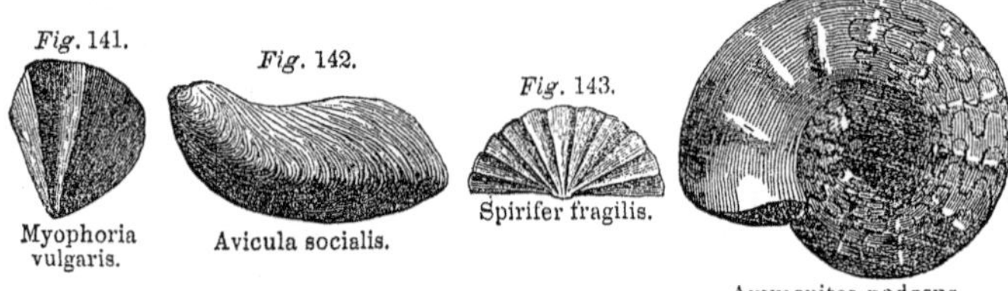

Fig. 141. Myophoria vulgaris. *Fig.* 142. Avicula socialis. *Fig.* 143. Spirifer fragilis. *Fig.* 144. Ammonites nodosus.

4 *Articulata.*—Some extinct genera of crabs and lobsters have been found in the muschelkalk.

5. *Fishes.*—Remains of more than sixty species of fishes have been discovered in the European trias, chiefly in the muschel kalk. In certain bituminous shales, which are associated with the red sandstone of the United States, fossil fishes are abundant. The most celebrated localities are at Sunderland, in Massachusetts, and at Middletown, in Connecticut. *Fig.* 145 represents the Catopterus gracilis, from Middletown.

Fig. 145.

CATOPTERUS GRACILIS.

What is said of the shells? of the articulata? of the fishes?

6. *Reptiles.*—Remains of lizards, of marine saurians, and of batrachians occur. One of the most extraordinary reptiles was the Labyrinthodon. This animal appears to have belonged to the frog tribe, although it was as large as an ox. The bones of the cranium and the jaws, and the bones of the hinder extremities, show that the genus was on the type of this order of reptiles, with affinities to saurians and to fishes. The bones of the legs and the teeth are quite anomalous.

Fig. 146.

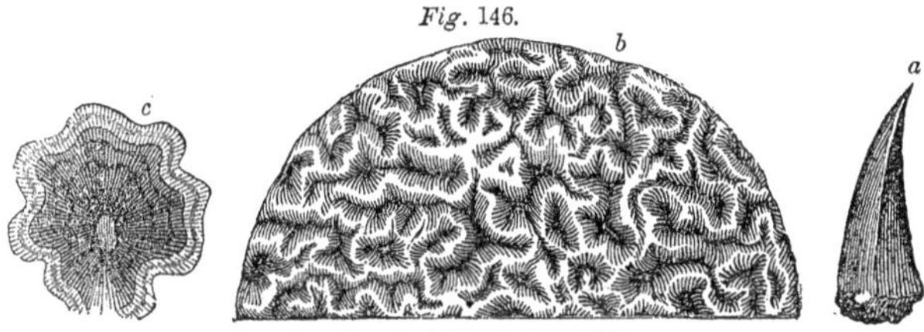

LABYRINTHODON JAEGERI.

a represents a tooth half of the natural size; *b* is half of a transverse polished section magnified twenty diameters; and *c* is one of the anfractuosities of *b* more highly magnified. The complicated structure of the teeth suggested the name *Labyrinthodon.*

In the beds of the same formation are very singular quadrupodal footmarks, which resemble a human hand, *Fig.* 147. The name Cheirotherium (hand-beast) is therefore given to the ani-

Fig. 147.

TRACK OF THE CHEIROTHERIUM.

What is said of the reptiles? of the Labyrinthodon? of the tracks of the Cheirotherium?

mal. The tracks of the hind feet are much larger than those of the fore feet.

Fig. 148 exhibits the manner in which those tracks are supposed to have been made.

Fig. 148.

The Cheirotherium and Labyrinthodon are now supposed to have been the same animal.

In the red sandstone of the western part of New Jersey, the bones of a large saurian reptile have recently been discovered.

In the red sandstone of the United States, and mostly in the Connecticut River valley, numerous footprints have been found, of which some are referable to reptiles and others to *birds;* while many so combine the characters of both of these classes, or are otherwise so anomalous, as to render the precise character of the animals which made them very doubtful. These tracks had been noticed for forty years by persons who were unacquainted with their nature and importance. In 1835 a specimen attracted the notice of Dr. Hitchcock, who subsequently discovered a great number of species, and who has investigated their character and origin with great success. The results obtained by him were at first received by geologists with more or less skepticism, but afterward with admiration, as one of the most extraordinary and interesting chapters in American Geology.

These footmarks prove the existence of about fifty species of animals, of which twelve were quadrupeds; of these, four were probably lizards, two were tortoises, and six were batrachians. Thirty-two were made by biped animals, of which eight were

What is said of the discovery of footmarks in the Connecticut River valley? of the number of species? of the kinds of quadrupeds? of the kinds of biped tracks?

probably birds with three thick toes; fourteen were birds with three or four slender toes; and eight may have been biped reptiles! Of the rest, some were made by invertebrate animals, as worms or molluscs, and others are wholly doubtful. All but two of this list occur in the valley of the Connecticut River.

The most gigantic of these animals was a bird, which may have had some resemblance to the ostrich tribe. It has been named Brontozoum giganteum. The feet were 14 to 20 inches long, and the step was usually four feet long, sometimes six feet. The nails of the toes were $1\frac{3}{4}$ inches long. Some of the tracks have

Fig. 149.

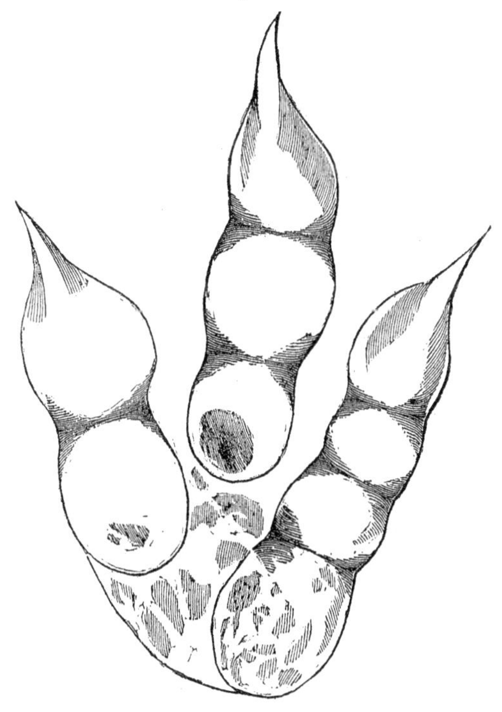

BRONTOZOUM GIGANTEUM.

What is said of the Brontozoum giganteum?

part of the impression of the skin well preserved: it seems to have been papillose and striated. That the bodies of these birds were of great bulk, may be inferred from the shape of the foot and from the depth of the impression made in the mud. Several parallel series of tracks have been found in lines oblique to the shore margin. The footprints in these series are opposite to each other, and thus show that the animals walked in company and were gregarious.

Fig. 149, p. 247, represents a track one fifth of the natural size.

Fig. 150.

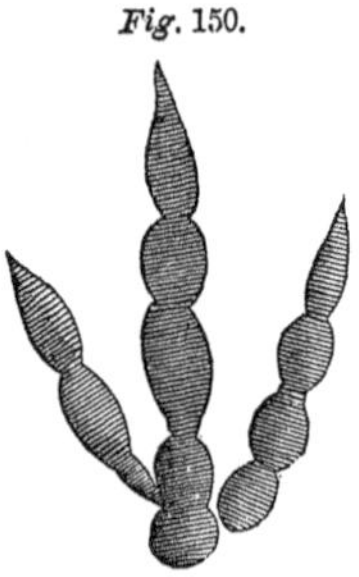

BRONTOZOUM GRACILLIMUM.

Fig. 150 ($\frac{1}{2}$ natural size) represents the track of Brontozoum gracillimum, a small species of the same genus. The step was six inches long.

Another large and remarkable biped species was the Steropezoum ingens, *Fig.* 151. This animal had a heel, or an appendage to the heel, which left a track resembling that of a brush. The surface must have been covered with ridges and striæ, rather than with a brush, as might be supposed from the figure. The animal was inferior in size only to the Brontozoum giganteum and Otozoum Moodii. The length of the foot was 23 to 25 inches, of the middle toe, 10$\frac{1}{2}$ inches, and of the step, four to six feet.

A very anomalous track (Ornithopus Adamsanus, *Fig.* 152) possibly may have belonged to a quadruped of large size. The foot was 13 inches long. It had a small fourth toe directed backward.

A large and interesting species was the Polemarchius gigas, *Fig.* 153, a biped which had a small hind toe, situated far back on the heel, like a spur. The long slender toes and heavy heel suggest both great quickness and force of motion, for they indicate, in the general structure of the animal, long limbs and powerful muscles. If, as is probable, the foot was used as a weapon of offense and defense, it must have been a formidable weapon.

The most extraordinary, and probably the largest of all the

What is said of the Steropezoum ingens? of Ornithopus Adamsanus? of Polemarchius gigas?

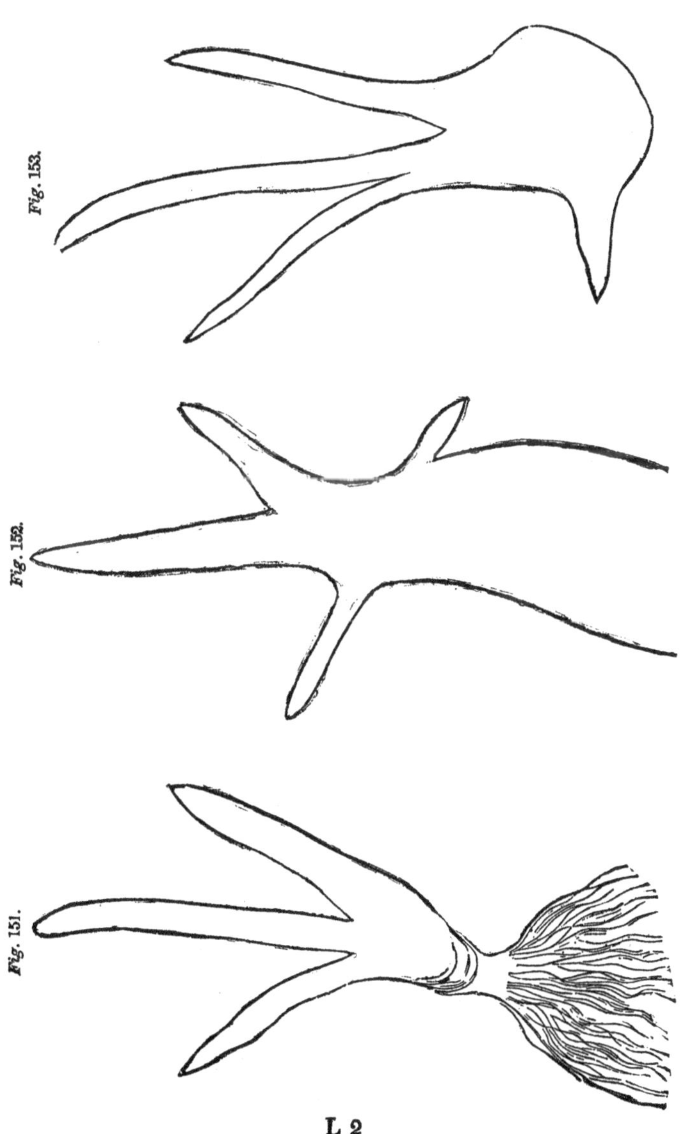

Fig. 153.

Fig. 152.

Fig. 151.

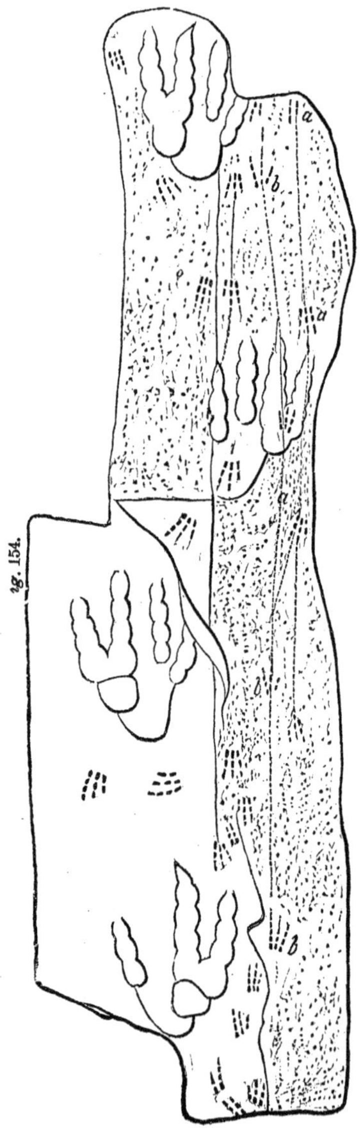

ig. 154.

animals of sandstone days, which Dr. Hitchcock has described, was probably a biped batrachian or toad, with a foot 20 inches long and 12 inches wide. It is obvious from the succession of steps, as seen in *Fig.* 154, that the animal was a biped with short legs. The foot resembled that of an embryo frog more than those of any other living animals. This form, in an adult state, indicates an inferior grade of a batrachian type. Yet this biped toad must have been as large as an elephant. It has been named Otozoum Moodii. The specimen represented in *Fig.* 154 ($\frac{1}{22}$d natural size) was found near Mount Holyoke, in Massachusetts. It contains also (*a a* and *b b*) tracks of a species of Brontozoum (B. parallelum), and has much of the surface pitted with rain-drops, as shown in the figure.

Fig. 155 represents the tracks of a small quadruped, Anisopus gracilis. The hind foot was .9 inch long, and the fore foot .55 inch long.

Fig. 155. *Fig.* 156.

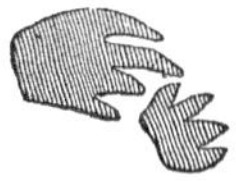

Fig. 156 ($\frac{1}{3}$d natural size) represents the tracks and step of another quadruped, Anisopus Deweyanus.

Numerous slabs have been found which are more or less covered with tracks. One of the best specimens, from Turner's Falls, Massachusetts, was sold to the British Museum by Dr. James Deane, of Greenfield. A portion of it is represented in *Fig.* 157, on page 252. It is about six by eight feet, and contains 75 tracks.

In connection with one kind of tracks, in Chicopee, Massachusetts, coprolites were found. They contain black carbonaceous grains, which are supposed to have been undigested seeds. Chemical analysis shows these specimens to have been the coprolites of omnivorous birds. Teeth ascribed to a mammal, *Microlestes antiquus*, have been found in Germany by Professor Plieninger.

What is said of the biped batrachian? of the coprolites from Chicopee?

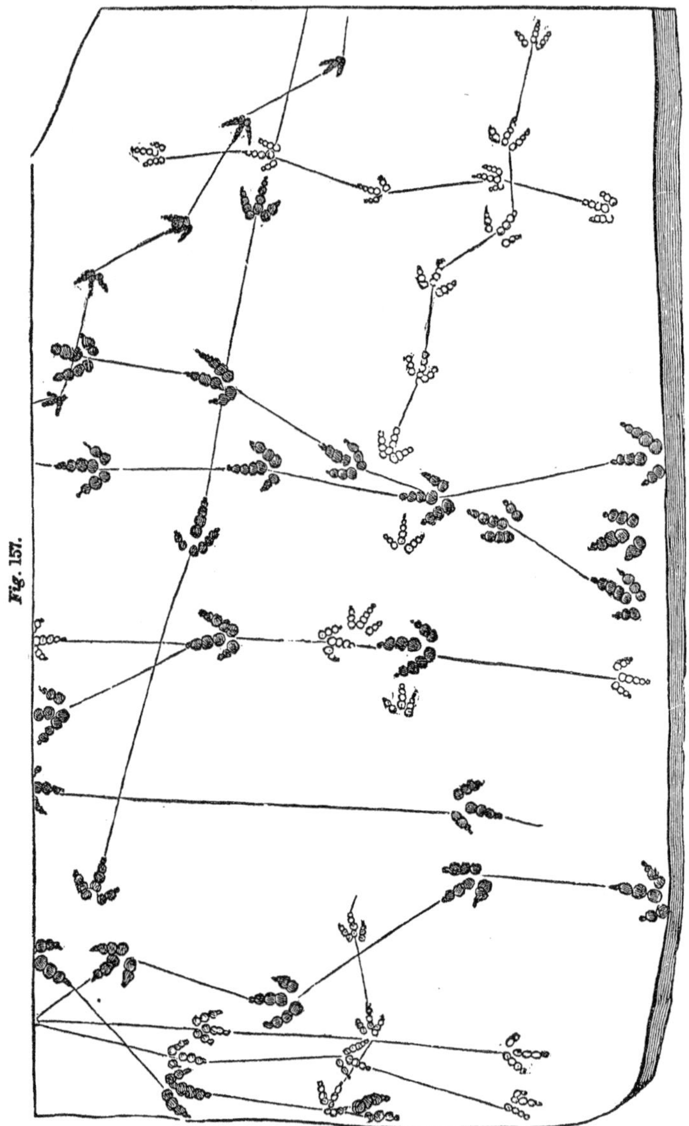

Fig. 157.

CHAPTER VII.

PALÆOZOIC PERIODS.

RECEDING yet farther into the remote ages of the past, we come to those formations which are called Palæozoic, because they contain the remains of the most ancient animals that are known to have existed. The passage from the Mesozoic to the Palæozoic formations is not marked by any abrupt transition in the lithological characters of the strata. Without fossils, it is even difficult to find the plane of division. Yet in the forms of animal life, we find a transition almost as abrupt as that from the Tertiary to the Mesozoic. Although the triassic strata (the oldest Mesozoic) usually repose conformably on the Permean rocks (the newest Palæozoic), yet the fossils are entirely different. Not one of the species which existed during the Permean epochs was continued into the trias.

Some of the Permean animals had existed through several of the previous Palæozoic periods, and most of the others were nearly allied to their more numerous predecessors. The species of the last Palæozoic period may, therefore, be regarded as the remnant of older types, which were gradually dwindling away. In the later Palæozoic periods, the dawn of a new order of things was seen in the introduction of a few species of saurians, a tribe of reptiles which, as we have seen, had an enormous development during the Mesozoic periods. Thus, as we go up the stream of time, we find all the genera, and even some entire families and orders disappearing, and others taking their place.

What is the meaning of Palæozoic? What is said of the transition from the Mesozoic to the Palæozoic formations? of the general character of the Permean fossils of the introduction of Saurians?

SECTION I.—PERMEAN SYSTEM.

I. *Geographical Distribution.*—This system derives its name from the ancient kingdom of Permea, in the eastern part of Russia in Europe, where the rocks of this period occupy a district 700 miles long from north to south, and nearly 400 miles wide. It is also well developed in England and France; but it is in Germany especially that it appears with a numerous series of well-marked subdivisions.

II. *Structure and Position.*—These rocks consist of many distinct strata of various characters. They are composed of white limestone, with gypsum and rock salt, sandstones with slates and copper ore, of magnesian limestone, conglomerates, &c.

Fig. 139 (page 242) exhibits the position of the system in the rocks of England. The following figure exhibits their subdivisions in Germany.

Fig. 158.

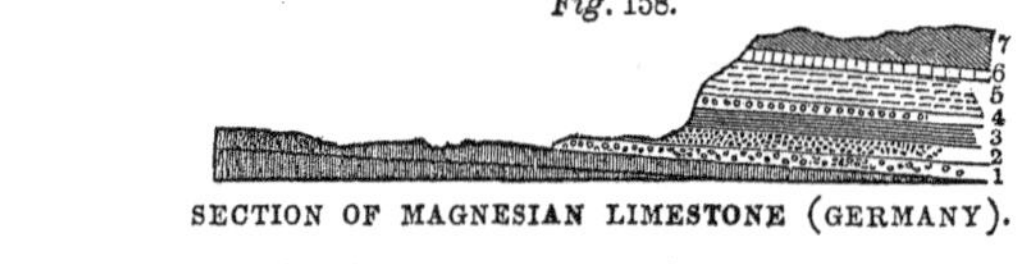

SECTION OF MAGNESIAN LIMESTONE (GERMANY).

Magnesian limestone series.	Zechstein, or mine-stone.	7. Letten; clay.
		6. Fetid limestone.
		5. Rauwacke, cellular magnesian limestone.
	Shaly series.	4. Argillaceous schist.
		3. Bituminous schist and copper slate.
		2. Arenaceous schist.
Lower new red series....		1. Rothe-todte-liegende.*

III. *Fossil Plants.*—The species are not numerous. A few of them are identical with the species of the carboniferous system, and the others belong to the genera of that system.

Fig. 159 represents a species, Odontopteris Strogonovi, which is peculiar to the Permean system.

IV. *Fossil Animals.*—One hundred and sixty-six species are

* Red-dead-lier, because of a red color, dead or worthless, not containing metals, and underlying the metalliferous strata.

What is said of the name and geographical distribution of the Permean rocks? of their structure? of their subdivisions in Germany? of the fossil plants? of the fossil animals?

Fig. 159.

ODONTOPTERIS STROGONOVI.

known, of which 148 belong exclusively to this system; the remaining 18 have been continued from older formations. Of these 18, ten are the shells of Brachiopods.

1. *Radiata.*—There are only 15 species of *corals*, and these are mostly rare The *Crinoideans* were represented by only one species.

2. *Mollusca.*—We select five species of shells which are characteristic of this system.

Fig. 160.

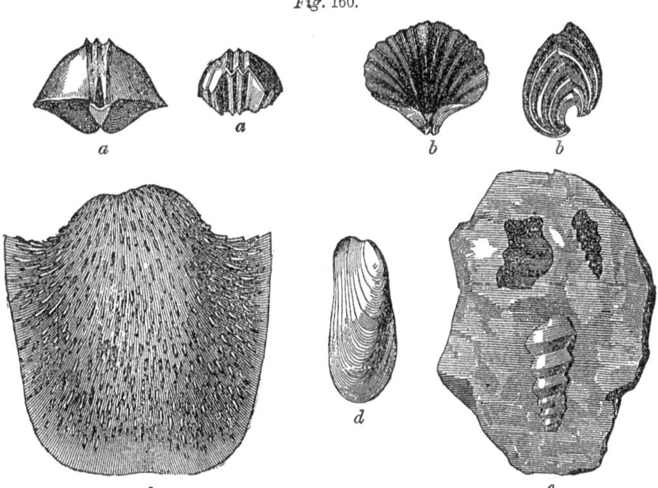

a a. Terebratula Schlotheimi. *c.* Productus horrescens.
b b. Spirifer Blasii. *d.* Modiola Pallasi.
e. Murchisonia subangulata.

3. *Articulata.*—In this period the existing genus, Limulus (horse-foot), was introduced, and the Trilobites, which had flourished abundantly during nearly all the older epochs, had disappeared.

What is said of the radiata? of the articulata?

4. *Fishes.*—The conditions of life were more favorable to this class of animals than to some others. Forty-three species have been found. The two most characteristic genera are Palæoniscus (*Fig.* 161) and Platysomus.

Fig. 161.

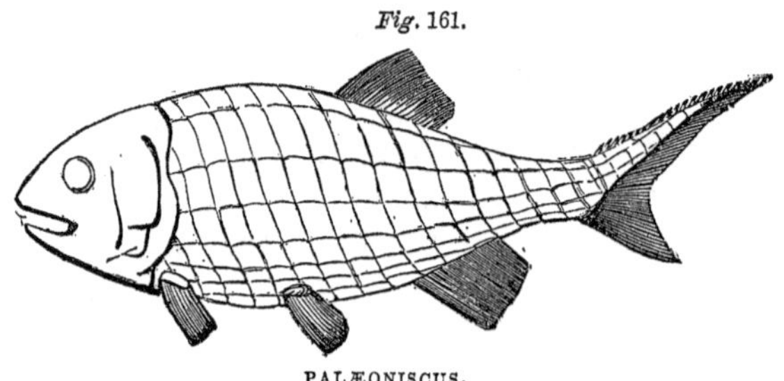

PALÆONISCUS.

5. *Reptiles.*—Some lizard-like reptiles are interesting, as the oldest representatives of this class with which we are acquainted, with the exception of some species, which are known by their footprints only, in the carboniferous system.

SECTION II.—CARBONIFEROUS SYSTEM.

I. *Geographical Distribution.*—This system is widely extended in both hemispheres. On this continent it furnishes a rich bituminous coal-field in Nova Scotia. In the eastern part of Massachusetts there is a small anthracite coal-field.

The largest explored coal-field in the world has its northeast extremity west of the Delaware River, in New York, and extends through Pennsylvania into Ohio westward, and to Alabama on the southwest. It covers more than 100,000 square miles, and contains more than one million of million tons of bituminous coal. A much less extensive but rich field of anthracite coal lies in the eastern part of Pennsylvania, east of the Delaware River. The central region of Michigan constitutes another large coal-field. Another covers most of Illinois, the southwest part of Indiana, and the adjacent part of Kentucky.

What is said of the fishes? of the reptiles? In what regions in North America does the carboniferous system occur? What is said of the largest coal-field?

The carboniferous system is found in Vancouver s Island and in New Mexico; and in South America, in Chili.

It occurs also in England, Scotland, Ireland, Central France, Spain, Belgium, Westphalia, and in Russia, from the White Sea on the north to Kaluga on the south. It is found in Java, in Borneo, and in China.

II. *Structure and Position.*—Sandstones, conglomerates, and shales, alternating, in some parts of the series, with beds of coal, constitute this system. In some districts, calcareous strata occur; but they are much less common in the Atlantic States of North America than the Western States or in Europe.

The accompanying section, *Fig.* 162, exhibits the relations of this system to the older rocks in the northeast extremity of the great Ohio coal-field in the northern part of Pennsylvania.

Fig. 162.

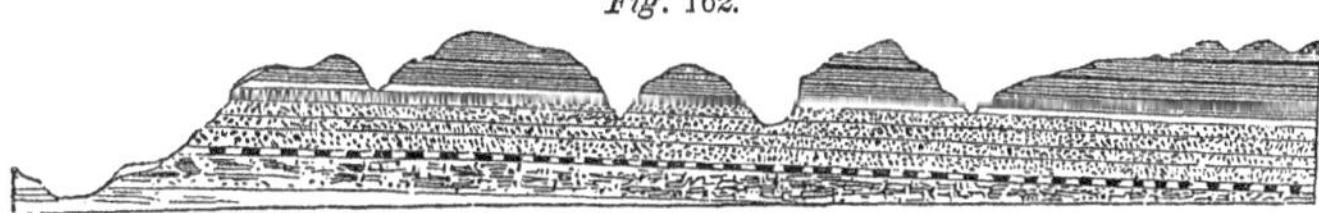

Here we see the effects of the extensive denudation which these rocks have suffered. Probably the contiguous parts of Pennsylvania and New York were once entirely covered with them. Now only a few insulated summits remain. *Fig.* 139 (p. 242) exhibits the relations of this system to the newer rocks in England.

Fig. 163, p. 258, exhibits details of structure in a section of the coal strata on Kenawha River, Ohio.

In what other countries does coal occur? What is the structure of the **rocks?** What is said of the effects of denudation?

Fig. 163.

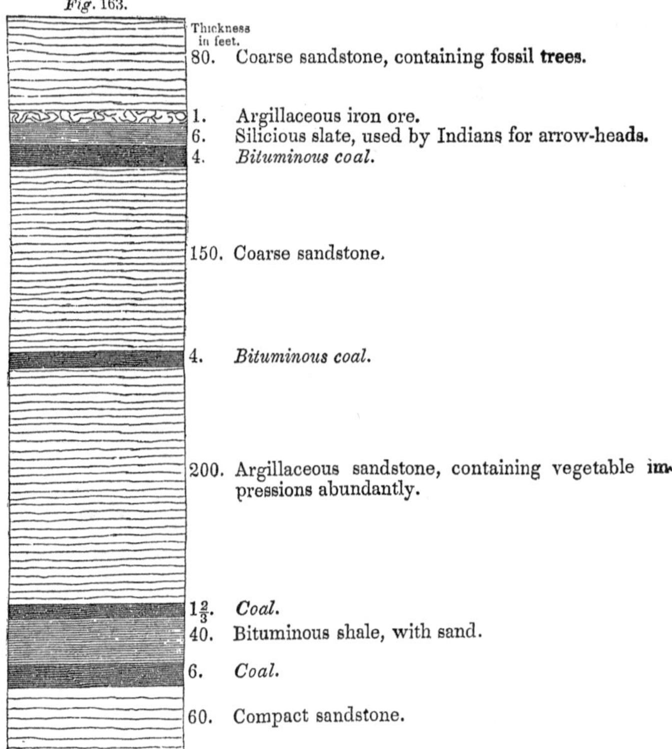

The continuity of coal and other strata is often interrupted by dislocations, of which usually no traces appear on the surface of the ground, in consequence of subsequent denudation of the surface. *Fig.* 164 represents a section of an English coal-field. The

Fig. 164.

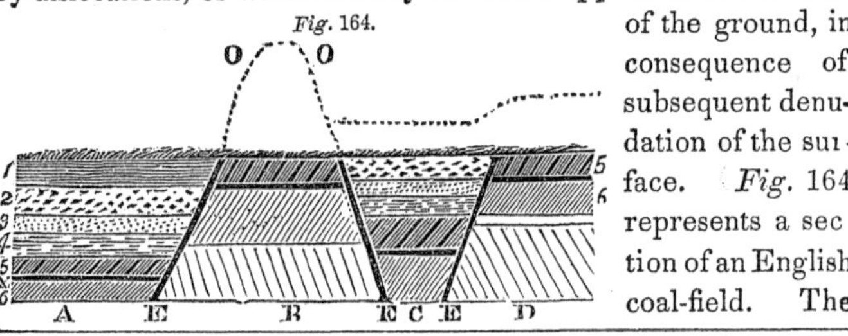

Describe the section at Kenawha River; the section of an English coal-field

strata are numbered in descending order. There are four different levels. In A B the coal X is 900 feet below the surface; in B C it is within 200 feet of the surface, covered by No. 5; in C D it is 700 feet below the surface, and No. 1 only is wanting; in D E it is somewhat higher than in B C.

III. *Fossils.*—1. *Plants.* This system is remarkable for containing not only a vast quantity of vegetable matter in the form of coal, but also for the great number and variety of the remains of plants. About 1000 species have been described, which is rather more than half of the entire number found in a fossil state. This is the more remarkable, since only a few species have been found in the older rocks. It is also remarkable that while few animal remains have been found in the coal formation, the plants were almost wholly of families which were not adapted for the food of animals. A large proportion of the vegetation of this period consisted of tree ferns, and of plants intermediate between the coniferæ and the endogens.

The wide distribution of the species is worthy of notice. Several have been found to be identical in Europe and in North America. The distribution of the species of animals was also on the same plan, not only in this but in all the earlier periods. It has, therefore, been inferred that the climate was then more uniform over the earth's surface than at the present time.

(1.) *Calamites* is the name given to a tribe of plants, which are sometimes mistaken by the ignorant for petrified snakes. Some botanists have supposed them to have been a kind of reeds allied to equisetum; but as they had a true bark, according to Professor Lindley, they could not have belonged to the class of endogens. They were branching plants, with hollow stems, of large size, and were very fragile. They are often found much compressed. The surface was covered with deep longitudinal furrows, which terminated at the joints. Some have been found three feet in diameter and thirty to forty feet long.

It is very rarely that the stellate sheaths of the joints are found

---

What is said of the species of fossil plants? of the general character of the plants? of their distribution? of calamites?

as in *Fig.* 165, *a*, of *Calamites radiata;* or that the remains of roots are found, as in *b*. *c* shows the curved upper extremity of *Calamites approximata*. *a* and *b* are $\frac{1}{2}$, and *c* $\frac{1}{5}$ natural size.

Fig. 166 ($\frac{1}{12}$ natural size) represents a calamite from Seekonk, Massachusetts, now in the Museum of Amherst College.

Fig. 166.

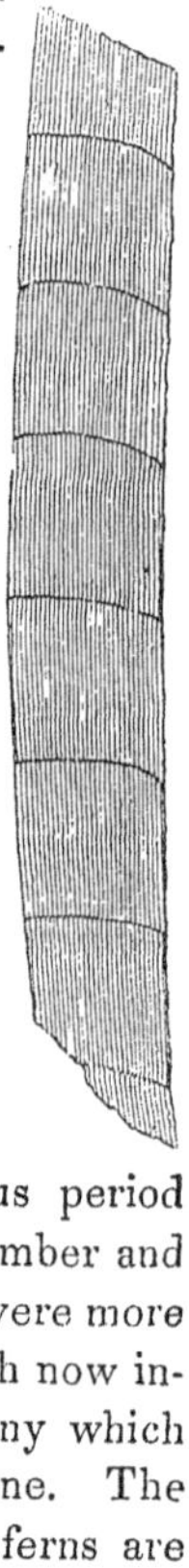

Fig. 165.

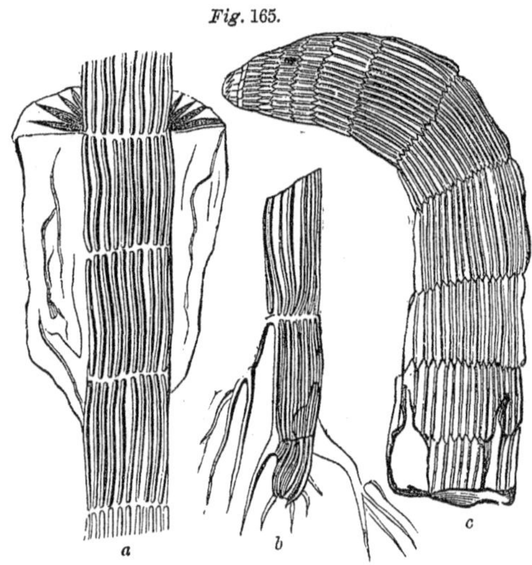

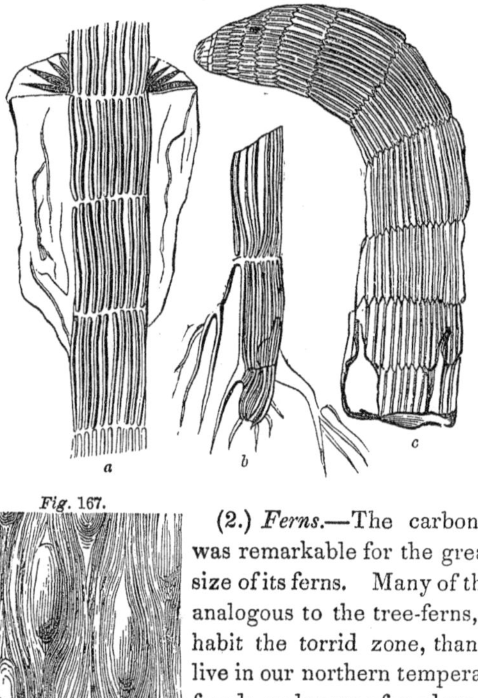

Fig. 167.

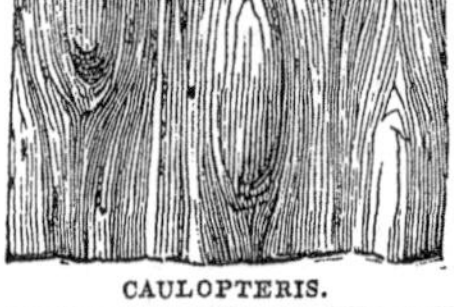

CAULOPTERIS.

(2.) *Ferns.*—The carboniferous period was remarkable for the great number and size of its ferns. Many of them were more analogous to the tree-ferns, which now inhabit the torrid zone, than to any which live in our northern temperate zone. The fronds or leaves of modern tree-ferns are often ten or twelve feet long, and are minutely and elegantly subdivided.

More than 200 extinct species have been found in the carboniferous system. The stems and leaves are almost invaria-

What is said of the general character of the ferns? of the species of ferns?

bly separate; and since the parts of extinct species and genera of plants can not be reunited with the same accuracy with which the comparative anatomist reconstructs an extinct animal, botanists have been obliged to give different names to the leaves and to the stems. The provisional name of Caulopteris has therefore been given to the stems (*Fig.* 167).

Fig. 168 represents a part of two leaves of *Pecopteris longifolia*, from Rhode Island; *Fig.* 169 represents a leaf and part of the stem of *Odontopteris Brardii*, also from Rhode Island.

Fig. 168 (PECOPTERIS LONGIFOLIA).

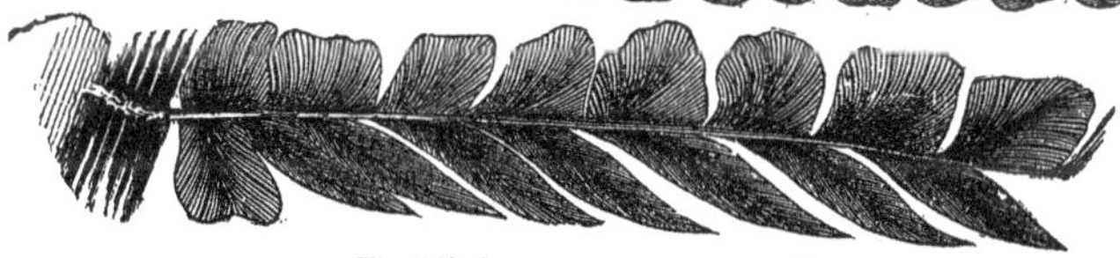

Fig. 169 (ODONTOPTERIS BRARDII).

Fig. 170. *Fig.* 171.

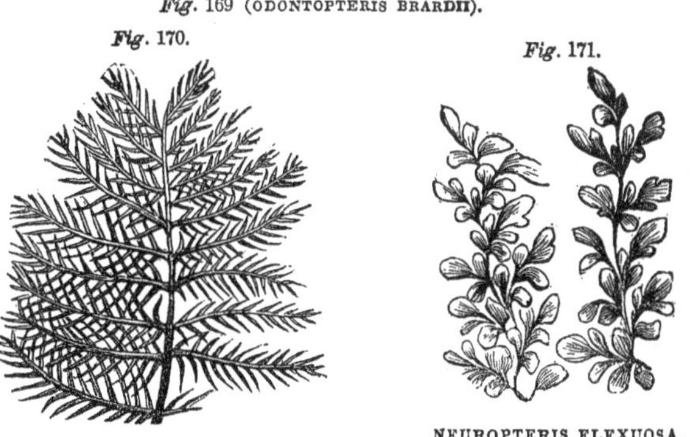

PECOPTERIS MANTELLI. NEUROPTERIS FLEXUOSA.

Fig. 170 is *Pecopteris Mantelli*, from England; and *Fig.* **171 is** *Neuropteris flexuosa*, also from England.

(3.) *Lepidodendron* (scaly tree) is the name given to an anom-

What is said of the Lepidodendron?

alous type of plants, which were much like the existing club-mosses (Lycopodiaceæ), in respect of texture, surface of the stem ramification, and foliage. But they were large trees. In an English coal mine one specimen was found which was forty feet long, and at the base thirteen feet in diameter; at the summit it had fifteen or twenty branches. Having a true bark, they could not have belonged to the same class with our Lycopodiaceæ. In the arrangement of the leaves they also resembled coniferæ.

Fig. 172 shows the ramifications of a Lepidodendron and the leaves of the extremities. This specimen was found in the coal shale of Newcastle, England.

Lepidostrobus is supposed by some to have been the fruit of the Lepidodendron. *Fig.* 173, *a*, represents the rare example of a young fruit at the end of a branch. *Fig.* 173, *b*, shows the form and structure of the ripe fruit, and its internal axis.

Halonia regularis, *Fig.* 173, *c*, from Coalbrook Dale, England, had a remarkably knotted surface. It was probably allied to the Lepidodendron.

(4.) *Sigillaria* and *Stigmaria.*—These were once supposed to be two distinct plants; but it is now well known that the former is the stem and the latter the root of a large tree. *Fig.* 174 represents a specimen found in a coal mine near Liverpool, England, with the roots proceeding from the stem.

In coal mines the roots of this tree are found abundantly in a deposit of clay, which invariably underlies the coal, and is therefore called underclay. The trunks are generally found in a horizontal position; but, in one instance, five stems were found erect, near Manchester, England, where they have been secured in their position. In another case, a group of forty trees was found, standing not more than three or four feet apart. The stems vary from a few inches to five feet in diameter, and five to sixty feet in length.

What is said of the Lepidostrobus? of the character of Sigillaria and Stigmaria? of their position?

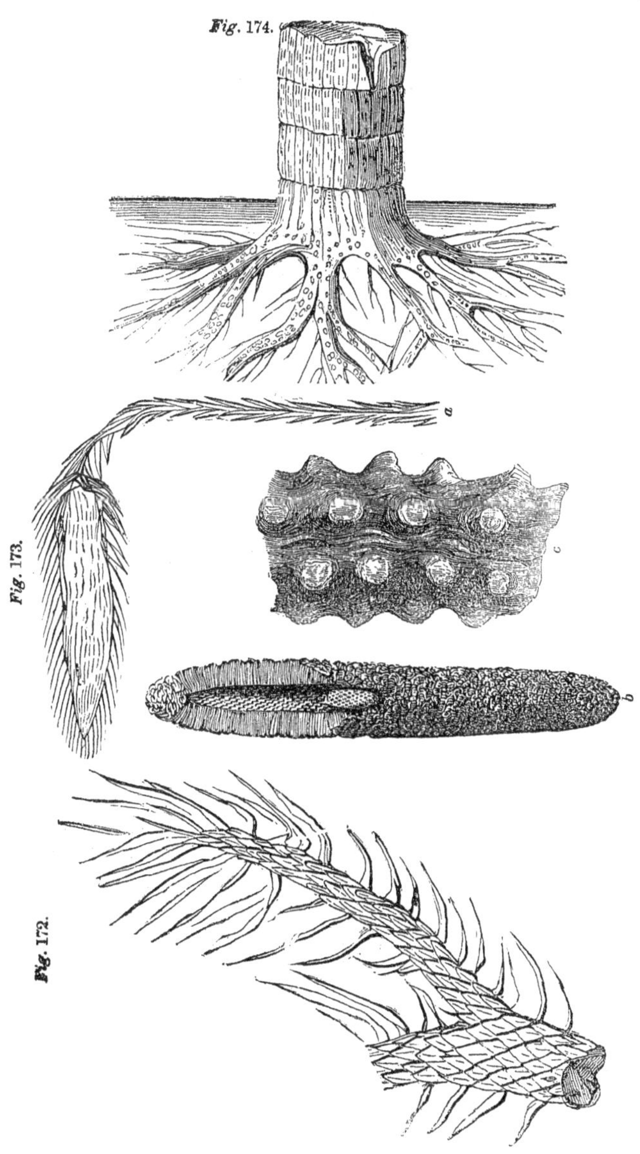
Fig. 174.
a
c
b
Fig. 173.
Fig. 172.

Fig. 175 shows the scarred surface of a species of Sigillaria.

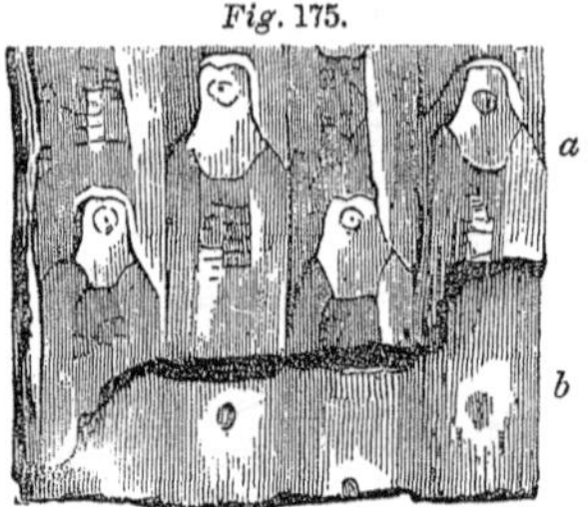

Fig. 175.

a. The scars left on the bark by the fallen petioles.

b. The surface beneath the bark, exposed by the removal of a part of the carbonized bark.

(5.) *Origin of Coal.*—(*a.*) The vegetable origin of coal might be inferred, with some probability, from the great abundance of fossil plants in connection with it. But it is more satisfactorily demonstrated from the microscopic examination of the structure of pieces of coal. The most convenient method is to take a piece of partially-burned anthracite from the fire. The vegetable cells, being more or less silicious, retain their form after the carbon is partly burned away, and may be seen with a microscope on the surface of the specimen. *Fig.* 176, *a*, represents several ducts in superimposed layers; *b* shows two ducts more highly magnified. The white spaces are the patches of silica, and the black lines are the unburned carbon. From such observations, Professor Bailey, of West Point, infers (1.) that the material of coal could never have been reduced to a homogeneous pulp; but (2.) that coal is composed of thin layers of vegetable bodies confusedly intermingled. This method of examination is less applicable to soft coal, on account of the partial fusion, and the swelling caused by the bitumen. But other methods have detected in it a vegetable structure.

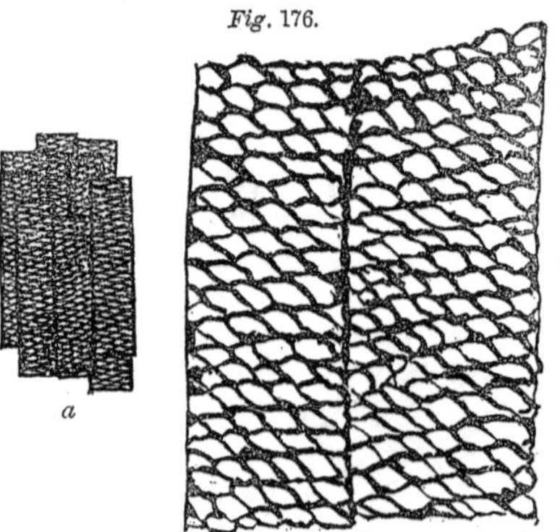

Fig. 176.

(*b.*) Different theories have been proposed to explain the man-

What is said of the origin of coal? of its structure?

ner in which the strata of coal may have been preserved. In some places, their alternation, in a thick series, with strata which contain marine shells, and with others which contain fresh water shells, is thought to indicate successive rise above and subsidence beneath the waters of the ocean. The subsidence accounts sufficiently for the covering and subsequent preservation of the vegetable matter. In other cases, perhaps more frequently, the vegetable matter may have been carried down large rivers into the sea, and becoming water-logged, have sunk to the bottom. But it has been said that this could not have been the origin of those coal-fields in which trunks of Sigillaria are found erect; although some stems might have this position from the settling of the roots, like snags in rivers.

2. *Animals.*—The number of species known is not far from one thousand. Most of them are shells. A large majority do not occur in the proper coal formation, but in the other members of the system.

(1.) Of *Radiated* animals the remains of fifteen to twenty species have been found, very few of which have been found in the other systems. *Fig.* 177 shows the corals, encrinites, &c., which

Fig. 177.

are exposed on a weather-worn slab of the carboniferous limestone of Iowa.

What is said of the theories of the preservation of the coal? of the species of fossil animals? of the Radiata?

Fig. 178

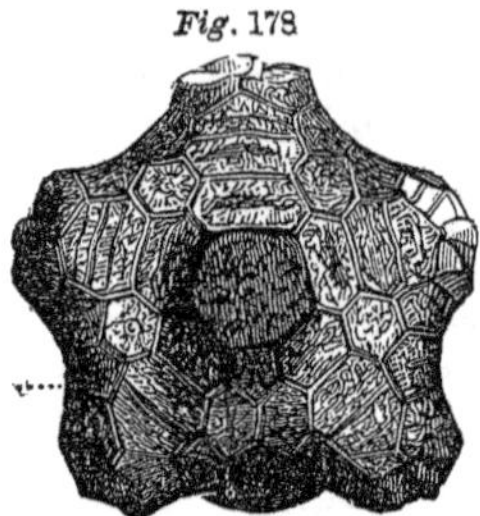

Actinocrinites triakonta-dactylus.

Fig. 178 shows the body of the thirty-fingered Actinocrinites of the mountain limestone of Europe. Each of the five branches was subdivided into two, and each of these into three. The figure shows the base of the branches.

Fig. 179.

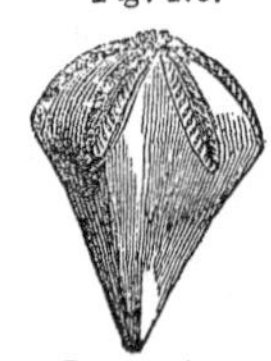

Pentremites inflatus.

Fig. 179 is Pentremites inflatus, from the same formation. Both of these animals had five-sided peduncles.

Fig. 180.

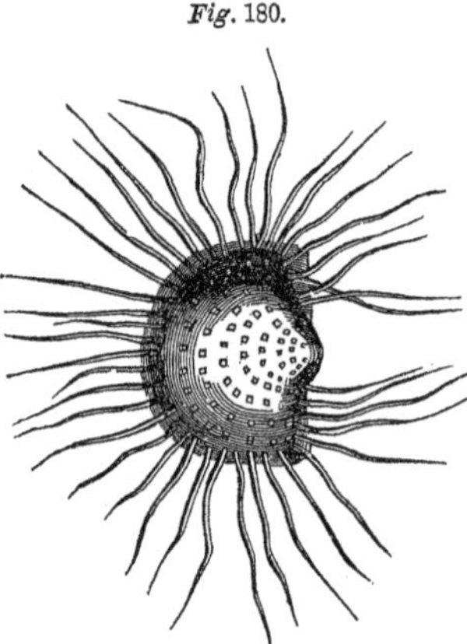

Productus spinulosus.

(2.) *Molluscs.*—The Foraminifera, a class of molluscs usually of microscopic size, and of an inferior grade of organization (see page 197), were represented by several species during this period. The class of Brachiopods was most abundantly represented. The shell of Productus spinulosus, *Fig.* 180, was remarkable for its long spines. There were thirty species of the genus Productus. Productus giganteus was a very large European species, five inches long and four inches high. Spirifer, another genus of Brachiopods, was represented by 25 species. *Fig.* 181 shows the parts of Spirifer striatus. *a*, the external surface. *b*, the interior, showing the spiral coil. *c*, part of the coil detached.

Fig. 181.

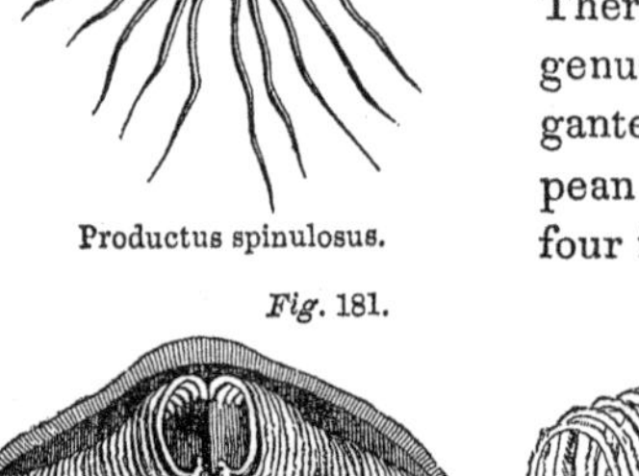

Spirifer striatus.

What is said of the Foraminifera? of the Brachiopods?

The Gasteropods were represented by several genera, many

Fig. 182.

of which now exist, while others became extinct in this period. Of the latter number was Euomphalus, which had been introduced in the earliest Palæozoic times. *Fig.* 182 represents a beautiful species, *E. equalis*, found in Russia. Although a Gasteropod, the animal had a chambered shell, for it formed im perforate partitions in its successive retreats from the older portions of the shell. *Natica Omaliana, Fig.* 183, has been found in Russia and in Belgium, and belonged to a genus which was introduced in the older Palæozoic periods, and is now abundantly represented in all climates.

Fig. 183.

Of the Cephalopods, Goniatites is a genus which became ex-

Fig. 184.

tinct in this period. *Fig.* 184 represents a remarkable species, G. Jossæ, which is found in Russia. The lines of the junction of the partitions with the exterior shell were very tortuous. Orthoceratites is another Cephalopod genus, which had been introduced in the earliest Palæozoic periods, and became extinct in this epoch. *Fig.* 185 shows the internal structure of a species. The genus Ammonites (see p. 219), although of early introduction, did not become extinct until the last of the Mesozoic periods. It was represented in this period by several species. Many of them occur in the Western States.

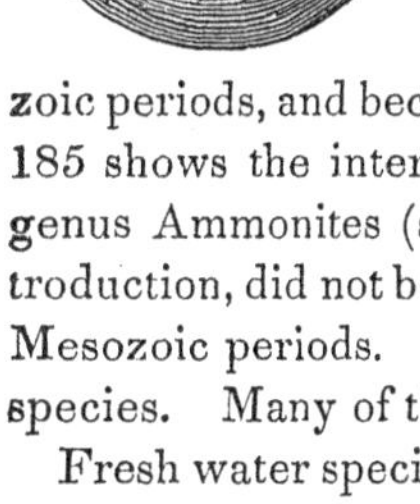

Fig. 185.

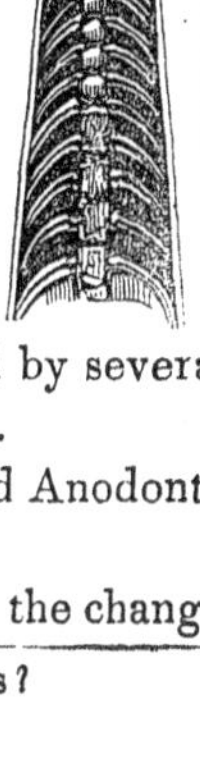

ORTHOCERATITES SIMMII.

Fresh water species of the existing genera Unio and Anodonta occur in the coal formation.

Although none of the species of this epoch survived the change

What is said of the Gasteropods? of the Cephalopods?

Fig. 186.

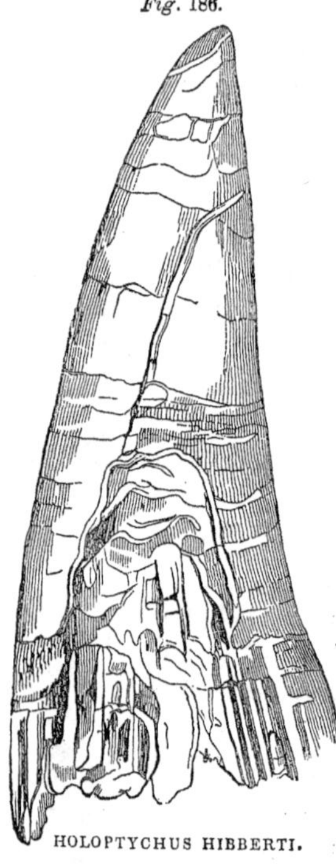

HOLOPTYCHUS HIBBERTI.

from Palæozoic to Mesozoic times, yet 23 genera of the shells have continued to the present time. They belong mostly to the Acephala and Gasteropoda.

(3.) *Fishes.*—The fishes of this period have been very abundantly preserved. Not less than 150 species have been found. Megalichthys and Holoptychus were gigantic sauroid fishes, whose remains are found with plants and crustaceans, which indicate a lacustrine or estuary formation. *Fig.* 186 represents a tooth of one of these fishes of the natural size. Their teeth are larger than those of any other fishes which have conical teeth.

(4.) *Reptiles.*—In the carboniferous rocks of Nova Scotia, the footprints of a reptile have been found. In those of Pennsylvania, the tracks of a reptile were discovered by Dr. A. T. King, who named the animal *Thenaropus heterodactylus*, *Fig.* 187. The hind foot resembles a human hand. The reptile may have been a batrachian.

Fig. 187.

THENAROPUS HETERODACTYLUS, KING.

In 1848, Dr. Isaac Lea, of Philadelphia, discovered near Pottsville, in Pennsylvania, certain reptilian tracks. There are six double impressions 13 inches apart. The animal which made them has been named by Dr. Lea *Sauropus primævus*. *Fig* 188

What is said of the genera of the shells of this period? of the fishes? of the reptiles? of the tracks discovered by Dr. Lea?

represents one pair of tracks about $\frac{2}{5}$ of the natural size. Dr

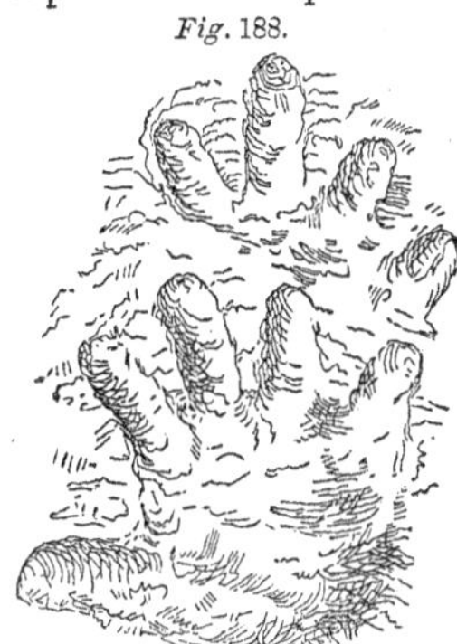
Fig. 188.

Lea refers their position to the Devonian system, but Professor H. D. Rogers places them above the middle of the carboniferous system, yet much below the tracks which were discovered by Dr. King. They are, therefore, very interesting as the oldest remains yet known of any reptilian or of any air-breathing animal.

IV. *Climate and Geography.*—From the vast quantity of the vegetable matter, and from the remains of fresh water shells, it may be inferred that there was no inconsiderable amount of dry land during this period. From the character of the plants, it has been inferred, also, that the land must have been low, and perhaps consisted to a large extent of islands. If, however, the beds of coal originated in vegetable matter brought down by rivers, there must have been large rivers, and consequently large continents.

Professor H. D. Rogers finds that the coal-beds of Eastern Pennsylvania were derived from a continent lying to the eastward, and occupying more or less of the space which is now filled by the Atlantic, while the ocean of this period occupied the region westward, and covered most of the continent of North America. The gradual retreat eastward of the Atlantic continent is indicated by the greater extension eastward of each successive coal-bed as we ascend the series. The retreat eastward of this continent must have been due to its subsidence beneath the ocean, as is obvious also from the thickness of the series of strata which intervene between the coal-beds.

It has been generally inferred, from the character of the plants, that the climate was tropical; and some have supposed that it was ultra-tropical. Most of the existing tree-ferns and other plants

What is said of the extent of dry land? of the continent occupying the are of the Atlantic? of the climate?

which resemble those of the carboniferous epoch are tropical. Yet the abundance of a similar vegetation in New Zealand, and the difference of the species and genera of that period from those which now exist, will not permit us to infer confidently that the mean temperature of the earth's surface was much higher than at present. But the temperate and northern regions may have been warmer, and it is almost certain that a very uniform climate prevailed over large areas.

During this period a great quantity of carbon was buried between the sedimentary strata. But since the carbon was of vegetable origin, and since plants derive their carbon from the carbonic acid gas in the atmosphere, it has been inferred that, up to this period, the atmosphere had been loaded with a much greater quantity of carbonic acid than it has since contained. The total amount of coal in the world has been estimated at 5,000,000,000,000 of tons. The quantity of carbon now in the carbonic acid of the atmosphere is 850,000,000,000 tons. Hence, previous to the coal vegetation, the atmosphere must have contained six times its present amount. Such an atmosphere, with a moist temperature, and a uniformly warm climate, would have fulfilled the conditions most favorable to an exuberant vegetation.

SECTION III.—DEVONIAN SYSTEM—OLD RED SANDSTONE.

I. *Geographical Distribution.*—In the United States this system has its principal development in a region which includes the Catskill Mountains of New York on the northeast, extending to the anthracite coal district of Pennsylvania on the south, while it is prolonged westward, in a narrow belt, around the great bituminous coal-field in Western Pennsylvania. Farther to the southwest, it extends in narrow belts along the Alleghany Mountains. But some geologists do not regard these rocks as belonging to the Devonian system.

In Russia, in Europe, this system covers a vast region 150,000 square miles in extent. Occupying most of the western part of

---

What is said of the atmosphere? of the distribution of Devonian rocks in the United States? in Europe?

Russia, it extends to the northeast in one long arm beyond Archangel, while another arm extends to the southeast beyond the River Don. It also occurs along the western flanks of the Ural Mountains. In Western Europe, this system is well developed in the Rhenish provinces; it has been most carefully studied in the southwest part of England, and in Scotland, where it is 10,000 feet thick. It occurs also in Ireland, France, and Spain.

II. *Structure and Position.*—The strata of the Devonian system are composed of dark red sandstones and conglomerates in the upper part, and of calcareous slates and flagging stones in the lower part, in England. In New York they consist of red, brown, gray, and green sandstones, and conglomerates, and of fissile sandstones and slates of the same colors.

Fig. 162 (p. 257), and *Fig.* 198 (p. 275) exhibit the position of this system in New York.

III. *Fossils.*—More than seven hundred species have been found, of which a majority are shells. Fishes are also preserved in great numbers.

1. *Plants.*—Very few species remain, and these are said to have been mostly marine.

Fig. 189.

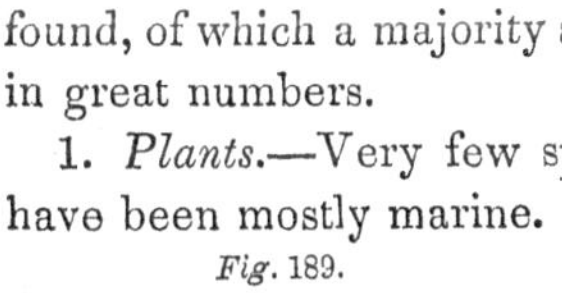

2. *Animals.*—(1.) *Radiata.* Corals and Crinoideans were very abundant during this period.

Fig. 189 is *Favosites polymorphus*, a coral of this period.

(2.) *Articulata.*—The existing genus *Serpula* makes its first appearance in these strata. These animals are marine worms, which are covered by a solid calcareous secretion. Their tubes are irregularly twisted, and are joined in large masses or attached to other marine bodies. Of crustaceans, trilobites occur much more rarely than in the older Palæozoic rocks. *Brontes flabellifer* was a crustacean of the order Macroura, somewhat like a lobster, but was four feet long. *Fig.* 190, page 272, represents the tail.

What is said of their structure? of their position? of the number of fossil species? of the plants? of the radiata? of serpula? of the crustaceans?

(3.) *Molluscs.*—Of the Brachiopods, the existing genus *Terebratula* was abundantly represented. The genus *Spirifer* was also well represented. The genus *Strigocephalus* was introduced during this period. *Fig.* 191 is *S. Bartini.*

Fig. 190.

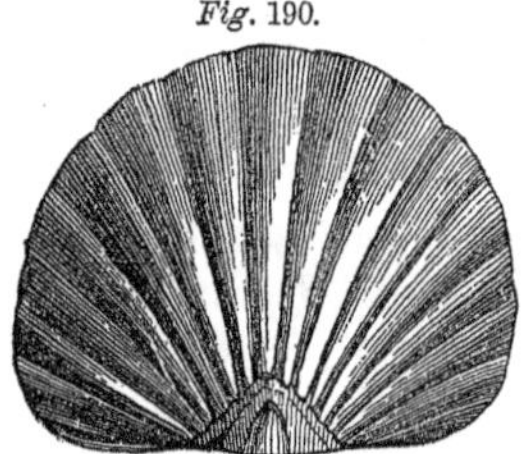

Fig. 191.

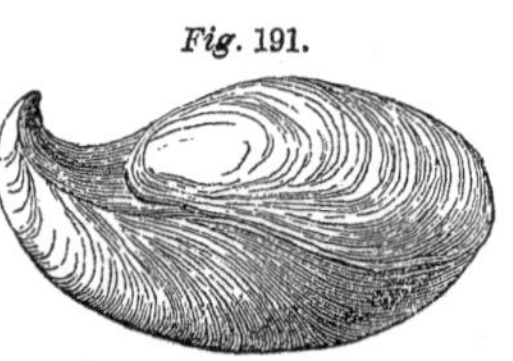

Strigocephalus Bartini.

Of the Acephala there were many species. *Fig.* 192 is *Cypricardites angustata,* from the Catskill region, in New York. *Murchisonia intermedia*, *Fig.* 193, is an example of the Gasteropods of this period. Of the Cephalopods, the genus *Clymenia*, *Fig.* 194, which had the siphuncle along the inner margin of the partitions, was introduced during this period. Thirty-five species of this genus have been found at one locality in Bavaria. This number embraces nearly all the known species.

Fig. 192.

Fig. 193.

Fig. 194.

Clymenia inæquistriata.

(4.) *Fishes.*—This period was remarkable for the number and extraordinary character of the fishes. Seventy-five species have

What is said of the Brachiopods? of the Acephala? of Clymenia? of the fishes?

been found. They occur chiefly in the sandstones of Scotland and Russia, while, in the same kind of rocks, shells are rare in both of these countries.

In one family of fishes the head was greatly developed. Such

Fig. 195.

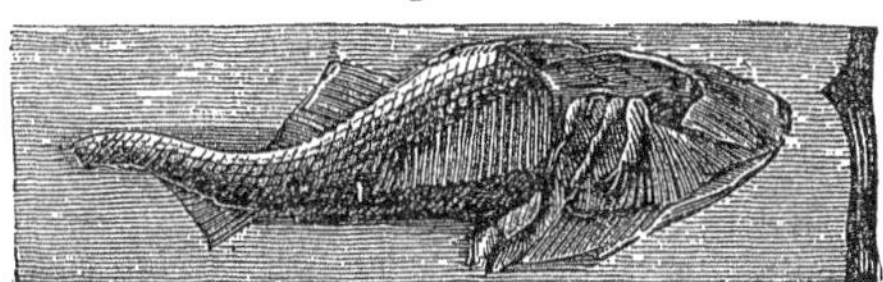

was *Cephalaspis Lyellis*, *Fig*. 195. *Pterichthys cornutus*, *Fig*. 196, was remarkable for its wing-like appendages. In other re-

Fig. 196.

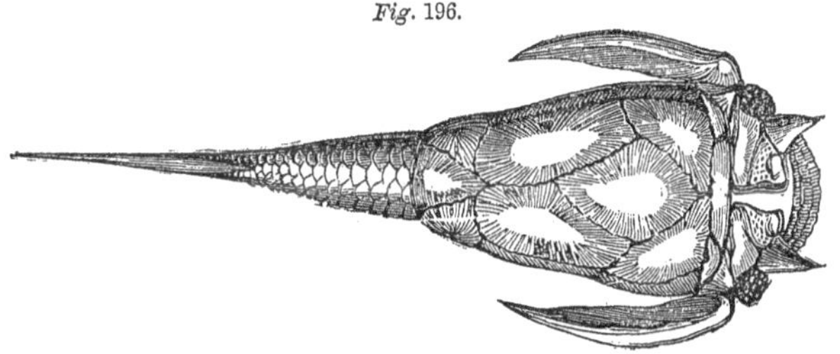

spects, the *Coccosteus oblongus* was allied to it, but had the tail needle-shaped, and much longer.

One of the most extraordinary genera was Holoptychus. It had very large, stout, enameled scales, which constituted a defensive armor of great strength. Some of the scales were nearly

Fig. 197.

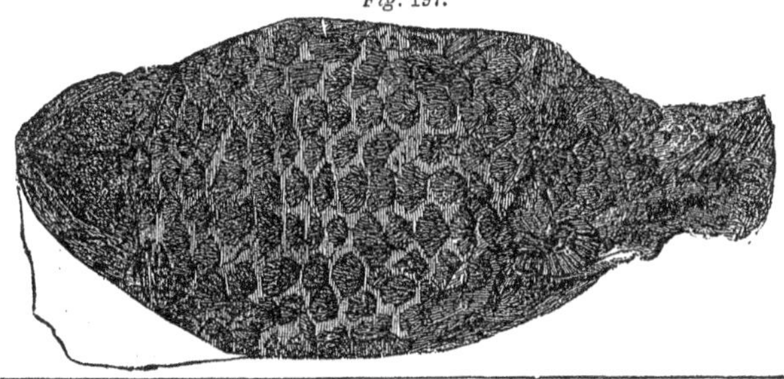

What is said of Cephalaspis? of Pterichthys? of Holoptychus?

three inches long and two and a half inches wide. *Fig*. 197 page 273, represents a specimen of *Holoptychus nobilissimus*, which is two feet four inches long and twelve inches wide. The animal appears to have sunk quietly, after death, to the bottom of the sea, lying on its back in the sandstone of Scotland. Some fish-bones in the Devonian strata of Dorpat, Russia, must have belonged to fishes 30 or 40 feet long. One bone is three feet in length. Scales of Holoptychus are common, also, in the old red sandstone of New York.

SECTION IV.—SILURIAN SYSTEM.

I. *Geographical Distribution.*—This system occurs in the basin of Hudson's Bay, in the valley of the River St. Lawrence, and in the valley of Lake Champlain. In the western part of New York, it is most fully developed with a great number of subdivisions, regularly superimposed, and with outcrops in favorable conditions for examination. In a southwest direction it extends along the Alleghany Mountains, which are mostly constituted of its members. In the Western States, the upper part of the system shares most of the surface with the carboniferous system.

In the Northern States, the Silurian system exists in a highly metamorphic condition, having been changed into crystalline rocks.

This system received its name from the place where it was first successfully investigated, the western part of England, which was anciently inhabited by a tribe called the Silures. These rocks occur also in Belgium, Germany, Norway, and Sweden. In Russia, they occupy a region around St. Petersburg, and are found in the Ural Mountains in a metamorphic condition. It is rendered probable, by fossils which have been collected by travelers, that the Silurian system exists in Terra del Fuego, in South Africa, and in New Holland.

II. *Structure and Position.*—A great variety of rocks is com-

---

What is said of the distribution of the Silurian system in North America? of its condition in the Northern States? of its name? of its distribution in Europe? of its structure?

prised in this system. Hard sandstones, fine slates, and flagging stones constitute a large part. Conglomerates are less common. Some of the formations are partly calcareous, and others consist chiefly of limestone, and furnish excellent marble. It is an extraordinary fact that, in the vicinity of St. Petersburg, this most ancient system contains beds of unconsolidated clay.

Fig. 198 exhibits the position of this system in a section extending from the north side of Lake Ontario across New York into Pennsylvania. (See Report by Professor Hall.)

Fig. 198.

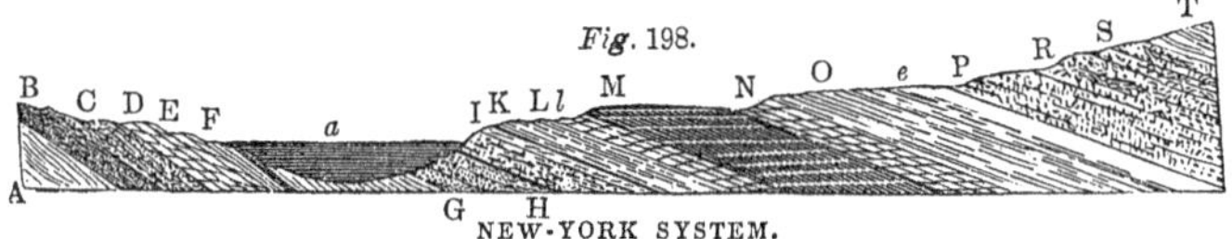

NEW-YORK SYSTEM.

1. A. Primary or metamorphic rocks.	10. K. Clinton group.
2. B. Potsdam sandstone.	11. L *l*. Niagara group.
3. C. Calciferous sandrock.	12. M. Onondaga salt group.
4. D. Black River limestone.	13. N. Helderberg series.
5. E. Trenton limestone.	14. O. Hamilton group.
6. F. Utica slate.	15. *e*. Tully limestone.
a. Lake Ontario.	16. P. Portage group.
7. G. Hudson River group.	17. R. Chemung group.
8. H. Gray sandstone and Oneida conglomerate.	18. S. Old Red or Devonian system.
9 I. Medina sandstone.	19. T. Conglomerate of the carboniferous system.

Fig. 199 exhibits the order of the Silurian rocks in England.

Fig. 199.

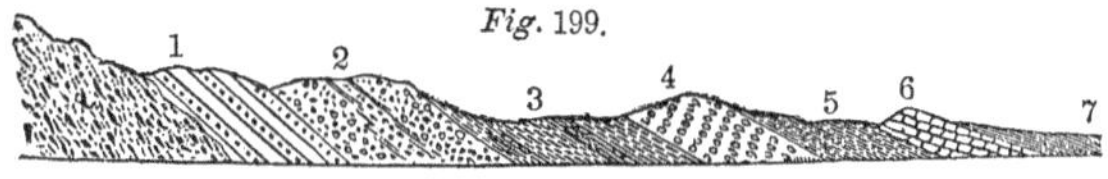

ORDER OF SUPERPOSITION OF THE SILURIAN ROCKS.

7. Upper Ludlow shale. 6. Aymestry, or Ludlow limestone. 5. Lower Ludlow shale.	Ludlow series.	Upper Silurian.
4. Wenlock limestone. 3. Wenlock shale.	Wenlock series.	
2. Caradoc sandstone. 1. Llandeilo flags.	....	Lower Silurian.

Fig. 200, page 276, exhibits the order of these rocks in the territory of Christiana, in Norway.

Describe the section across New York; the section in England.

Fig. 200.

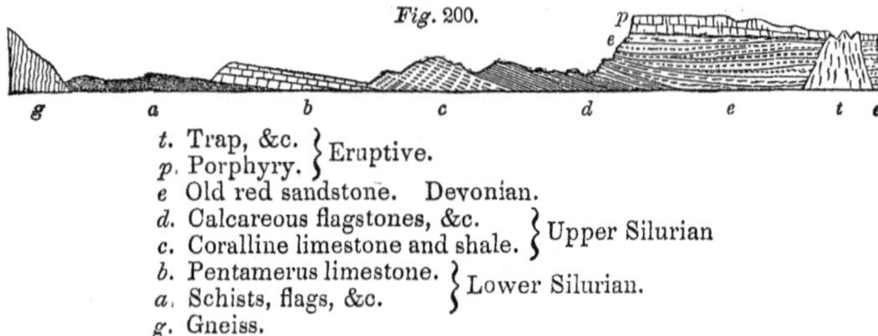

t. Trap, &c. } Eruptive.
p. Porphyry. }
e Old red sandstone. Devonian.
d. Calcareous flagstones, &c. } Upper Silurian
c. Coralline limestone and shale. }
b. Pentamerus limestone. } Lower Silurian.
a. Schists, flags, &c. }
g. Gneiss.

III. *Subdivisions of the Silurian System.*—On account of the number of regular formations, each characterized by some peculiar fossils, while other fossil species extend through and characterize entire series of successive formations, it is practicable, in most countries, to make convenient subdivisions of this system. In New York it has been divided into four divisions. They are,

1. Erie Division; R to O in *Fig.* 198.
2. Helderberg Division; N. to M.
3. Ontario Division; *l* to I.
4. Champlain Division; H to B.

The first of these divisions, perhaps, includes in its upper part some strata of Devonian age, the limit between the two systems not being clearly defined in New York.

If the same subdivisions could be identified in distant countries, they might be regarded as distinct systems. But they are not persistent over large areas. In the Silurian districts of the Western States some of the New York formations are wanting, while a few others are more fully developed. This may be seen

Fig. 201.

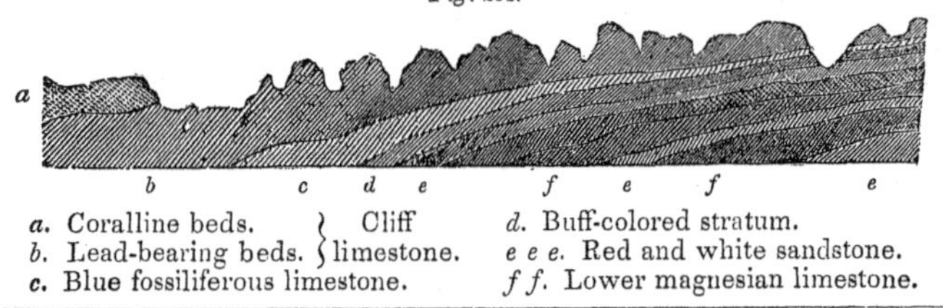

a. Coralline beds. } Cliff limestone.
b. Lead-bearing beds. }
c. Blue fossiliferous limestone.
d. Buff-colored stratum.
e e e. Red and white sandstone.
f f. Lower magnesian limestone.

Describe the section in Norway. What is said of the subdivisions? Why are they not distinct systems?

in the preceding section of the Silurian rocks of Iowa, *Fig* 201, from Mr. D. D. Owen's Report.

The cliff limestone is so called on account of the great number of cliffs composed of this rock. The cliffs have resulted from denudations of the surface. Its upper beds abound in corals and shells, and the lower part constitutes the great lead region.

In *Figs.* 199 and 200 we have seen the subdivisions of the system in England and in Norway. The simple division into Upper and Lower Silurian may be applied in most countries where these rocks occur.

IV. *Fossils.*—Very few species of plants or of fishes have been found. Except fishes and a chelonian (*fresh-water tortoise*), all the vertebrated classes are wanting. Several crustacea occur, and the family of trilobites was more fully represented in this than in any other period. The crinoideans and corals were also well represented. But a large majority of the fossils are shells.

More than one thousand species of Silurian fossils are known. Professor Hall has described three hundred and eighty-one species belonging to the Champlain division, and three hundred and forty-one belonging to the Ontario division, within the limits of New York.

1. *Radiata.*—(1.) *Corals.* One of the most beautiful of the Silurian fossils is the chain coral, *Catenipora escharoides. Fig.* 203, page 279, represents a specimen from Iowa, from the Upper Silurian, or cliff limestone. The same formation also contains several other species and genera of corals.

Fig. 202, *l*, page 278, represents a solid coral, *Favistella stellata*, from the Hudson River group of New York. *n* and *p* represent other solid corals: *n* is *Stictopora fenestrata*, from the Chazy limestone of New York; and *p* is *Chætetes Lycoperdon*, from the Trenton limestone of New York. In *Fig.* 207 (p. 281), *j* and *l* also represent solid corals: *j* is a species of *Cyathophyllum*, from the Hamilton group in New York; and *l* is *Astræa rugosa*, from the Onondaga limestone of New York.

Describe the section in Iowa. What is said of the kinds of fossils? of their number? of the chain coral? Mention some examples of solid corals.

Fig. 202.

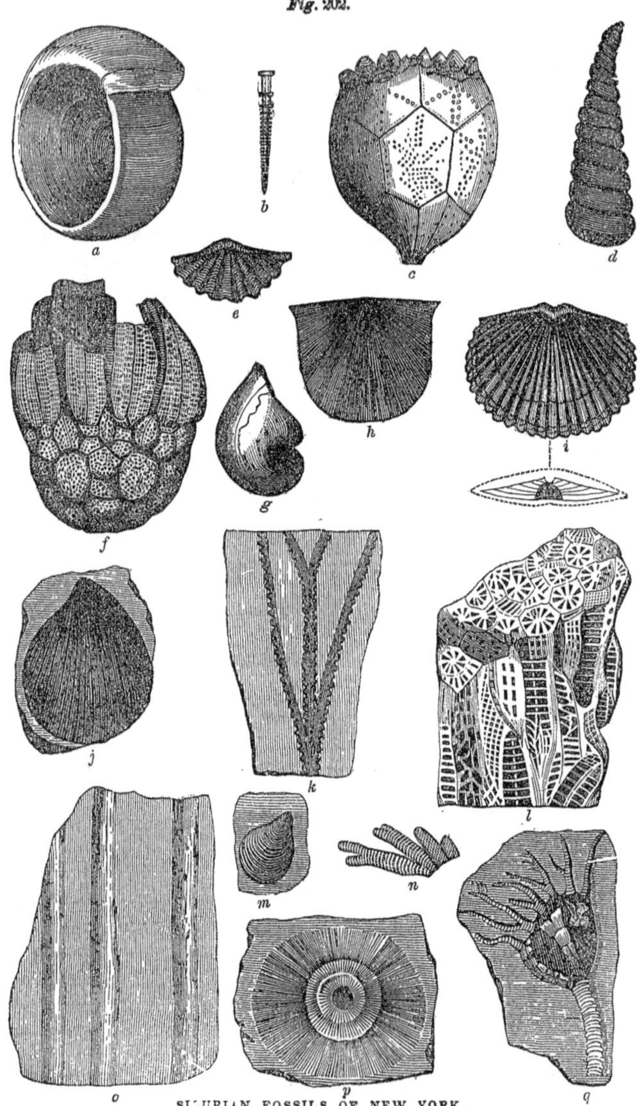

SILURIAN FOSSILS OF NEW YORK.

Fig. 203.

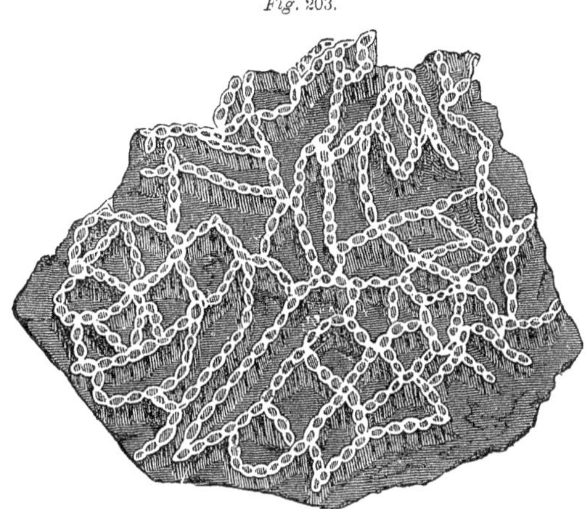

Fig. 202, *b*, represents the *Tentaculites ornatus*, from the Water Lime group of New York. It is doubtful whether this genus was more nearly allied to corals or to crinoideans.

(2.) *Graptolites* were probably zoophytes. Their character has been much discussed, and it is yet uncertain to what family in this class of animals they should be referred. Some have supposed them to have resembled the sea-pen, Pennatula. They are very common in some argillo-calcareous slates (*Fig.* 202, *k*).

(3.) *Crinoideans.*—The following figures exhibit some of the Silurian types of this elegant family of starfishes. *Fig.* 204 is a basal view of the body of *Platycrinus discoideus*, the peduncle having been attached to the central piece; the ramifying arms were jointed to the side-pieces. This species is from Iowa.

Fig. 204.

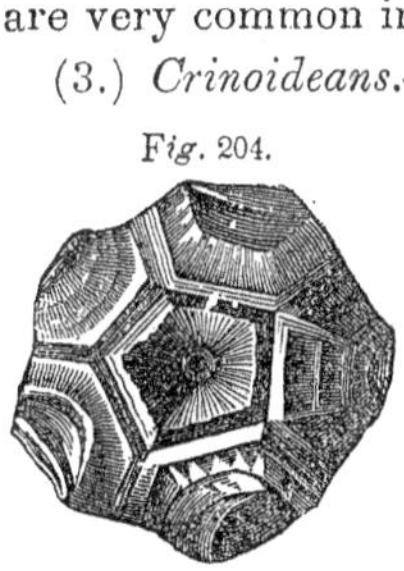

Fig. 202, *f*, represents the body and part of the arms of *Hypanthocrinites cœlatus*, from Lockport, in New York. *q* represents the body, arms, and a part of the stem of *Schizocrinus nodosus*, from the Trenton limestone at Glenn's Falls, New York. *Fig.* 207, *g*, represents all the

What is said of Graptolites? Mention some examples of Crinoideans.

parts of *Cyathocrinus ornatissimus,* from the Portage group, at Portland, in New York. *Fig.* 202, *c,* represents the body of *Caryocrinus ornatus,* from the Niagara group, in New York.

2. *Articulata.*—The crustaceans were represented chiefly by the family of trilobites. These animals were divided lengthwise into three lobes, and transversely into numerous segments. They were able to roll themselves up, like the wood-louse. Many were entombed, and are now found in this condition, as represented in *Fig.* 205, *Calymene Fischeri,* from Russia. Their eyes were prominent and compound, consisting of many tubes, which were less numerous and larger than those which we see in the compound eyes of insects (*Fig.* 207, *k k*). Although multitudes of specimens have been carefully examined, no legs have been discovered. Probably they had small and fragile legs, which were not much used for locomotion. *Fig.* 207, *k,* represents *Calymene bufo,* from the Hamilton group, in New York: this specimen is partially coiled.

Fig. 205.

3. *Mollusca.*—The class *Brachiopoda* had its greatest development in the Silurian periods. *Fig.* 202, *e,* represents *Dethyris de cemplicata,* from the Niagara group, at Lockport, in New York. *Fig.* 202, *h,* is *Strophomena striata,* from the same locality. *i* is probably *Orthis flabellulum,* also from Lockport. *Fig.* 207, *a,* represents *Atrypa hystrix,* from the Chemung group, at Bath, in New York. *b* represents *Delthyris acuminata,* from the Chemung group, at Ithaca, in New York. *f* is *Strophomena setigera,* from the Marcellus shale, at Avon, in New York. *Fig.* 206 is a European species, *Pentamerus Knightii.*

Fig. 206.

Pentamerus Knightii.

What is said of the structure of Trilobites? Mention some examples of Brachiopoda.

Fig. 207.

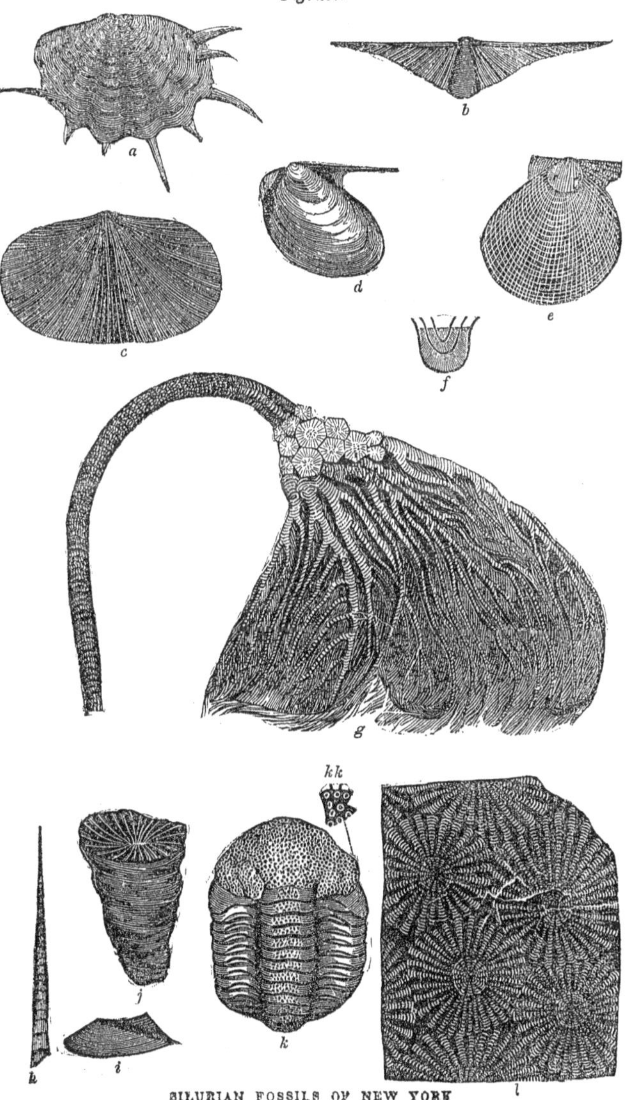

SILURIAN FOSSILS OF NEW YORK

Fig. 207, *c*, represents *Orthis interlineata,* from the Chemung group, of New York. *Fig.* 202, *m*, represents *Lingula antiqua,* of the Potsdam sandstone, in New York. This is the oldest of the fossiliferous rocks; yet the same genus inhabits the tropical seas of the present time.

Although the *Acephala* are now far more numerous than the *Brachiopoda*, yet in the Silurian periods they were less abundant. *Fig.* 202, *j*, represents a species of a genus long since extinct, *Pterinea carinata,* from the Hudson River group, in New York. *Fig.* 207, *d*, is *Avicula spinigera,* from the Chemung group, at Painted Post, in New York. This extinct species belongs to the existing genus, which includes the pearl oysters. *e* represents an extinct species of the existing genus of scallop shells, *Pecten cancellatus,* from the Chemung group, at Phillipsburgh, in New York. *i* represents *Cypricardia truncata,* from the Hamilton group, at Cayuga Lake, in New York.

The *Gasteropoda,* now by far the most numerous of all the classes of mollusca, were also rare. *Fig.* 202, *a*, represents the *Euomphalus profundus,* from the Pentamerus limestone, in New York. *g* is the *Bellerophon bilobatus,* from the Trenton limestone, in New York.

The *Cephalopoda,* now extremely rare, were common, and were represented by a great variety of generic forms. *Fig.* 202, *d*, is *Cornulites arcuatus,* from the Niagara group, in New York. *Fig.* 208 represents a remarkable shell, *Lituites Ordini,* from the Silurian rocks of Russia. *Fig.* 207, *h*, represents a remarkably

Fig. 208.

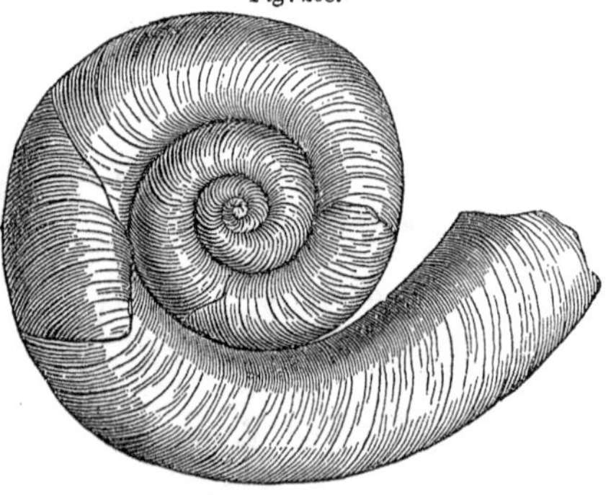

Mention some examples of Acephala; of Gasteropoda; of Cephalopoda.

slender, chambered shell, *Orthoceras aciculum*, from the Portage group, at Cashaqua Creek, in New York.

4. *Fishes and Reptiles.*—Fragments of eight or ten species of fishes have been found, and the tracks of at least one reptile have been discovered in Canada by Mr. Logan.

5. *Plants.*—The fossil represented in *Fig.* 202, *o*, is supposed to have been a marine plant, and has been called *Scolithus linearis.* It is found in the Potsdam sandstone, in the valley of Lake Champlain.

V. *Climate and Geography.*—The farther we recede from the present time in the history of the earth's surface, the less definite is our knowledge of the distribution of land and water. Although no remains of land animals have yet been found, we can not infer from this negative fact, in the present imperfect state of knowledge, that such species did not exist. On the contrary, many of the fossils indicate shoal water, and it is not credible that the surface of the earth should have been so adjusted as to have been one shoreless ocean, yet abounding over extensive areas with shoals. There must have been archipelagos, and it is probable that there was a great continent occupying the area of the Atlantic. This is inferred from the manner in which some formations thin out westward, where also they are of finer materials. Both of these facts indicate a continent on the east as the source from which the materials were derived.

From the great number of species of shells in the Silurian system within the State of New York, it may be inferred that the climate was tropical, or, at least, much warmer than it is now; for, at the present time, it is only in warm regions that such a profusion of species is found in such limited areas. From the wide distribution of many of the species of fossils, it may also be inferred that the climate was more uniform over the earth's surface.

CUMBRIAN AND CAMBRIAN SERIES.

In the north of England and in Wales, there exist certain slates, and other more or less metamorphic strata, on the precise

What is said of the fishes? of the Scolithus? of the probable existence of and animals? of dry land? of the climate?

relations of which to the Silurian system there has been a dif ference of opinion. It is generally believed that they belong to the older part of the great Silurian period, since the fossiliferous strata contain species of fossils, many of which occur also in the typical Silurian rocks. Some geologists, however, suppose that a part of these rocks, although belonging to the same system, are older than the Caradoc sandstone and the Llandeilo flags (see p. 275), while others consider them as the equivalents of those rocks. English geologists, therefore, describe them geographically, and the names above given are to be understood geographically. They do not indicate well-established systems of formations, having peculiar fossils and occupying a distinct period.

In Pennsylvania, also, there is a series of shales lying beneath the Potsdam sandstone.

TACONIC ROCKS.

Certain slates and limestones in the western part of Vermont and Massachusetts, and in the adjacent parts of New York, have been called Taconic rocks. They have been supposed by Professor Emmons to constitute a system of formations more ancient than the Silurian (New York) system. A few very rare fossils occurring in the slates were thought to characterize them. On page 117 we have given a section of Snake Mountain, in Vermont, where these rocks are most perfectly exhibited. Here they are found to contain abundantly the same species of fossils which belong to the Champlain division of the New York system.

In most places the Taconic rocks have been so much disturbed and altered, that their proper place in the series is quite obscure. They are semi-crystalline and much jointed, and the fossils are mostly obliterated. It is convenient, therefore, to retain this name as designating a group of strata more or less metamorphic, but of which a part certainly and probably all belong to the Silurian system.

Where do the Cumbrian rocks occur? what is said of their age? what difference of opinion respecting them? Where are the Taconic rocks? what is said of the fossils in them? of the character of the rocks?

CHAPTER VIII.

METAMORPHIC AND PRIMARY STRATA

We have frequently had occasion to speak of *metamorphic* rocks. In the present chapter we propose to give a brief and general account of them, of their origin, structure, and varieties.

The word metamorphic literally signifies changed in form, and is applied to such stratified rocks as have lost, in a greater or less degree, their original structure. The mere change from the condition of sand or mud into that of solid rock is not metamorphism, since solid strata often retain their original structure with their fossils in a good state of preservation. But when the original mechanical structure of the strata is superseded by a crystalline structure, and the fossils have been nearly or wholly obliterated, the rock is said to be metamorphic.

As might be expected, all degrees of metamorphism occur. Some rocks are slightly crystalline; their fossils are indistinct, but not obliterated; and their stratification remains quite distinct, but a jointed structure has been superinduced. Other rocks have lost nearly all traces of their origin in sand, or mud, or corals; their fossils are mostly obliterated; and they contain, in few localities, only indistinct traces of organic bodies. Such are most of the Taconic rocks. But there are extensive regions, as in the New England States, in which the rocks are wholly crystalline in their structure, and are entirely destitute of organic remains retaining no evidence of their original deposition beyond that of stratification. Even this has disappeared in many rocks: some, like the fine slates, assume a conspicuous slaty cleavage, which is often independent of and in a different direction from their original lamination, which is now indistinct. Other strata, as

What is the meaning of metamorphic? What is the lowest degree of metamorphism? a higher degree? What is the most complete metamorphism?

gneiss, appear to have been more or less perfectly fused, and pass, by insensible gradations, into unstratified masses.

Nearly all the metamorphic rocks were once supposed to be older than the Palæozoic formations. They were included in the class of *primary strata*, and were supposed to have been deposited in periods anterior to the existence of living beings on the surface of the earth. But most of the so-called primary strata have successively been identified with the fossiliferous strata. The origin of others is yet doubtful; and some lying beneath the oldest Palæozoic formations are unquestionably more ancient. Yet we are not to infer from the absence of fossils that they were not originally fossiliferous, for we know that extensive regions consist of rocks of the same structure and characters, which are destitute of fossils only because they have been subjected to intense metamorphic action, while most of the oldest rocks which retain their original structure contain fossils.

The true primary strata, therefore, are rocks which are more ancient than the oldest fossiliferous strata, which had probably a similar origin, but which retain no evidence of the existence of organic beings during the periods of their deposition. They are the Ultima Thule of our geological knowledge. They are the limits, not of the facts of the earth's history, but of our means of knowledge. As the most ancient profane history of man is lost in the dimness of ancient myths, so is the geological history lost in the obscure teachings of these most ancient metamorphic strata.

The principal *cause of metamorphism* is the power of crystallization, aided by heat and by pressure. It is well known that long-continued intense artificial heat may change the structure of stones. Slabs of sandstone in furnaces have assumed a crystalline structure without fusion. But the strata of the earth's crust, being prevented by the mass of superincumbent materials from losing the heat which is imparted by adjacent volcanic agencies

What was once supposed to be the age of metamorphic strata? What do we know of some of them? What is said of those of doubtful age? What are primary strata? What are the causes of metamorphism?

are thus for a long series of ages subjected to this metamorphic agency. In many substances the power of crystallization acts with extreme slowness. It is probable, therefore, that the long continuance of intense heat has produced important changes which would not otherwise have occurred.

It is well known, too, that great pressure with heat produces remarkable effects. By this means chalk has been converted into marble. A degree of heat which, without pressure, would decompose the substance, with pressure merely enables the crystallogenic power to give it a new structure. The ancient and the metamorphic limestones were once in the condition of corals, of chalk, of tufa, and of other calcareous deposits; but by heat and pressure they have been converted into more or less crystalline marble.

The agencies of heat and pressure also exalt the power of chemical affinity, and enable substances to enter into combinations, which could not otherwise have been formed. Hence the crystalline strata contain a great number of minerals, which are never found in unaltered sedimentary rocks. If, therefore, metamorphism has robbed the palæontologist of fossils, it has enriched the mineralogist with some of the choicest treasures of his science.

Metamorphism not only obliterates fossils and induces a crystalline structure, but it also is accompanied by the introduction of *joints* and *a slaty cleavage.*

Joints are smooth planes of division, which are entirely independent of stratification. They usually occur in two or more sets of planes, which cross each other so as to divide the rock into prismatic forms. But these forms are incapable of subdivision.

The planes of *slaty cleavage* are nearly or quite parallel, and produce forms which, by subdivision, may be reduced to thin laminæ. Slaty cleavage is sometimes identical with the lamination which results from the original deposition of fine sediment, but

---

How has time aided in producing changes of structure? What is the effect of pressure? What effect of heat and pressure on chemical affinities What are joints? What is said of slaty cleavage?

is very commonly quite independent of it. In the latter case it is often extremely difficult to detect the planes of deposition, and geologists have been sometimes deceived in the structure of such rocks.

The cause of joints and of slaty cleavage is yet obscure. That joints are not exclusively the effect of igneous agency, appears from their occurrence, although rarely, in unconsolidated clays of the latest (the pleistocene) formation. They are, however, far more common in metamorphic rocks. It is probable that the agency is somewhat analogous to that of crystallization; in other words, that joints are the effect of an attraction between the particles of mineral matter.

The joints of conglomerate rocks often divide evenly the pebbles which lie in the planes of division in the partially metamorphic conglomerates of Roxbury and Dorchester, in Massachusetts. The quarrymen have exposed, often to the extent of several rods, smooth surfaces, which have been formed by such joints passing evenly through all the various materials of flinty and slaty pebbles, and of coarse and fine sand, of which the rock was originally constituted.

The principal varieties of rocks in the metamorphic and primary strata are, clay slate (argillaceous slate), crystalline marble, hornblende slate, quartz rock, mica slate, and gneiss.

Clay slates were originally deposits of clay, the odor of which may be perceived when they are moist. As roofing and graphic slates they are well known. They are more or less free from a gritty composition, as the clay was originally more or less free from sand. They are among the softer rocks; are usually dark-colored, but some varieties are of light colors, as green, drab, &c.

Crystalline marble, as we have before said, was originally coral, or chalk, or tufa, and has been crystallized by heat under pressure. When pure, it is white; but many metamorphic regions abound with silicious, micaceous, or argillaceous strata, which

---

What is said of the cause of joints? of joints in conglomerate rocks? Mention the principal varieties of metamorphic rocks. What is said of clay slates? of crystalline marble?

are also calcareous. Originally such strata were composed of sand or mud, with more or less calcareous matter.

Hornblende slate consists of quartz and hornblende, and is abundant in some parts of the New England States.

Quartz rock is composed chiefly of quartz. It originated probably in sandstone. It should not be confounded with veins of pure quartz, which intersect other rocks, and which are not of sedimentary origin. Quartz rock is often found in thick beds, which consist of nearly pure quartz, and retain no trace of lamination. It is then a very hard and indestructible rock. But it occurs also in thinner strata, with more or less mica in its composition, and then passes into mica slate.

Mica slate consists of layers more or less thin, that are composed of quartz and mica, with the scales of mica, for the most part, lying nearly or quite in the planes of lamination. Well-crystallized minerals, as garnets, often enter more or less into its composition. When feldspar in considerable quantity is added, it becomes gneiss.

Gneiss is, therefore, composed of the three minerals, quartz, mica, and feldspar. Its composition is precisely the same with that of granite, but it differs in being stratified. Of some mountain masses one part is gneiss and another part is granite, this part having been entirely melted, while the other was only crystallized.

One of the most interesting examples of metamorphic agency may be seen in the crystalline strata of the western part of the New England States. Commencing in Canada, they occupy most of Vermont and the western parts of Massachusetts and of Connecticut. These rocks have been much disturbed and elevated, and constitute the mountain ranges of the regions which they occupy. For many years they were supposed to be primary strata, until at length the frequent identification of other supposed primary strata with Palæozoic, and even with Mesozoic formations, led to a more diligent search for fossils. In Canada these

What is said of hornblende slate? of quartz rock? of mica slate? of gneiss? of metamorphic rocks in the western part of New England and in Canada?

rocks are further removed from the great source of metamorphic agency, which was in New England. While, therefore, the search for fossils was unsuccessful in these states, in that province it was rewarded with the discovery of a few specimens. By these the rocks in question were shown to be of Silurian age.

The Alleghany Mountains constitute another belt in which metamorphic agency has been exerted. They consist mostly of Devonian and Silurian formations, which are partially metamorphic.

The source of metamorphic action may usually be discovered in the greater or less proximity of unstratified rocks. The influence of those which are called hypogene (see description of the first class of unstratified rocks in the next chapter) has pervaded extensive regions, while the effects of the Plutonic rocks (the second class in the same chapter) have usually been more local. The influence of trap rocks has generally been limited to a few rods or a few feet, while that of granite has been felt for many miles; but in the island of Jamaica, enormous mountain masses of Plutonic rocks have partially altered limestones to a distance of thirty or forty miles.

CHAPTER IX.

UNSTRATIFIED ROCKS

In our history of the changes to which the crust of the globe has been subjected, and by which the stratified rocks have been formed, we have briefly noticed those irregular masses which appear to have been forced in among the strata by igneous action during the successive geological periods. These masses are called *unstratified* or *igneous rocks.*

We have now reached a point in the history of the earth beyond which the succession of events is not distinctly recorded. If stratified rocks older than those already described ever exist-

---

What is said of fossils in metamorphic rocks? of the Alleghany Mountains? of the source of metamorphic agency? What point in the history of the earth has now been reached?

ed, as most geologists believe, all known traces of them have been obliterated by igneous agency, but how long a period intervened between the oldest known strata and the time when deposits of strata commenced it is impossible to determine. We know, however, that there was a time when the whole crust of the earth consisted of unstratified rocks, and that these rocks formed the base on which all the strata repose, and out of which the primitive strata were formed.

This mass or crust was probably granite, a compound of quartz, mica, and hornblende. Granite is so various in composition and structure, and has been erupted at so many different periods, that we can ascend the stream of time no further to note the successive changes in the mineral masses, and designate the epochs of geological history. It only remains, therefore, to give a short description of the various kinds of unstratified rocks.

There are about eight varieties of unstratified rocks. They may be arranged in three classes, the *Granitic*, the *Trappean*, and the *Volcanic* rocks

SECTION I.—GRANITIC ROCKS.

In this class of unstratified rocks may be included *Granite*, *Syenite*, and *Quartz*.

I. *Granite.*—True granite is a triple compound of quartz, feldspar, and mica. The proportions and arrangements of these simple minerals are indefinitely varied. There are very coarse granites, in which large crystals of the constituent minerals are united to each other; and there are very fine-grained granites, in which it is difficult to distinguish the different ingredients. Between these extremes there is every variety of structure and composition, although feldspar is generally found in the largest proportion.

What was the condition of the earth before stratification commenced? Which is probably the oldest igneous rock? What is said of it? Why can we not ascend the stream of time further? How many kinds are there of unstratified rocks? How may they be classed? What rocks belong to the first class? Describe true granite.

1. *Varieties of Granite.*—When the rock is nearly or quite destitute of mica, the quartz and feldspar present a surface which resembles written characters, on which account it is called *graphic granite.* The following figures represent the appearance of the surface. *Fig.* 209 is a section in the same direction with the laminæ, and *Fig.* 210 is a portion of the surface of a rock found

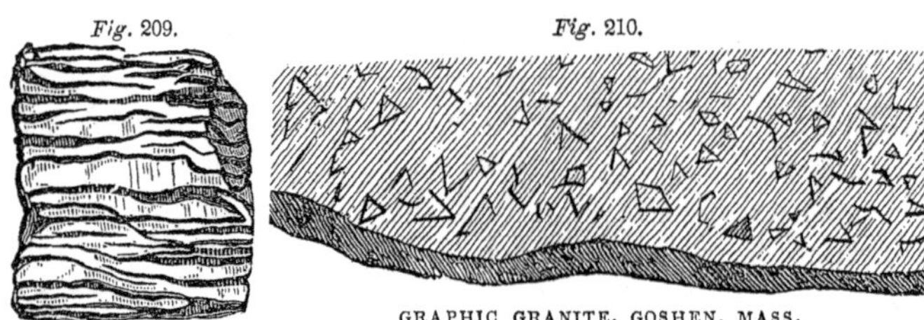

Fig. 209. *Fig.* 210.

GRAPHIC GRANITE, GOSHEN, MASS.

in Goshen, Mass. (Hitchcock's Report), and is a good example of this variety of granite.

2. When the crystals of feldspar are very distinct, the rock is called *porphyritic granite.*

3. Talc sometimes takes the place of the mica; the rock is then designated as *talcose granite,* and is called by the French *protogene.* By the decomposition of this rock there is formed the celebrated *porcelain clay,* so extensively used in pottery.

4. When the rock is mostly feldspar, it is called *feldspathic granite.*

All these varieties, however, often gradually pass into each other, and also into the other kinds of unstratified rocks.

The color of granite is also very various, although it is usually grayish white or flesh colored.

II. *Syenite.*—Syenite is similar to granite in structure, but differs from it in composition. Hornblende takes the place of the mica, and imparts to the rock a darker color. It is generally a fine-grained rock. Mica is, however, often present in this rock, and it is then called *syenitic granite.*

Describe graphic granite. What is said of the three varieties of granite? Describe syenite, its composition and varieties

The term syenite is derived from Syene, in Upper Egypt, whence most of the rocks used in the construction of ancient Egyptian monuments were obtained. This rock, however, is found to be a red granite, containing black mica and minute portions of hornblende, while Sinai, in Arabia, is composed of a true syenite.

III. *Quartz Rock.*—The mineral quartz often exists in such masses as to be entitled to a place among the unstratified rocks. It sometimes occurs in thin seams, or in regular layers, in which cases it probably had an aqueous origin; but large dikes of quartz not unfrequently occur among the strata in such forms as to indicate an igneous origin. Large mountains sometimes consist mostly of quartz. It is distinguished as the rock with which gold is usually associated.

IV. *Geographical Distribution of Granitic Rocks.*—The rocks of this class are widely distributed over the surface of the earth They frequently form extensive mountain ranges, but are generally associated with the non-fossiliferous or metamorphic strata.

Granite occurs in various parts of New England. It is found in bands and patches extending across the eastern portion of Massachusetts. Commencing at Andover, a belt of granite passes, in a southwest direction, to Rhode Island, having the mica and clay slates on the west, and porphyry and syenite on the east. Another belt extends from Buzzard's Bay, in a northeast direction, to Cape Ann.

Granite and syenite are quarried, for building purposes, in many places near Boston. The most celebrated localities of the latter are at Quincy and Braintree.

West of these belts, in various parts of New England, granite is found in patches, some of which are of great extent, as in Essex county, Vermont, and in many portions of New Hampshire, Maine, and along the range of the Green Mountains. It is also often met with in the Middle, Southern, and Western States.

From what is the name derived? How has quartz originated? For what is it distinguished? Where are granitic rocks found in New England? in other portions of the United States?

On the western coast of the United States granitic rocks extend along the whole range of the Rocky Mountains, and throughout the Andes, in South America. In some places among the Andes, granite rises to an elevation of 12,000 feet.

Spires of granite among the Alps, in Europe, are of great height, as in the case of the Aiguille de Dree, which is a solid shaft elevated 4000 feet above its base.

Quartz rocks abound in California, and constitute the guanqua of the inexhaustible deposits of gold which have so recently been discovered in that state.

V. *Geological Position and Age.*—Granitic rocks are generally associated with the metamorphic strata, either existing in large belts or masses, or cutting the strata in the form of dikes. They are not, however, confined to the older strata, but often penetrate the coal-beds, and, in a few cases, extend as high up as the tertiary; hence these rocks have been erupted at different periods during the deposition of the stratified masses. This inference is confirmed by the chemical and mechanical effects which granitic dikes have produced upon each other and upon the strata.

The following figure of a bowlder found at West Hampton, Massachusetts (Hitchcock's Report), will illustrate the kind of proof which these rocks present of the truth of this statement.

Fig. 211.

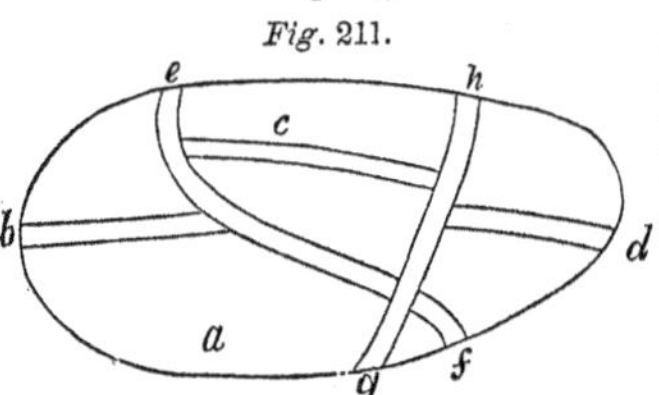

a represents the bowlder, which is a genuine granite. *b c d* a vein of granite passing directly through it. *f e* is a second vein, which cut off the first, and removed it upward. *g h* is a third vein, which cut off the other two. Each of the veins is of different varieties of granite. In this case we have four distinct periods of ejection. Similar examples are found among all the metamorphic rocks. It is hence inferred that, although some of the granites are the oldest

---

Where are granitic rocks found in South America? in the Alps? What valuable metal is usually associated with this class of rocks? What is the geological position of granite? What evidence of different periods of eruption?

of the rocks, others have been formed at different epochs, and can not, therefore, be assigned to any particular period of the earth's history.

SECTION II.—TRAPPEAN ROCKS.

Under this division we include porphyry, greenstone, trachite, basalt, and amygdaloid.

These rocks are composed mostly of feldspar and hornblende, or augite. The term *trap* is derived from the Swedish word *trappa*, a stair, and applied to this class because they often consist of blocks arranged like steps.

I. *Porphyry.*—The term *porphyry* is applied rather to the *structure* of rocks than to their *composition.* The ancient porphyries have a base of compact feldspar, through which are distributed crystals of feldspar. This formed a very hard and durable rock, capable of receiving a very high polish. Any rock, however, with a homogeneous base containing crystals of some of the simple minerals disseminated through it, is called porphyry. Hence we have *greenstone porphyry*, which has a base of green stone, *claystone porphyry*, *pitchstone porphyry*, *trachytic porphyry*, &c. The ancient porphyries were purple, hence the name; but each variety has a different color.

II. *Greenstone.*—This term includes the compact varieties of the class in which hornblende predominates, and imparts to the rock a greenish color. The structure is often porphyritic, or very coarse grained, in which case it is designated as *porphyritic greenstone* and *syenitic greenstone.*

III. *Trachyte.*—This rock consists of glassy feldspar, hornblende, mica, and titaniferous iron ore. It has a rough surface, on which account it has received its name. Its structure is porphyritic, and its color of a grayish white.

IV. *Basalt.*—This is a compact, fine-grained variety of the

What is the inference respecting the age of granite? What is the composition and structure of the trappean rocks? How is the term porphyry used? What is said of the ancient porphyries? of greenstone? of trachyte? What is basalt?

trap family, and is similar in composition to greenstone, although it contains the variety of hornblende called augite, with distinct grains of olivine. Its color is black, or grayish black. The rock can not easily be distinguished from greenstone.

V. *Amygdaloid.*—This term, like porphyry, is applied to the structure of certain rocks of the trap family. It abounds in rounded cavities, which are filled with quartz, chalcedony, calcareous spar, and some other minerals. This gives a peculiar texture to the rock, so that it appears like paste filled with almonds, hence its name. The softer base is called *wacke*, which is sometimes destitute of vesicles.

VI. *Columnar Structure of the Trap Rocks.*—Rocks of the trap family and some others are often distinguished by their columnar structure. They are divided into prisms, which are more or less regular, with from three to eight faces, and from a few feet to 200 feet in length. These columns are generally divided by joints, and the sections are usually concave at the top and convex at the bottom; although the form is sometimes reversed, as at Titan's Pier, Mount Holyoke, Massachusetts.

The following view of the Giant's Causeway, Ireland, *Fig.* 212,

Fig. 212.

THE GIANT'S CAUSEWAY.

shows the position of these columns. The columns are pentagonal, and from one to five feet in diameter.

Fingal's Cave, on the island of Staffa, is another example of the same structure (*Fig.* 213).

These columns sometimes stand perpendicularly to the hori-

What is amygdaloid? What is said of the structure of the trappean rocks?

Fig. 213.

FINGAL'S CAVE, STAFFA.

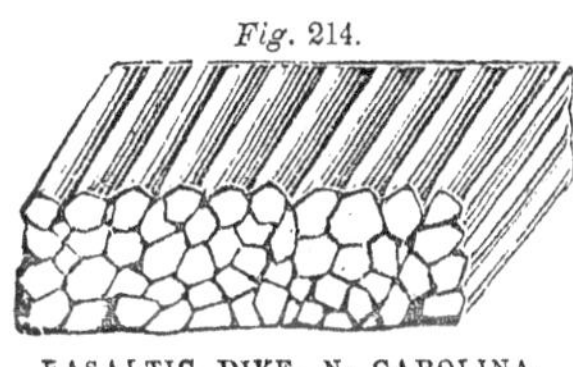

Fig. 214.

FASALTIC DIKE, N. CAROLINA.

zon, and, when broken and disintegrated by the action of the atmosphere or of the waves, present the appearance of ruined castles. When they are horizontal, they appear like walls, and have often been mistaken for ancient fortifications (*Fig.* 214).

The cause of this peculiar columnar structure is believed to be a tendency of the matter, when cooling, to assume a globular form, and this view is confirmed by the fact that the same structure is found in recent lavas and in the clay of furnaces, but it has been more satisfactorily verified by the experiments of Watt, who melted 700 pounds of basalt and allowed it slowly to cool. In this process the matter gradually separated into globular masses, and these, by pressing against each other, were finally formed into regular columns.

VII. *Geographical Distribution.*—Trap rocks are found in mountain ranges and in the form of dikes, which have cut througn the strata in nearly every part of the earth's crust.

In the U. States, trap rocks occur on the Kennebec River, in Maine, where they rise in mountain masses from 200 to 300 feet in height, and trap dikes are found in every portion of the state.

What is the position and appearance of the columns? What is the cause of the columnar structure? How is this proved? Where do trappean rocks occur in New England?

North and south of Boston, Massachusetts, there is a belt of greenstone associated with porphyry, which makes its appearance at Nahant, Lynn, Charlestown, Roxbury, and many other places. The ridges are sometimes 500 feet in height. Another ridge of greenstone passes along the Connecticut River valley from West Rock, in New Haven, nearly to Vermont. It includes Mounts Tom and Holyoke, which are elevated a thousand feet or more in height. In the valley of Lake Champlain numerous naked dikes occur. A few miles above the city of New York greenstone trap forms a series of elevations called the *Palisades.* Three ridges pass through the State of New Jersey, and beds and high summits make their appearance at least as far south as North Carolina.

But the most extensive ranges of trap are found west of the Rocky Mountains. The Columbia River passes through mountains of trap from 400 to 1000 feet high. The following figure exhibits the appearance of the rocks.

Fig. 215.

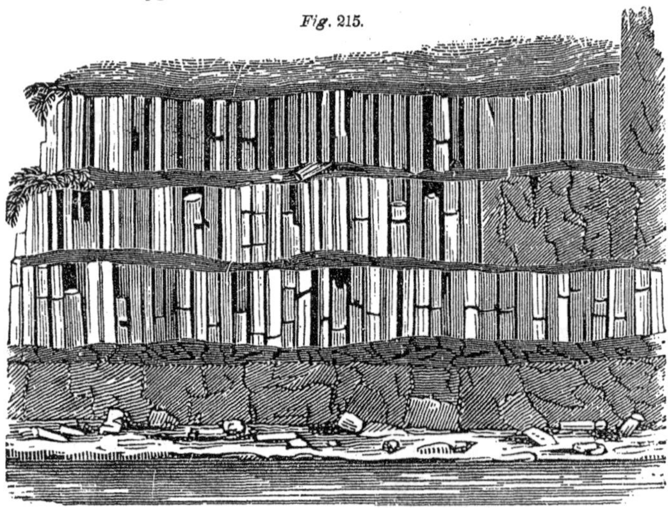

BASALTIC COLUMNS, COLUMBIA RIVER.

Where do trappean rocks occur in New York and New Jersey? What other localities?

A belt of trap about three miles wide and one hundred and thirty miles long extends along the Bay of Fundy, Nova Scotia. The action of the waves has exposed vast columns of greenstone, from 300 to 400 feet in height. Many beautiful minerals, such as quartz crystals, agate, chalcedony, amethyst, and specular iron, abound in this range. Some valuable ores are associated with greenstone dikes, particularly the ores of copper, as those of Lake Superior, Connecticut, and New Jersey.

Trap rocks are very extensively developed in the Andes, on the Eastern Continent. In New South Wales, basaltic mountains rise to the height of four thousand feet, and present phenomena similar to those of Staffa. These rocks, however, are found in so many localities, that our limits forbid a further enumeration.

VIII. *Geological Position of the Trap Rocks.*—Trap rocks are protruded through all the strata from the earliest times down to the close of the tertiary period. There appears to have been two periods, however, during which they were erupted in the greatest abundance. The greenstone, which is associated with the new red sandstone, and the basalt of the tertiary, form the most extensive ranges or masses of this class of unstratified rocks.

SECTION III.—VOLCANIC ROCKS.

This division of the unstratified rocks includes all the mineral substances ejected from volcanic craters. The term *lava* has been applied to this matter from the fact of its flowing out in a liquid form.

Lava is composed mostly of feldspar and augite, although a large number of the other simple minerals are found distributed through it, one hundred species having been found in the products of Vesuvius.

Lavas are of several kinds, which differ in structure, composition, and color. In the dark-colored lavas, *augite* constitutes a

What is the geological age and position of the trappean rocks? During what periods were they erupted in the greatest abundance? What rocks are included in the volcanic class? Describe the different varieties of lava.

considerable portion; and in the light-colored lavas, *feldspar* is the principal ingredient.

1. When feldspar preponderates, the lavas are called *trachytic*, and resemble *trachyte*. The structure depends upon the mode of cooling. The interior of the mass, or those portions which are cooled under pressure, are compact; but the surface, which is exposed to the air, is exceedingly porous, and, when ejected into water, so light as to swim upon the surface in the form of *pumice*.

2. When augite is the principal ingredient, *basaltic lavas* result. These can hardly be distinguished from basalt; in fact, when cooled under pressure, they form *compact* basalt, but, when cooled in the open air, they become filled with vesicles, and are generally termed *scoriæ*.

3. A variety of lava called *graystone lava* is intermediate between the trachytic and basaltic lavas. *Vitreous lavas, obsidian*, and *pitchstone*, are similar to melted glass.

4. The materials which are ejected from volcanoes during an eruption, in the form of fine powder and angular fragments, fall down upon the neighboring land or sea, and there becoming mixed with sand or shells, form a peculiar rock called *tuff*. The varieties are denominated *trap tuff, volcanic tuff, volcanic breccia, tufaceous breccia*, &c. Many other substances are ejected from volcanoes, but they are of little geological interest. We have given on p. 67 the geographical distribution of volcanic vents.

EXTINCT VOLCANOES.

Extinct volcanoes are very numerous, and the materials they have ejected are similar to the lavas and basalts of the present period. The evidence of their existence during the tertiary period has been presented p. 181. From their size and appearance it is inferred by many geologists that volcanic agency was exerted on a much more extensive scale than at present, and that it has been diminishing from the earliest geological times.

What is said of extinct volcanoes?

SECTION IV.—ORIGIN OF THE UNSTRATIFIED ROCKS.

We now proceed to give a summary of the evidence which is relied upon to prove the igneous origin of the unstratified rocks. This evidence may be presented under the following heads: I. Their structure and composition; II. Their position; and, III. Their effects upon the strata.

I. The *structure* and *composition* of the unstratified rocks prove their igneous origin. Commencing with the lavas of existing volcanoes, of whose igneous origin there can be no doubt, we notice the same structure, and often the same constituents in the extinct volcanoes, in all the trachytes and in basalt. So complete is the evidence derived from physical and chemical characters, that most rocks of the trap family, when isolated from their position, can with difficulty be distinguished from many varieties of lava. This resemblance is not so perfect in rocks of the granitic class, and yet granite and syenite often assume the trappean structure. The crystals are more distinct, a fact which is accounted for on the supposition that they were formed under the ocean or deep in the earth, and were more slowly cooled. There is also considerable difference in their composition, but all the unstratified rocks by almost insensible gradations pass into each other, the lavas into trachyte, trachyte into greenstone, and greenstone into porphyry and granite.

II. The *position* of the unstratified in relation to the stratified rocks shows that at the time of their formation they were in a melted or plastic state. Dikes of each variety not only pass directly through the strata, but they spread out laterally, filling up the cavities and flowing over the surfaces of the strata. The appearance of the dikes is precisely similar to those in volcanic mountains, where rents are often made for miles in extent, and filled with melted lava. The principal difference is, that the trap and granitic dikes are on a much larger scale, and of much greater extent. They occupy every position which would be

What was the origin of the unstratified rocks? What is the first proof of this? the second proof?

assumed by matter in a state of fusion, ejected among the strata by an upward force, as seen in the following figure.

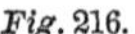

Fig. 216.

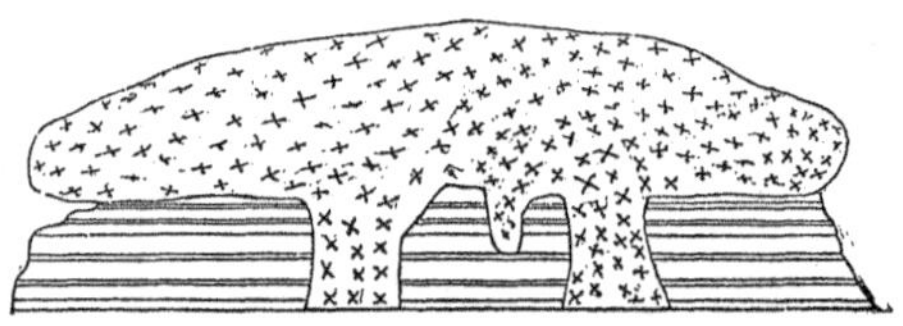

GRANITIC DIKE IN LIMESTONE, DERBY, VT.

III. The mechanical and chemical effects of the unstratified rocks upon the strata and upon each other prove their igneous origin. The strata are lifted up, broken, and plicated on each side of dikes, greenstone, and granitic ridges, showing that the latter were forced up from beneath; and there are many examples in which portions of the stratified rock through which granitic and trap dikes pass are inclosed in their substance. This seems conclusive as to the igneous condition of the matter when it was erupted. *Fig.* 217 is an example of this kind of agency (Hitchcock's Report).

Fig. 217.

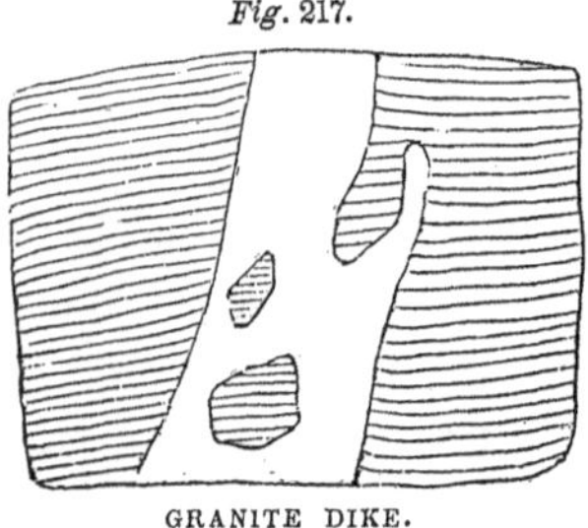

GRANITE DIKE.

The strata in contact with the dikes, and often to a considerable distance from them, are altered in texture and composition in a manner precisely similar to rocks which have been subjected to volcanic action. Dikes of granite passing through beds of chalk have converted the chalk into crystalline marble. The same effect has been produced upon chalk by heating it confined in an iron vessel. Copper ores through which greenstone dikes have been thrust have been smelted and converted into pure metal. Examples of this effect are met with in Nova Scotia, Lake Superior copper mines, and in many other places where copper ore is found.

What is the third proof of the origin of unstratified rocks? Describe the chemical and mechanical effects of dikes upon chalk, coal, and other strata through which they pass.

Similar chemical effects have been produced upon the coal-beds. In the eastern portions of the great Appalachian coal-field, the anthracite coal is found in the vicinity of the unstratified rocks; and we know that heat will drive out the bituminous matters from soft coal, in which case it resembles anthracite. That the anthracite was originally bituminous is proved by its structure, and by the fact that, in passing in a northwest direction across this coal-field, the bituminous matter gradually increases, until, at a considerable distance from the unstratified rocks, its original bituminous character prevails.*

Generally, near the dikes the strata are more crystalline, and this is a well-known effect of heat.

We find, also, many examples where one dike has cut off another. A horizontal dike has been cut off and lifted up by one which is vertical. This is so frequently the case with granite (page 294) and greenstone, that no doubt can be entertained by any one who has witnessed the phenomena of the igneous origin of the whole class of unstratified rocks.

SECTION V.—ORIGIN OF THE STRATIFIED ROCKS.

The stratified rocks, as we have abundantly shown, received their structure and regular arrangement through the immediate agency of water. But the materials of many of them, especially of the older strata, were evidently derived directly from the igneous rocks. The composition of gneiss, which is believed to be the oldest stratified rock, is identical with that of granite. Granite often passes into gneiss so gradually as to show that in early times the same materials were alternately acted upon by fire and water, and that they existed at one time in a stratified, and at others in an unstratified condition.

The various classes of slates, as mica slate, talcose slate, and argillaceous slate, appear to have been derived from granitic rocks which had been subjected to chemical and mechanical

* This is doubted by some geologists.

How have the stratified rocks been formed? In what state were their materials before stratification commenced? What proof of this?

agencies. In different parts of the stratified series we often find conglomerates which are composed of pebbles of igneous and aqueous rocks, cemented together by finer materials.

From the above facts, and from those stated in the preceding section, it is an obvious inference that *the whole crust of the globe accessible to observation was once in a melted state.*

We have already presented the evidence for believing that the interior of the earth is at the present time subjected to intense heat (p. 70); so that we have now traced back the history of our planet to a period when its entire mass was in a state of igneous fluidity.

In confirmation of this condition of the earth at some former period, the supposed igneous state of other bodies of the solar system has been frequently urged. The moon appears to have been subjected, at no very distant period, to intense igneous agency. The sun, and some of the planets, have also been supposed to be pervaded with intense heat at the present time. This, however, is merely an hypothesis: their actual condition is unknown.

SECTION VI.—NEBULAR HYPOTHESIS.

It has been a question much discussed by astronomers, whether the earth, the solar system, and the whole material universe, were once in a state of vapor. The subject is designated the *nebular hypothesis.* The hypothesis supposes that the matter of which the universe is composed existed as vast nebulæ, out of which, by the laws of gravity and of centrifugal force, under the molding hand of God, the present *system of the world* has been formed.

If we suppose that God created matter and diffused it throughout space, and that from time to time he gradually molded it into suns and planets, under the laws of attraction which he has impressed upon material atoms, we shall have all the conditions required to account for the present arrangement of the universe.

What inference in respect to the early condition of the earth? What analogies are urged in confirmation of this? What question has been much discussed by astronomers? Give a description of the nebular hypothesis.

There are many arguments which are relied on to establish the truth of this hypothesis. An important one is furnished by the condition of the earth's crust.

If we assume that the matter of the solar system existed at first as a nebula, there would be a tendency, from the action of gravitation, in all the atoms to move toward the center, a solid nucleus would be formed by condensation, and the latent heat would render it fluid.

By the meeting of opposite currents a revolution would soon be established, and the matter on the circumference of this vast nebula would become so dense that the centrifugal force would overcome the cohesion, and a ring would be separated from the mass. This ring, as the nebula left it, by its further condensation would be collected into a globe, which would be fused by the development of its latent heat. The same process would be repeated until all the planets were formed, and by the same law the planets would throw off their satellites.

Now we have shown that the earth was once in the precise condition which such a process would produce, that is, in the state of fusion from heat.

All the facts and phenomena connected with the present condition of the earth and of the solar system are either accounted for by, or consistent with, this hypothesis. The planets all move in the same direction in their orbits. They lie very nearly in the plane of the ecliptic. They are all compressed at their poles and enlarged at their equators, a condition which would result from the action of the centrifugal force. Many bodies, as the comets, exist as nebulous matter, and, in various portions of the heavens, the telescope has revealed to us numerous objects which are believed by some to be matter in a nebulous condition.

This hypothesis is adopted by many astronomers, and many learned treatises have been written in its support, the most celebrated of which is the Mécanique Céleste of La Place.

It should be observed, however, that many nebulæ have lately been resolved by powerful telescopes, especially by the large re-

What are the proofs of the truth of the nebular hypothesis?

flector of Lord Rosse, and by others, and shown to be clusters of stars. These discoveries have somewhat weakened the evidence in support of the hypothesis; but, on the other hand, the discoveries of Kirkwood in respect to the relations between the number of rotations of the planets and the diameters of their spheres of attraction, would show that, at some remote period, there must have been a physical connection of the different bodies of the solar system; and if "his analogies," as they are termed, are fully established, they will offer the strongest proof hitherto furnished of the former nebulous condition of the universe.

CHAPTER X.

ANTIQUITY OF THE EARTH.

THE antiquity of the earth is a subject which has excited much interest both among geologists and theologians. The idea of long periods of time previous to the introduction of man has been supposed by many to conflict with the Mosaic history. Before we attempt to reconcile the records of geology with this history, it is desirable to establish, by physical evidence, the great age of the world. We have already given, in the preceding parts of the work, the principal facts which are the basis of this argument, and it only remains to gather into one view the chief points which are relied on to establish the truth of the proposition,

That the earth existed for long periods, and passed through many changes previous to the introduction of the human species.

The evidence upon which the argument for the antiquity of the earth is based is derived from the structure, character, and position of the rocks, the remains of animals and plants, and the mutual relations of the organic and inorganic portion of the earth's crust.

What has weakened the evidence in favor of the nebular hypothesis? What has strengthened it? What is the object of this chapter? From what sources is the evidence of the great age of the world derived?

SECTION I.—THE ANTIQUITY OF THE EARTH, AS INFERRED FROM THE STRUCTURE, COMPOSITION, AND POSITION OF THE ROCKS.

I. *Basis of the Argument.*—In our examination of the rocks which are now forming at the bottom of lakes and the ocean, we have noticed that their *structure* indicates the mode of their formation. They are arranged in regular layers, which may be easily separated in one direction, and, as we see the process actually in progress, we find no difficulty in understanding it. The sand, mud, and organic bodies which are borne from the continents by the rivers into the ocean subside to the bottom in thin layers, inclosing marine animals and plants. These layers, by means of the mechanical pressure of water, and by the chemical agency of lime and iron, aided sometimes by heat, are gradually consolidated. In many cases these layers have been broken up and variously inclined by volcanic agency.

If now we take a rock thus formed and compare it with those rocks which we have described as *stratified*, we shall notice a very great similarity of *structure*. The resemblance is so perfect in many cases, that it would be difficult to decide which was formed first. The legitimate inference is, that not only the rocks thus compared, but the whole class of strata which possess this structure, wherever they may be found on the surface of the earth, have been formed by the same agencies and by a similar process; that is, that their materials were brought by the agency of water from the then existing continents, and deposited in layers at the bottom of ancient seas and lakes; that these strata have been hardened into solid rock, and, finally, that the strata have been broken and tilted up on their edges by a force acting from beneath, or by the agency of heat. The argument is based on the analogy between the present and the past changes in the crust of the globe.

II. *Character and Position of the Strata.*—The character and position of these strata prove that a series of *successive changes* have taken place which require *long periods of time.*

What is the basis of the argument by which the great age of the world is proved? From what is the first argument derived?

The separate layers which constitute a formation in each of the geological epochs are, as we have seen, conformable to each other, but the formations themselves are often inclined or uncon formable. The *several groups, however, always maintain the same uniform order of position.* The Palæozoic rocks in the United States, from the Potsdam sandstone, the base of the Silurian system, to the top of the coal-beds, are generally conformable to each other, but each of the separate formations always occupies the same relative position.

The coal strata in England are often much inclined, and strata of sandstone lie directly across their edges. It is obvious, therefore, that the coal-beds were deposited and tilted up before the sandstones were formed. The tertiary of the London and Paris basins is found in basin-shaped depressions of the chalk, and lies unconformably to it. In penetrating through their beds, we find London clay, plastic clay, chalk and green sand, and this order is never reversed wherever these rocks are found in junction on the surface of the earth. The tertiary is always above the cretaceous, and never below it; the cretaceous above the oolitic, the oolitic above the liassic, triassic, &c. The position of each group shows one uniform order of formation, and this indicates their relative age.

This conclusion is further sustained by the fact that the rocks in the higher groups are often *composed* in part of fragments of those in the lower groups. Thus the lower tertiary beds in the London and Paris basins are filled with pebbles of chalk. The new red sandstone of the Connecticut River valley is composed of the metamorphic Silurian rocks of Vermont. The conglomerate rocks are composed entirely of materials derived from rocks which occupy a lower position, and examples of a similar char acter are found throughout the strata, all tending to show that the group which is lower in relative position has been formed previous to the one which lies above it.

What order is observed among the several formations? Give examples. By what is the relative age of the formations indicated? By what fact is this conclusion confirmed?

Observations on this subject have been conducted with the greatest care by the ablest geologists in Europe and in this country, and this uniform order of arrangement has been satisfactorily demonstrated.

The position and composition, then, of the stratified rocks require a series of successive changes in the materials of the earth's crust. This fact is fatal to the hypothesis which attributes the formation of the strata to sudden and simultaneous deposition from water or to the agency of deluges, for one group in many cases must not only have had time for its deposition, but also to be hardened into rock from the condition of fine sand or mud, then to be elevated and exposed to the agents of degradation before the formation of succeeding groups could have commenced. Now when we consider the fact that two thirds of existing continents are covered with stratified rocks to a depth of six to ten miles, when, from their structure, composition, and position, we are compelled to believe that their formation must have proceeded gradually, and in most cases in tranquil waters, we are furnished with a powerful argument in support of the proposition that the series of changes required for their deposition must have extended through *long periods of time.*

III. The relation of the unstratified or igneous rocks to the stratified strengthens the above conclusion. We have seen that the igneous rocks have penetrated the strata at different and distant periods in every direction, lifting them up, and plicating them in such manner as to prove a series of successive changes (see page 302). "Each individual movement has contributed its share toward the final object of conducting the molten materials of an uninhabitable planet through long successions of change and of convulsive movements, to a tranquil state of equilibrium, in which it has become the convenient and delightful habitation of man, and of the multitudes of terrestrial creatures that are his fellow-tenants, of its actual surface."—*Buckland.*

Where have observations been conducted, and with what result? To what hypothesis are these facts opposed? What is the conclusion? What other facts confirm this conclusion?

SECTION II.—THE ANTIQUITY OF THE EARTH AS INFERRED FROM THE REMAINS OF ORGANIZED BEINGS.

The character and position of the mineral masses which constitute the crust of the earth render the supposition of its having been molded into its present form in six days, or in 6000 years, in the highest degree improbable.

But when we examine the remains of animals and plants, and notice their character and position, we derive an argument for the antiquity of the earth which is wholly unanswerable, while the opposite doctrine is rendered not only incredible but absurd. No honest mind can believe it, unless it is prepared to believe that the Deity has furnished us with reasoning powers for the purpose of deceiving us.

I. *The distribution and abundance* of the remains of organization in the earth's crust evince a high antiquity of the earth.

We have seen that the remains of animals and of plants are found in all the stratified rocks from the base of the Silurian system through all the successive deposits up to the present period

Each group contains a peculiar *type.* The families and genera often extend through two or more formations, but the species are generally confined to a single geological period. As the condition of the earth was changed, and the old species died out, either new species of the same genera were introduced, or an entirely new race were brought upon the stage of life.

In the older rocks, the Silurian, the animals and plants belonged to the lower types of organization. The plants were mostly marine, and the mollusca, radiata, articulata, and a few fishes and reptiles were the only types of animal existence.

As we rise in the series fishes are more abundant, and reptiles make their appearance in greater numbers near the close of the Palæozoic period. In the trias we have the first evidence of the existence of birds, in the numerous impressions of their feet upon

---

What is said of the argument derived from the remains of organization? of the distribution of fossils? What types prevailed in the lower rocks? What changes are observed as we proceed from the earlier to the later geological times? When do reptiles make their appearance? birds?

sandstones, as those of the Connecticut River valley. The saurian tribes, of singular and monstrous proportions, succeed the birds in the lias, and extend to the chalk.

At the close of the cretaceous period the ancient races were all exterminated, and, with the commencement of the tertiary mammals, with but three exceptions (page 235), were for the first time introduced. Cetaceans, pachydermata, and carnivora, were the prevailing forms. They were allied to existing families, but the species are all extinct. In this general distribution we may discover a succession of animal forms which were adapted to the condition of the mineral masses, and which, in their character, gradually become more and more allied to existing species.

The same succession in the forms of vegetable life may be observed in the successive periods of the earth's history.

As these fossil bodies are found in all parts of the rocky strata, they reveal to us the conditions of the earth when they existed as living creatures, and the changes by which they were entombed in its rocky beds. Their character and position form the most perfect chronological charts by which to determine the great age of the world.

The *abundance* of these remains sets aside every hypothesis which would account for their existence by any other processes than those which we have described. "It must appear almost incredible to those who have not minutely attended to natural phenomena, that the microscopic examination of a mass of rude and lifeless limestone should often disclose the curious fact that large proportions of its substance have once formed parts of living bodies. It is surprising to consider that the walls of our houses are sometimes composed of little else than comminuted shells, that were once the domicile of other animals at the bottom of ancient seas and lakes. It is marvelous that mankind should have gone on for so many centuries in ignorance of the fact which is now so fully demonstrated, that no small part of the present surface of the earth is derived from the remains of animals which

---

When do saurians make their appearance? mammals? What succession in the types of vegetable life? What is said of the abundance of these remains?

constituted the population of ancient seas. Many extensive plains and massive mountains form, as it were, the great charnel-house of preceding generations, in which the petrified exuvia of extinct races of animals and plants are piled into stupendous monuments of the operations of life and death during almost immeasurable periods of past time."—*Buckland.*

II. But the argument for the antiquity of the earth, as inferred from the remains of organization, may be presented with greater distinctness by specific examples. The vegetables of the carboniferous period furnish the most convincing proof of the great age of the world.

1. This evidence appears in the peculiar character of the vegetation and present position of the fossil trees (p. 264), indicating a condition of the earth so unlike the present, that we are forced to believe that a long period of time was consumed in their growth and deposition in the strata of the earth. Most of them are tropical plants, and yet they are buried in the rocks beneath the polar snows.* Their position often shows that they grew in the localities where they are found, and were not drifted from warmer climes. Large trees are found buried 1000 feet beneath the earth's surface, and geologically they are at the depth of several miles. How can their position be accounted for? It is possible that a few trees may have fallen into rents of the earth's crust produced by volcanic agency; a larger number might have been submerged by a sinking down of the earth's crust during earthquakes, covered up, and then elevated again by the same force. In these cases we should expect to find them of the same species with existing plants, but this is not the fact. The coal plants are *all extinct species.*

2. But when we examine the *character* and *position* of the coal-beds, and the *extent* of the coal-fields, such an hypothesis is wholly

* Vegetables are found in the rocks as far north as Baffin's Bay and Melville Island, in latitude 75 degrees.

What is said of the character of the vegetables of the carboniferous period? What does their position in the rocks show? How came they to be buried in the rocks? What hypothesis is suggested?

inadequate to explain the phenomena. The Derbyshire coal-field, in England, furnishes an example in point. Immediately under the coal are strata of shale from 300 to 600 feet in depth, then strata of the mill-stone grit filled with the remains of vegetables. Above the grit are regular coal strata, consisting of sandstones in thin sheets, iron stones containing numerous fossils, and soft argillaceous beds called *shale*. Two strata contain fresh-water shells. The whole thickness is 3930 feet. The series includes thirty different strata of coal, varying from 6 inches to 11 feet in thickness. The whole depth of the coal seams is 78 feet.

The Scotch coal-field in some places consists of 10 beds, whose united thickness is 100 feet. The South Welsh coal basin contains 23 beds of coal, whose total thickness is 93 feet. These fields extend over hundreds of square miles.

But the largest coal-field in the world is in the United States. The great Appalachian coal formation extends over a portion of the States of Pennsylvania, Maryland, Virginia, Alabama, Tennessee, Kentucky, and Ohio. "Its length from northeast to southwest is rather more than 720 miles, and its greatest width about 180 miles. Upon a moderate estimate, its superficial area amounts to 63,000 square miles."—*H. D. Rogers.*

The coal strata consist of alternate layers of limestone, sandstone, shales, and seams of coal, and are from a few hundred to three thousand feet in thickness. The coal itself appears to have been formed mostly from some peat-creating plant, which grew on extensive marshes upon the borders of the sea, as very few large trees are found in the coal-beds, although their leaves and branches are so numerous as to lead to the inference that extensive forests flourished upon the margins of those "vast marine savannahs."

Now the quantity of coal in all the coal-fields in the world has been estimated at five thousand billions of tons, and this immense

---

What objections to this hypothesis? What examples are mentioned? What do these examples prove? What is said of the extent of the Appalachian coal-field? of the character of coal? What estimate has been made of the quantity of coal in the world?

quantity of carbon must have been abstracted from the atmosphere by the process of vegetation, for by microscopical examinations of thin slices of coal the vegetable structure has been so distinctly seen, that in many cases the families and species of plants which formed the coal have been determined.

The vegetables, then, for each of the 30 or 60 seams in the several coal-fields, must have had time to grow, just as plants grow at the present day—this is proved by the concentric rings of annual increase—time to be torn away by storms and sea-waves, produced, no doubt, by earthquake-pulsations, which elevated and platted the strata, and brought the sea ooze or mud over the peat marshes; there must have been time for the vegetable matter thus covered with mud to be carbonized (which is a slow process), and after each bed was thus slowly formed, there must have been time for the whole to have been broken up and elevated by volcanic agency to its present position. Considering the character of the coal strata and the extent of the coal-fields, is it possible to believe that all has been deposited since the creation of man?

It should be observed that, above the coal strata, the remains of vegetables exist in all the rocks, but most of the species are different from those of the coal period. In the tertiary beds of Germany are found extensive beds of brown coal, containing trees from 9 to 13 feet in diameter, in one of which there were counted 792 rings of annual increase. The tree must have required nearly eight centuries for its growth, and afterward converted into coal.

But if any are disposed to doubt the evidence of antiquity derived from these fossil bodies, let them go to Senegal, in Africa, and examine the baobab-tree (*Adansonia digitata*), which is 30 feet in diameter, and has been estimated by the concentric rings of annual increase, to be five thousand one hundred and fifty years old; and, if they are still in doubt, let them visit Central

---

Whence has the coal been derived? What proof of this? Mention the changes to which the coal strata must have been subjected. What is the conclusion from this examination? In what other rocks are the remains of vegetables found? What is said of the size and age of existing trees?

America, and notice the celebrated Toxodium, of Chapultepec which is 117 feet in circumference, and has been estimated by De Candolle to be still more aged. Professor Henslow estimates its longevity from 4000 to 6000 years. Then reflect that these antediluvian trees, that flourished in the days of the patriarchs—that were a thousand years old when Noah was born—that were cotemporary with our First Parents—reflect that these venerable trees stand upon the loose soils which overlie all the preceding geological formations with their organic contents. The soil was formed there before the trees were planted, and, before the soil was formed, the entire series of rocks were deposited, and similar trees, in countless numbers, had their long periods of growth and decay.

III. We might, if our limits would permit, bring many specific examples of the character and position of the fossil animals, which would fully confirm the truth of the proposition we are considering. Few are aware of the immense numbers and of the peculiar character of the animal remains which modern researches have brought to light. Their existence in the rocks is of itself ample demonstration of a higher antiquity than is generally assigned to the earth.

The earliest animals were principally radiata, mollusca, and fishes. They prevailed almost exclusively until near the close of the carboniferous period, when we find the first evidence of the existence of reptiles. Throughout the Palæozoic deposits, their remains form a considerable portion of many strata of rocks.

With the triassic period birds were, for the first time, introduced, and reptiles became more numerous. The evidence of the existence of these animals is presented to us by their tracks and fœcal remains. These have already been described. Their distribution through the strata, their gigantic dimensions, indicate not only conditions of the earth unlike the present, but vast periods of time during which they flourished. Having left the

---

What changes occurred before those trees existed? What is said of the character and abundance of animal remains? What was the character of the earliest animals?

indisputable marks of their existence in the rocks, they were, at the close of this period, all exterminated, and were succeeded in the next group by lizards, saurians, and tortoises.

From the commencement of the liassic down to the close of the cretaceous period, we find some of the most extraordinary organic forms which have ever existed; so unlike to preceding types, and to those that succeeded, as to impress upon us the belief of a peculiar condition of the earth's surface during the period of their existence. The remains of huge saurians, of gigantic reptiles, and of flying lizards, are distributed through strata more than 4000 feet in thickness, and are associated with innumerable generations of other inhabitants that "swam in the waters or basked on the shores of the primeval lakes and seas." But, in one of those successive changes which had swept away preceding races, they in turn were all destroyed, and were succeeded by entirely new and distinct types of animals more nearly related to existing races.

During the tertiary period the land quadrupeds (*Mammalia*) were introduced in great numbers, and continued to flourish down to the close of the Pleistocene, when the species were mostly destroyed.

Last of all, in the most recent deposits are found the remains of man and his cotemporary races; and the fact that human relics are not found lower in the rocks is positive proof that all those other races lived and passed away previous to the appearance of man on earth.

It appears, then, as the result of geological investigations, that innumerable generations of organized beings have peopled the earth during its past history. In such countless numbers, in such infinite variety, in such great abundance are their remains stored in the rocky strata, forming almost the entire soil on which we tread, that we may properly inquire with Young for the "dust that has not been alive;" but it is impossible to believe that they have all lived within the last 6000 years.

What types succeeded the earliest animals? What is the evidence which the successive races furnish of long periods of time? What conclusion is drawn from the examination of animal remains?

SECTION III.—ANTIQUITY OF THE EARTH AS INFERRED FROM THE MUTUAL RELATIONS OF THE ORGANIC AND INORGANIC PORTIONS OF THE EARTH'S CRUST.

If now we combine the two branches of this argument, viz., that derived from the remains of organic beings, and that from the structure and position of the rocks, and notice the changes which must have been effected by the three agents, water, fire, and vitality, we can not fail to perceive that the physical evidence of the great age of the world is perfectly conclusive. These relations have been partially presented in the preceding sections. Let us select a few examples.

I. In Auvergne, in central France, there occur a series of the older tertiary deposits, which of themselves appear to settle the question of the age of the world. This district embraces an area of 1600 square miles (see page 194). It consists of a great basin filled with fresh-water limestones, sandstones, clays, marls, and volcanic products, to the depth of from 700 to 1000 feet. The strata rest on the primitive rocks, which must have been elevated from the bed of the primeval ocean, forming large fresh-water lakes, before the tertiary deposits could have commenced.

In some places the limestones and marls alternate with each other. The marls consist of thin leaves about the thickness of paper. Each leaf contains multitudes of the crustaceous coverings of the cypris and of small shells.

The cypris moults its shell annually, and, as the coverings fall down upon the bottom of the lake, they are covered with mud, and become cemented into sheets. Each marl leaf was then just one year in forming. The leaves are not more than $\frac{1}{20}$th of an inch in thickness. For the deposition, then, of one inch, it required 20 years; for one foot, 240 years. Now the vertical thickness of the marl strata is from 60 to 100 feet, and this would require from 14,000 to 24,000 years! The other deposits were

In what way is the argument for the antiquity of the earth presented in section third? What facts in Auvergne evince a high antiquity of the earth? Describe the manner in which the marl beds were formed. How long a time is required to form them?

not, probably, formed in a more rapid manner than the marls as they are filled with the remains of innumerable shell-fishes, and the bones of land quadrupeds.

In the upper beds are found strata of volcanic ashes, which indicate the commencement of the volcanic period of this region, during which more than 60 volcanic mountains were built up, and now remain with their craters several hundred feet in depth. The lava currents are easily traced from their vents in different directions, passing across the streams, forming new systems of lakes, and changing the whole face of the country.

During this period the mastodon, the elephant, and large herbifera were created, and inhabited the neighboring land. They were covered up by showers of volcanic ashes and lava streams, or drifted into the lakes, where their bones are preserved in a state of great perfection. After a long period, the volcanoes cease their activity, and the new lakes are filled with fresh-water deposits. Bears, hyenas, and various carnivorous tribes are associated with the mastodon and elephant, and, finally, these are all exterminated, and the existing races succeeded. If now we estimate these changes by the present operation of natural laws, it must have required millions of years for them to have taken place.

II. But in other parts of the earth, as in the north of Italy and Germany, extensive deposits occur which are more recent than the marls of Auvergne; and if we examine parts of Sicily south of Ætna, we shall notice strata, 2000 feet in thickness, still more recent. These rocks are so filled with shell-fishes as to show that the whole was beneath the sea during their formation, and that the island was not elevated until the last shell was deposited. If now we pass to Ætna, we may notice that the whole mountain, which is ninety miles in circumference, and eleven thousand feet in height, has been sent up since the deposits south

---

What other changes occurred in the marl beds? What animals existed? What has become of them? What other deposits of this period are referred to as indicating long periods of time? What is said of Ætna, and what proof of antiquity derived from its character and history?

of the mountain were formed. And yet the mountain has a venerable age. It has existed nearly as at the present time since history gives us any account of it. But "of the eighty most conspicuous cones which adorn its flanks, only one of the largest, Monti Rossi, has been produced since the times of authentic history."—*Lyell.* If we were to remove these cones, it would but slightly diminish the height of the mountain, and we should doubtless find a similar series of cones beneath. If, then, but one cone has been formed within a period of 2000 years—if, on this supposition, 1,600,000 years were consumed in the formation of the 80 cones—who will undertake to measure the time which transpired during the formation of the whole mountain?

III. Passing from the tertiary group to the older formations, we find first the chalk, which is about 1000 feet in depth, and is composed of small shells and corals; these must have been formed by animals previous to their deposition in regular strata. Under the chalk are beds of green sand 480 feet in depth, and, having penetrated through these, we come to the Wealden, 900 feet in vertical depth. The latter is a fresh-water deposit, liying between strata which were formed beneath salt water. In this group we notice the remains of whole forests, the trees standing in the position in which they grew, as in the "Portland dirt bed." Here the land must have been above the water a sufficient time for trees to grow two, three, and four feet in diameter, then to have been submerged, and upon it all the other groups deposited, and, finally, the whole must have been elevated to its present position several hundred feet above the level of the sea.

IV. If any still doubt the great age of the world, let them penetrate still deeper into the oolitic and liassic groups, nearly 3000 feet through series of strata filled with the remains of saurians, pterodactyls, and huge iguanodons, and in the same manner continue their observations through the triassic system to the depth of 1500 feet, and become acquainted with the gigantic frogs and birds that left their footprints on the now hardened sandstones.

What is the argument derived from the cretaceous rocks? the Wealden? the oolitic and liassic formations? the red sandstone?

"Solving a problem,
Man never has to leave a trace on earth
Too deep for time and fate to wear away."

Thence proceed to the coal-beds and view the ancient forests which supplied the materials for fossil coal, 10,000 feet deeper; thence downward three miles to the base of the Potsdam sandstone, until all traces of organization disappear, from which point they may descend to unknown depths, until they reach the original igneous rocks and the internal fires. And if these observations do not convince them, then leave all these stupendous monuments of antiquity, and come down to modern times, to the formation of terraced valleys and coral islands, processes which are now in progress, and not, therefore, liable to deceive them.

V. The terraces upon the river valleys have evidently resulted from the erosive action of the rivers by which their beds have been lowered. Three terraces or more often occur one above the other. Cities have been built upon the lowest of these terraces for more than 2000 years, during which period the action of the streams has not lowered their bottoms but a few feet or inches. The time required for rivers thus to lower their beds 40, 50, and 60 feet, must have been more than 6000 or 10,000 years. In many places the streams have cut through rocks to a great depth, as the Niagara River, which also proves that the action must have been long continued.

VI. Many of the coral islands indicate very long periods of time. We have described (p. 78) the manner in which these islands have been formed. Their exact rate of increase is not known, but, from all the observations which have been made, it is exceedingly slow.

Ehrenberg, who has examined the corals of the Red Sea, doubts whether channels and harbors have been filled by these animals during the historical period. The single corals which he saw of a globular form, and about nine feet in diameter, he

What is the argument derived from the coal and rocks below? What evidence is furnished by terraces and channels? What evidence of age do the coral islands furnish?

supposes might have been seen by Pharaoh in nearly the same state as at present. If they increased half an inch in a year, it would take 6000 years to build a reef 250 feet in height; but many of them are much thicker, for outside of the reef the waters become suddenly very deep, "no soundings having been obtained at the depth of 150 fathoms," even in the channels which lead into the lagoons.

VII. Each of the several groups of rocks which form the crust of the earth affords ample proof of the great age of the world; but when we take into view the whole series of changes, organic and inorganic, the argument for the antiquity of the world is so perfectly overwhelming, that those who have carefully attended to the evidence on which it is based are not only convinced of its truth, but are unable to express by numbers the long periods which must have passed since the first creation. The exact age of the world can not be determined; but, to aid our conceptions, we may perhaps be allowed to make a few numerical calculations. These estimates must be based upon the rate at which rocks are now forming in lakes and seas. The observations on this subject are few, but it is said that the lakes of Scotland have shoaled about a foot in two hundred years, and that the mud which the rivers transport would not more than raise the general floor of the ocean one foot in a thousand years. On this basis, if we take the estimated depth of the Silurian rocks (p. 121) at 8700 feet, their deposition would require a period of 8,700,000 years. The old red sandstone, or Devonian system, estimated at 10,000 feet, would consume 10,000,000 years. The carboniferous system is estimated also at 10,000 feet, and, regarding the deposits as made in lakes, they would require 2,000,000 years.

The Permean system	(oceanic)	800 feet	. .	800,000	years.
" Triassic	" "	900 "	. .	900,000	"
" Liassic	" "	700 "	. .	700,000	"
" Oolitic	" "	1400 "	. .	1,400,000	"

What effect is produced upon those who carefully examine the evidences o. age? Can the exact age of the world be determined? In what way may our conceptions be aided? Upon what must the estimates be based?

The Wealden system	(lake dep.)	900 feet	.	180,000 years.
" Cretaceous "	(oceanic)	1480 "	.	1,480,000 "
" Tertiary "	(lake dep.)	3500 "	.	700,000 "

The average thickness of the whole strata has been estimated at 10 miles. On this supposition it would require, at the present rate of lake deposits, 12,560,000 years; but if the rate of oceanic deposits is taken as a standard of comparison, they must have occupied a period of 52,800,000 years!

These estimates may serve to give us some definite impression of the great age of the world, and yet it is evident that they do vast injustice to the antiquity of our venerable planet. Astronomers tell us that the telescope reveals the existence of stars so deeply sunk in space, that light moving at the rate of twelve million of miles in one minute would be half a million of years in reaching the earth, and if these stars were blotted from the face of the heavens they would be seen for half a million of years to come. It is not improbable, therefore, that our earth, the whole universe around us, is many million times older than would be inferred by these calculations.

At what point in the past eternity this mighty system was called into being we can not determine, nor can we know whether at any time in the future it may be blotted out.

We should infer, however, from the history of the past, that our earth is destined to pass through other changes which will sweep away the existing races, and, if man be not the highest possible type of physical existence, that the renovated earth will be peopled by new and more wonderful forms of organic beings, who will be endowed with more exalted, intellectual, and moral powers.

How rapidly do rocks form in lakes? in the ocean? On these data, what length of time was occupied in the deposition of the several classes of rocks? How long time for the whole class of stratified rocks? Do these estimates truly represent the actual age of the world? From the history of the past, what changes are to be expected?

SECTION IV.—OBJECTIONS TO THE ANTIQUITY OF THE EARTH CONSIDERED.

I. In order to set aside the evidence which the rocks afford of the great age of the world, it has been urged that the agents of nature may have operated with much greater energy in the earlier than in the later periods of the earth's history; and that the strata, with their imbedded fossils, were produced between Adam and Noah, and chiefly by the waters of the deluge.

This hypothesis requires so great activity of the agents concerned in the production of animals and plants, as well as of those by which the rocks were formed, that there are no analogies founded upon the present operation of physical laws to sustain it. The only question is, whether there are good and substantial reasons against it. From the many reasons which oppose this theory, a few only are needed.

1. Many of the fossil trees have the concentric rings of annual growth, and these yearly additions are not thicker than those in the trees of existing forests; but if vegetation grew previous to the flood at a rate which the hypothesis requires, we ought to find some evidence of it in their structure; for if all the plants in the earth's crust grew and were inhumed between Adam and the close of the deluge, the antediluvian vegetation must have grown, according to the preceding estimates (page 322), 7.602 times as rapidly as the plants of our own era.

2. The same must have been true of animals; but we have the most convincing evidence in the structure of animal relics that their growth was not increased at a rate required by this hypothesis.

3. The character and position of the rocks demonstrate that the laws which governed their formation did not essentially differ in activity from their present rate. But in case the rocky beds were deposited by the Noachian deluge, we should be under the necessity of believing that the agents of change operated during the year of the flood 52,800,000 times as powerfully as at

---

What hypothesis has been suggested to set aside the evidence of the antiquity of the earth? What is said of this hypothesis? What reasons against it derived from the plants? from the animals? from the rocks?

present; and, even if we allow the whole time from Adam to Noah (1650 years), then the agents must have exceeded 32,000 times their present activity. This would produce a state of agitation sufficient, we should suppose, to preclude the existence of animals on the earth's surface.

4. But suppose that this is not a correct view of the case, and that animals and plants could have flourished; if it is maintained that any considerable portion of the changes which the rocks indicate took place before the flood, as the growth of animals and plants, then the antediluvians must at least have had rather turbulent times, and, as frogs grew as large as elephants, men would doubtless have attained to a colossal magnitude, "giants indeed in those days." We would suggest, however, to the believer in the hypothesis, why some of those giants' bones have not been found buried with his cotemporaries in the rocks?

There are about 30,000 fossil species found in the rocks, and only a few, in the upper beds, identical with those that exist at the present day. What has become of the remains of man and the animals which now inhabit the earth? Was there a new creation after the flood? If man existed during the deposition of the tertiary and secondary rocks, we might reasonably expect to find some evidence of it; some cities or implements of husbandry would have been preserved. It is reasonable to conclude that the human skeleton would be as durable as the bones of other animals, or the delicate leaves of plants; but no such remains have been found below the alluvial deposits. This fact appears to settle the question; for if the waters of the Noachian deluge had been continued sufficiently long to have deposited the stratified rocks, the simple fact that the bones of man are not associated with the fossil races is of itself a perfect confutation of the hypothesis under consideration. For a fuller discussion of this subject the student is referred to Hitchcock's Religion of Geology, which is the most able work on the connection of science and revelation which has hitherto appeared.

What reasons against this hypothesis derived from the absence of the bones of man?

II. In order to account for the present appearances in the earth's crust, it has been suggested that the laws of nature were altered after the fall of our First Parents. This hypothesis does not differ essentially from the preceding, so far as it depends upon natural laws, but simply assigns a reason for the supposed alteration. But we find no evidence either in the Bible or in the rocks of such a change. The character of the fossils proves that no such change has taken place since the commencement of organized life.

III. The last hypothesis which we will notice supposes that the rocks, with all their imbedded fossils, were created exactly as we find them by the direct power of God.

A naturalist who is ignorant of geological doctrines commences an examination of the rocks, and is surprised to find the remains of several thousand species of animals and plants different from those which exist upon the earth's surface. He finds them imbedded in rocks to the depth of several miles, and preserved in such perfection that he is able to assign them their position in the grand scale of organization. He inquires of a believer in the above hypothesis how these relics became buried so deeply in the solid rocks. The answer is, "They were created there;" "the earth was so made;" for he is told, "It is evident that when God made trees he would make them with the concentric rings of annual growth, so that they might appear to be several hundred years old on the day of their creation. The same principle applies to the making of worlds." "He made the earth as it would have appeared if it had existed millions of years." "The fossils and rocky strata are but *types* of what is now in progress." Would such an explanation satisfy any reasonable mind?

We would not doubt the power of God to form the earth in accordance with the above hypothesis; but what he has done and what he has power to do are quite different propositions.

If the Deity formed in the rocks *types* of the animals which were afterward to inhabit the earth, why did he not give us a

---

What is the second hypothesis, and what objections to it are mentioned? the last hypothesis? How is this hypothesis answered?

type of man? There are types of elephants, bears, whales, birds, monkeys, shell-fishes, and even insects and worms; why not a type of the human species? It is at least impossible to conceive of the reason unless the inventor of the hypothesis considers the monkey tribe as the proper type; and if so, it would seem that, with regard to some of the race, the type has not been greatly improved upon; for we can not conceive how any rational mind can give its assent to an hypothesis so purely fanciful as this. And yet, as it is the most plausible of any, we will proceed seriously to develop its logical and philosophical bearings.

As the hypothesis sets aside all secondary agents and laws, we have nothing to guide us in the present order of nature. In attempting to refute it, therefore, we can only appeal to analogies, and inquire whether such a view is consistent with what we know of the ways of the Divine Being? whether, in entertaining the hypothesis, we must not adopt principles which will subvert all reasoning on subjects which are not the immediate objects of personal observation?

The hypothesis is based on the assumption that the uniform course of nature must be set aside so far as to allow of a beginning—a time when natural laws were established—so that with regard to the original production of man and other animals, the analogy of present laws is of no logical force. This doctrine we fully admit, for we have positive proof that organic beings were first *created* by the direct power of God. But all the processes and laws of nature are not primary institutions, but many of them result from those that are. It is an axiom in reasoning that each law must be regarded as having been the same in all past time, unless there are good reasons for the contrary belief. Thus, for example, it is a uniform law that animals and plants are preceded by similar animals and plants. How do we know that this has not always been the case? We know that it has not, first, from the evidences of design and of a designer, which are

What does this hypothesis set aside? upon what is it based? What is the present law in respect to the origin of animals? Has this law ever been different? What reason for it?

not accounted for by tracing the series backward even to infinity and, secondly, from a revelation which particularly describes the creation of man and the present races of animals and plants. By the first reason we settle it as a fact that this uniform law of parentage must have commenced in some period of the past, and by the second reason the time for the present races is assigned. We can not, therefore, affirm that every animal was preceded by one of its kind; there is one case, at least, in which the present order can not be made the basis of reasoning.

Are there the same or similar reasons for setting aside the present order of nature in our reasonings upon the formation of the rocks, and of the remains of animals and plants which they contain? It is now a uniform law that the bones of men and other animals belong to living beings before they are entombed in the earth. Is there any reason to believe that this law has ever been different? It is obvious that there is no necessity in the case, for the creation of living animals and plants will perfectly account for their existence in the rocks, and we have no historical record or revelation of any kind that in any way implies that such relics were created as we find them.

The admitted fact, then, that the ordinary course of nature in the production of organic beings must be set aside both by the necessity of a first creation and by revelation, can not be urged either logically or philosophically to dispense with the law in respect to the fossil races, because the creation of living beings fully accounts for the existence of their remains in the rocks.

On the other hand, we may urge the analogy of the present system of nature to prove that the fossil species were first created as living beings.

The true state of the argument is this: In the rocks which are now forming we find the remains of man and other relics of organization. We know that they once formed parts of living beings. We next inquire when the latter were first introduced,

---

Are there similar reasons for setting aside the law of the fossil bodies? Why not? What analogy may be urged to prove that the fossils once belonged to living beings?

and find a revelation which fixes the time to be about 6000 years since. Beneath these relics, deeper in the rocks, are found the remains of a different series of animals and plants, such as have no living representatives, and we infer that these also were once in a living state, although we have no account of their creation; but we infer it from the necessity of the case, and from the fact that revelation has given us an account of the creation of one race which we know exist as living beings before they are buried in the earth.

The analogy, then, of the creation *before* the inhumation of existing races may be urged in proof of the creation and subsequent deposition in the rocks of all preceding races; in other words, it is logically inferred from the present law of the creation of animals and plants before their inhumation in the rocks, that the fossil races not derived from existing species were created as living creatures before they were buried in the earth's crust; hence *they were not created as we find them.*

We have presented this argument in different forms, because it has been so frequently and boastfully said that "the reasonings of geologists are destitute of sound logic." "The earth *may* have been created so, and therefore no man can prove that it was not." "The most that can be said is that the antiquity of the earth is somewhat probable!"

We will not attempt to confute this hypothesis further by direct argument, but will allow the believers in it the full benefit of their logic.

The mode of reasoning which the hypothesis adopts would enable us to arrest all inquiries and settle all difficulties in respect to the whole history of the past with the same facility with which it enables us to solve geological problems. The Pyramids and mummies of Egypt, the ruins of Pompeii and of Rome the cities of Central America, can not be proved to have been the works of man, or to have been built at all, because "they might have been created so." They are merely types of what is now in

---

What would be the result of adopting the mode of reasoning which the hypothesis requires?

progress. But suppose we have histories of their origin and decay, ought we not also to have types of histories? It is vastly more reasonable to believe that all things whose origin we did not see—cities, works of art, and men—were created just as they now appear, than to believe that the remains of animals and plants were created deep in the crust of the earth; for we can see some object in giving existence to living beings, but what possible reason can be assigned for the creation of comminuted shells, bones of land quadrupeds, saurians, lizards, and crocodiles, in a more or less petrified condition; some of the bones appearing as if they had been fractured and then grown together during the life of the animal? It is infinitely more reasonable to believe that all we behold is no evidence of creative power, than to believe that such power has been exerted in the manner required by the hypothesis.

In the language of a distinguished cultivator of science, Professor Silliman, "The man who can believe that the iguanodon, with his gigantic form of [30] feet in length, [5] in breadth, and 15 in girth, was created in the midst of consolidated sandstones, and placed down one thousand or twelve hundred feet from the surface of the earth, in a rock composed of ruins and fragments, and containing vegetables, shell-fish, and rounded pebbles, such a man can believe any thing, either with or without evidence. He must be left to his own reflections, since he can not be reached by sound argument; with such men discussion is useless, for the foundation of all conviction or persuasion is removed."

Let us then banish such fancies from the works of God, and with our Bibles in our hands meet this question of the antiquity of the earth, and show, by comparing record with record, that both may be received, because both proceed from the same great fountain of light and truth.

CHAPTER XI.

CONNECTION OF GEOLOGY WITH NATURAL THEOLOGY AND WITH REVELATION.

HAVING presented the physical evidence of the antiquity of the earth, there are two other subjects which are so intimately connected with the doctrines of geology, that a brief discussion of them seems necessary to the completeness of the present work. These subjects are the evidences furnished by geology for the existence and attributes of God, and the apparent discrepancies between the records of geology and the Mosaic history.

SECTION I.—GEOLOGY AND NATURAL THEOLOGY.

The evidence furnished by geology in proof of the existence and attributes of God, is the same in kind with that furnished by the earth and its inhabitants at the present time. The same evidences of *design*, and therefore of a designer, meet us in every period of geological history.

But most of the arguments derived from the existing condition of the earth are greatly extended, and their truth more variously and fully confirmed.

I. The *changes* to which the earth has been subjected require the exertion of *infinite Power*.

This is seen in the frequent changes of sea and land, the long-continued action of aqueous and igneous forces by which the earth has been molded into its present form, but more especially in the *creation* and *distribution* of many distinct races of organic beings. We can trace the remains of organization in the rocks of the earth's surface, until we reach a period of its history when the condition of its mineral masses precluded the existence of

---

What two subjects are discussed in this chapter? What is the nature of the evidence furnished by geology in proof of the existence and attributes of God? What evidences that organic beings had a beginning?

life. This fact entirely sets aside the hypothesis of "an eternal succession of animals and plants." They had a beginning, and this required *creative power.*

This power has been exerted at successive periods, as is shown by the new families which have from time to time been introduced; and hence the records of geology furnish numerous additional examples of the exertion of creative power, which greatly enlarge our ideas of the extent of this attribute of the Divine Being.

II. The *structure* and *adaptation* of the successive races of organic beings which have peopled the earth during its past history to the conditions of its surface show that power was guided by *intelligence* and *wisdom.*

The evidence of this appears in the anatomical structure and habits of the particular types of organization which prevailed during each of the geological periods. Every change in the mineral masses, in the climate and surface of the earth, was attended by corresponding changes in the animal and vegetable races. Previous to the coal period, the atmosphere was so charged with carbonic acid as to preclude the existence of air-breathing animals, and to favor a luxuriant tropical vegetation. Immediately after the coal-beds were deposited, the atmosphere became deprived of a large portion of its carbon, and birds are for the first time introduced. During the succeeding periods, down to the tertiary, the condition of the earth was fitted to reptiles, and they were the prevailing forms of life. At the commencement of the tertiary the surface had become suited to land quadrupeds, and they are then introduced.

III. In all the varieties of form and structure in the organic and inorganic kingdoms of nature there is perfect *unity of design.* From the earliest geological times to the present period the same physical and vital laws have prevailed. The rocks of every period show, by their composition, structure, and physical properties, that they were under the influence of the same atomic,

---

What proof is this of power? What evidence does geology furnish of inteligence and wisdom? What of the unity of God?

chemical, and electrical attractions, the same gravitating forces as at present.

The four great divisions of the animal, and the prevailing types of the vegetable kingdoms, were established at the commencement of organic life, and amid great variety of form and structure, in successive periods, the general plan is always preserved. The laws of anatomy and physiology are ever the same. Each group of rocks, and each family of animals and plants, have an intimate and fixed relation. One uniform plan, the same great *thought* of the divine mind, pervades the whole. Each served its appropriate end, and contributed to the final result of preparing the earth to become the fit habitation of intellectual and moral beings.

The unity of the divine intelligence, as evinced in the character of the fossil races, is most forcibly presented by Dr. Buckland in his Bridgewater Treatise. "There is," says he, "such a never-failing identity in the fundamental principles of their construction, and such uniform adoption of analogous means to produce various ends, with so much only of departure from one common type of mechanism as was requisite to adapt each instrument to its own especial function, and to fit each species to its peculiar place and office in the scale of created beings, that we can scarcely fail to acknowledge in all these facts a demonstration of the *unity* of the intelligence in which such transcendent harmony originated; and we may almost dare to assert that neither atheism nor polytheism would ever have found acceptance in the world had the evidences of high intelligence and unity of design which have been disclosed by modern discoveries in physical science been fully known to the authors or abettors of systems to which they are so diametrically opposed. It is the same handwriting that we read, the same system and contrivance that we trace, the same unity of object and relation to final causes which we see maintained throughout, and constantly proclaiming the unity of the great divine original."

IV. In all periods of the earth's history, the power, wisdom, and unity of the Divine Being have always been attended with *be nevolence*, that is, with the design of imparting happiness to living creatures, and to make provision for higher developments of the divine goodness in the future beings which were to be introduced.

The changes which preceded the first animals had direct reference to their enjoyment. They were adapted in form and habits to the condition of the mineral masses. The successive sedimentary deposits, and the convulsions which have thrown the surface into ridges, mingling those ingredients which were necessary to support vegetable life, contributed each to increase the aggregate of animal enjoyment. The introduction of carnivorous races even had the same tendency, by restraining a too excessive multiplication of any particular species, and by preventing the miseries connected with the infirmities of age and of a natural death.

But we see more particularly the Divine benevolence in those *provisions* which were made for the benefit of man ages before his appearance on earth.

These are seen in the prolific vegetation of the carboniferous period during which the materials for fossil coal were preparing, were stored up in the rocks, and, through the long ages that succeeded, were gradually wrought into their present condition, so that when the earth should be disrobed of its forests, man should have his wants supplied from these ancient stores of vegetation, and be reminded constantly of the prospective benevolence of the Creator.

The mines of valuable ores and minerals, in their distribution and abundance, are examples of the same benevolent provision.

The whole series of volcanic and aqueous agencies by which the primitive materials have been molded into their present forms, giving rise to that almost infinite variety of scenery which the surface of the earth presents, and of productions to which it has thus been fitted, evince foresight and benevolence in the Creator which challenges our profound gratitude and love. It is true that the past as well as the present condition of the earth

Mention the several facts which prove the benevolence of the Deity? What instances of prospective benevolence?

has been such that the Divine benevolence has been obscured; good and evil have been mingled in the development of the Divine plan, but benevolence prevails in all, and if we were able to look through the whole system, pain and apparent evil would no doubt be seen to result directly from infinite goodness.

In addition to what was absolutely necessary to maintain the existence of man and other animals, there is much that is intended to gratify his taste and regale his senses. The scenery of the earth, its varied surface of mountains and valleys, its rivers and oceans, its ever-changing views, its gorgeous colorings, these are ever filling the soul with grateful emotions. The thousand voices of the earth, its wonderful harmonies, its silences even, all tend to increase the happiness of sentient beings. This condition of the earth has been slowly maturing during the long ages of the past, and most beautifully and strikingly illustrates the prospective benevolence of the Creator. The records of geology, when fairly interpreted, all unite in proving and illustrating the attributes of God; they show that he is a God of power, wisdom, and benevolence, and that, amid all the changes of material forms, he is unchangeably the same.

V. But the greatest aid which geology has rendered to natural theology is in enlarging our ideas of the *system of the world* and of the *plans of God.*

Until the time of Copernicus, the space in which the universe was supposed to be contained was very limited, confined within a crystalline hollow globe of inconsiderable diameter called the firmament; but so soon as the telescope was directed to the heavens, this firmament was found to have no existence, and the idea of boundless space, filled with the display of God's power dawned upon the minds of men.

In the same way geology has enlarged our ideas of the time during which the universe has existed. Until within a very recent period, the idea was almost universal that the heavens and

---

Does not pain and evil in the world militate against the benevolence of God? Why not? What other illustration of benevolence? In what has geology rendered the greatest aid to natural theology?

the earth were created about six thousand years since; but an examination of the rocks brings to view a series of changes, which stretch backward through periods of such amazing length, that the imagination is not able to fix the time of the beginning, and from the analogies which come up from the past, we are pointed forward to indefinite ages during which man and his cotemporary races will work out the grand purposes of their earthly existence; and not only so, but from the changes to which the earth has been subjected, we may believe that other changes will take place which will sweep away the present races of animals and plants, and that new and higher forms of organic beings will inhabit the renovated earth.

It is thus that the records of geology carry us backward far beyond the time when human history began, and point us to indefinite ages after it shall have been completed. How does it exalt and enlarge our ideas of the plans of God! How great is his power, how unsearchable his wisdom, how wonderful his benevolence! Surely "His understanding is infinite!"

SECTION II.—CONNECTION OF GEOLOGY WITH REVELATION.

The material universe gives abundant proof and illustration of the being and attributes of God. Its purpose appears to have been to exhibit the divine character to finite intelligences.

First, we have presented the attribute of power in the creation of matter. Secondly, we find in addition to power, wisdom and intelligence in its arrangement. Thirdly, to these are added benevolence and providential care in its properties, uses, and government. And as each successive development was more and more glorious, as was shown particularly in the higher orders of beings which were formed, we should expect, after the creation of man the most exalted of all, an exhibition of other attributes which pertain to his nature as spiritual and immortal. The creation of an intellectual and moral race would lead to the expectation that God would make a revelation suited to their wants.

From the history of the past what change may be expected? After the creation of man, what kind of a revelation was to be expected?

Such a revelation has been given in the Bible. This manifestation of his higher attributes—his moral character—harmonizes with all his other revelations. In fact, each record confirms the truth of the other, and proves both to be from the same divine intelligence and wisdom. From the infinite nature of God, however—from the imperfection of language, and the difficulty of rightly interpreting these records, we might anticipate apparent discrepancies; an expectation which has been abundantly confirmed in the history of the past. At the present time, the most important differences refer to the age of the world, and the introduction of death. We shall therefore conclude the present work by a very brief discussion of these topics.

I. *There is an apparent discrepancy between the Mosaic and geological records in respect to the age of the world.*

We have presented the reasons for believing that the earth has existed for a much longer period than six thousand years, and the question is, whether the language of the first chapter of Genesis definitely fixes the time of the creation of the universe. "In the beginning God created the heavens and the earth," is the majestic and sublime language with which the sacred record commences. It should be noticed,

1. That the phrase "in the beginning" does not refer to time absolutely, but to the commencement of any period, process, or series of events, and is always limited by the subject to which it is attached, as the beginning of a kingdom, of the reign of a king, the beginning of life, &c. It undoubtedly, therefore, in this case, refers to the time when the universe was created. The revelation commences by ascribing the origin of all things to the power of God, in opposition, no doubt, to the views of heathen nations, which referred the creation of the universe to inferior deities, to chance, or to the belief that it was eternal. The heavens and the earth were absolutely created by the God of the Hebrews. There can be no doubt, therefore, but that the first verse refers

---

Has such a revelation been given? What apparent discrepancies are noticed? What does the phrase "in the beginning" mean? Does it fix the exact time of the creation?

to the time when the material universe was first brought into being.

2. The simple inquiry now is, whether this first verse is a general caption to the remainder of the chapter, or whether it is an independent proposition not connected with what follows, but intended merely to ascribe the origin of all things to God. That is, whether the time of the beginning was about six thousand years since, or whether it is left *undetermined*. It is obvious that either interpretation may be adopted, so far as the language itself is concerned. It has been quite generally considered as a general assertion, the particulars of which are stated in subsequent verses, although many commentators have regarded it as an independent statement, irrespective of the facts of geology.

By what means now can the true interpretation be settled? The Bible is to be interpreted like any other ancient record, that is, by considering the *object* of the record, the *character*, *knowledge*, and *condition* of the writers, and of those to whom the writing was addressed. By considering these points and others, we may be guided to a true interpretation, and to a solution of the difficulties which the believers in revelation have often found in the records of science.

(1.) *The object* of a revelation was to instruct men in respect to their duty and destiny as *moral* beings, and not to teach them *physical* science. The reason of this is very obvious. God had given the evidences of his existence and his natural attributes in the material universe. He had endowed man with powers capable of investigating and of reading this revelation; these were sufficient for him as a physical being; but he needed more; he needed instruction in respect to his spiritual and immortal nature, and this he could not find in the works of God. To supply this knowledge, a revelation direct from God must be given, and such is the object of the Bible, as appears in every page from the beginning to the end. Had the Bible undertaken to teach science, we could not have received it as a revelation from God, in-

What is the next inquiry? How is the Bible to be interpreted? What was the object of a revelation?

asmuch as a revelation, in order to be credible, must make known that which our natural powers are otherwise incapable of discovering.

(2.) The people to whom the revelation was first given were just emerging from a state of slavery, of semi-barbarism, and were wholly ignorant of the principles of science, for the sciences did not then exist. They had no idea of secondary forces, but ascribed every change to the direct power of God. The heavenly bodies were moved by his hand; they heard his voice in the thunder; he opened the windows of heaven in the solid firmament but just above their heads, and poured out rain to water the earth; consequently, all natural phenomena were regarded according to *apparent* rather than actual truth. As revelation was not intended to controvert these views, but to communicate spiritual truth, its language must be adapted to the necessities of the case. The people could not, without miraculous agency, have understood it, if, when it referred to the material universe, it had adopted language which should be scientifically accurate; and hence, in condescension to the views of men in a rude state of society, God wonderfully adapted his revelation in style and manner to their necessities. It was no part, therefore, of his purpose to teach them the chronology of the earth—neither Moses nor the people could have understood him if he had made the attempt—but to give the origin of the human race, and its relations to himself. Considering, then, the object of revelation, and the condition of men at the time, we are at perfect liberty to regard the first verse in Genesis as separate from the remainder of the chapter. The Hebrew particle which connects the next verse with it, rendered *and*, is well known to be generally used simply to connect the language, and not the *sense*. It is often rendered *but* and *afterward*, and may be rendered in this case, "Afterward the earth became waste and desolate." How long

---

What was the character of the people to whom the revelation was given? What were their ideas of natural phenomena? Could they have understood a scientific explanation? How is the first verse of the first chapter of Genesis to be understood?

afterward we are not informed. It may have been millions of years, during which all the changes required by geology may have taken place.

(3.) As the language of the first chapter admits of two interpretations, it is obvious that this is a case in which the facts of science may be brought forward to determine which shall be adopted. The evidence of the great age of the world derived from geology and astronomy is sufficient to decide this question, and we not only are at liberty, but are compelled to believe that between the beginning and the six days of creation a period intervened of which revelation gives us no account.

If we adopt this interpretation of the first chapter, viz., that between the original creation of the universe and the six days' work a long period of time intervened, the difficulty in respect to the great age of the world will be wholly obviated.

3. The only remaining inquiry is, whether such an interpretation is consistent with succeeding verses, and with other portions of the sacred narrative? Keeping in view the principle that the language of this chapter is descriptive of the appearance as viewed by a human spectator, let us examine the remaining verses.

"And the earth was without form, and void; and darkness was upon [brooded over] the face of the deep." "*Without form, and void,*" means *laid waste,* not what is generally understood as a *chaos.* The language does not describe a *chaos*, but a wasted, agitated condition of the earth. It had experienced a revolution. The whole, or, more probably, portions of its surface, had been agitated by volcanic agency, and the darkness was due to vapors and volcanic ashes, which usually accompany eruptions, and which often become so dense as to obscure the light of the sun (see page 50).

"And the spirit of God (that is, the wind) moved over the

Are we at liberty to make use of the facts of science in interpreting the Bible? How is the apparent discrepancy of the two records reconciled? What is the meaning of the phrase "without form, and void?" How was this state produced? What does the phrase "spirit of God" mean?

face of the waters." This drove away the clouds and mists, so that the light shined through the dark and murky atmosphere. To the Hebrew mind whatever appeared was described as produced directly by the command of God. "And God *said*, Let there be light: and there was light."

This completes the first day's work. On the second day the firmament was made, that is, the clouds ascended above the earth. The language is fitted to the idea which was entertained in the time of Moses, that there was a solid firmament, above which rain, hail, and thunder were stored, and through which windows were made, to allow them to descend to the earth. The same idea prevailed until within a recent period.

On the third day the land and water were separated. This was doubtless brought about by elevations and depressions of land, and appears to have been the termination of the agitations of the earth, and the vegetable kingdom was created. It is not necessary to understand this description as having reference to all the species that now exist, but to those which were in the region of this disturbance, and such as were necessary for the use of the animals that were to be formed.

On the fourth day the sun and moon were *appointed* for signs and seasons. The language here, "And God *made* two great lights," does not mean that they were then created, but appointed for signs and seasons to the new order of beings which were to follow. The atmosphere had now become so cleared of vapors that the sun and moon were distinctly visible to a spectator upon the earth. The sun is the greater light, and the moon the lesser. "He made the stars also." This statement shows that the language was adapted to human observers, who were ignorant of the extent of the universe; for we now know that the sun, moon, and earth, the whole solar system, constitute but a point com-

---

In what way was the light created? why this interpretation? What ideas had the Hebrews of a firmament? How was the land and water separated? Is it probable that all the vegetables now existing were created at that time? How is the language respecting the creation of the sun, moon, and stars to be interpreted?

pared with "the stars," and yet a single sentence is all that is devoted to them.

On the fifth day animals were created. This representation does not necessarily imply that the whole animal kingdom was then created, for there is no violence done to the language of the Bible by restricting the meaning of terms which describe a large quantity or number. In these cases universal terms are usually employed: *all*, the *whole*, are employed to describe a large part. The reader will find numerous examples of this use of language in the Sacred writings, particularly in describing the plagues of Egypt: "*All* the cattle died," "*all* the vegetation was destroyed," &c., and yet we find cattle and vegetables spoken of immediately afterward.

Last of all, on the sixth day, man, the most exalted creature of earth, was created in the image of his Maker, endowed with a spiritual and immortal nature, and fitted to have dominion over all the other works of God. This whole scene was one of amazing grandeur. It was no doubt witnessed by the heavenly intelligences, and, as the first formed pair walked erect in the sacred garden, in the vigor and grace of their new life, in holy communion with each other, and with their heavenly Father, the choirs of heaven assembled, the morning stars sang together, and all the "Sons of God shouted for joy."

The theory, then, which obviates the difficulty which appears to exist between the Mosaic and geological records, is simply that, after the creation of the universe, the earth existed for an unknown period, during which the geological changes which we have sketched in this work took place, to describe which it was not the purpose of a revelation, and that it was afterward thrown into a desolate state, and then fitted up in six literal days for the habitation of man and his cotemporary races.

It should be observed, however, that several other theories, to

---

Is is necessary to suppose that all the species of animals were created on the fifth day? why not? What was the last act of creation? What other theory has been proposed to reconcile the language of revelation with the facts of science?

reconcile these two records have been proposed, the most important of which was proposed by Jameson, and supposes that the six days of the creation were long periods of time, during which geological changes were in progress; and many still adhere to this view. But there are certainly very strong objections to it.

The Mosaic history represents the vegetable kingdom to have been created on the third day; but an examination of the rocks shows us that animals were created as early as vegetables; they are found in the lowest rocks which contain any remains of organized beings. We should expect to find the remains of the plants which flourished during that long third day, but there is at least no evidence that the coal vegetation answers to this description.

Another objection to this hypothesis is, that the vegetables and animals that now exist are, with a few exceptions, specifically different from those found in the rocks, and different families have been created and destroyed at several different and distant epochs, facts which can not be reconciled with the idea that the six days of creation were long periods, for during nearly every period both animals and vegetables were created, and peopled the earth and the waters.

The fact, too, stated in the second chapter, that no rain had taken place until about the time of the creation of man, is certainly inconsistent with the idea that the days were long periods, for we have positive proof on this subject: the *impressions of rain drops* have been preserved in many of the rocks. But if, during the time these rocks were forming, the earth was watered by a mist, it is impossible to account for their existence.*

In order to substantiate the views suggested in this section, a few extracts from distinguished writers on this subject are added:

"A philological survey of the initial section of the Bible (Gen., i., 1, to ii., 3) brings out the result,

"1. That the first sentence is a simple, independent, all-com-

* The reader should consult on this subject President Hitchcock's Religion of Geology, Dr. J. Pye Smith's Lectures, and Dr. Buckland's Bridgewater Treatise.

What objection to this theory?

prehending axiom to this effect: that *matter*, elementary or combined, aggregated only or organized, and *dependent sentient and intellectual beings*, have not existed from eternity, either in self-continuity or succession, but had a *beginning;* that their beginning took place by the all-powerful will of ONE BEING, the Self-existent, Independent, and Infinite in all perfections; and that the date of that beginning is not made known.

"2. That, at a recent epoch, our planet was brought into a state of disorganization, detritus, or ruin (perhaps we have no perfectly appropriate term), from a former condition.

"3. That it pleased the Almighty, Wise, and Benevolent Supreme, out of that state of ruin, to adjust the surface of the earth to its now existing condition, partly by the operation of the mechanical and chemical causes (what we usually call *Laws of Nature*) which Himself had established, and partly—that is, whenever it was necessary—by His own creative power, or other immediate intervention; the whole extending through the period of six natural days.

"It has been, indeed, maintained, that the conjunction *and*, with which the next sentence begins, connects the succeeding matter with the preceding, so as to forbid the intercalating of any considerable space of time. To this we reply, that the Hebrew conjunction, agreeably to the simplicity of ancient languages, expresses an annexation of subject or a continuation of speech, in any mode whatever, remote as well as proximate. For denoting such different modes of annexation, the Greek and other languages have a variety of particles; but their use is, in Hebrew, compensated by the shades of meaning which the tone in oral speech, and the connection in writing, could supply. To go no further than the first two leaves of the Hebrew Bible, we find this copula rendered in our authorized version by *thus*, *but*, *now*, and *also*.

"This interpretation is what I have been laboring to diffuse for more than thirty years, in private and in public, by preaching, by academical lecturing, and by printing. But it is not my interpretation, though I believe that I originally derived it from the sole study of the Bible-text. Clemens of Alexandria, Origen Basil, Chrysostom, and Augustine, among the fathers (though not in a truly philosophical way, which was not to be expected), departed from the vulgar notion; and some judicious interpreters of the sixteenth and seventeenth centuries have done the same; in particular, Bishop Patrick and Dr. David Jennings Of modern Scripture critics I say nothing; for prejudice, justly or unjustly, may lie against them. Not that the question is to be settled by human authority. Our only appeal for decision is to the Bible itself, fairly interpreted. But the mention of venerable names may be useful to allay the apprehensions of some good persons, who only hear obscurely of these subjects, and have not the means of forming an independent judgment on solid grounds.

"I therefore, with many, feel greatly obliged to Dr. Buckland for having come in aid of this, which I believe to be *the true sense and meaning* of the sacred writers. I am framing no hypotheses in geology; I only plead that *the ground is clear*, and that the dictates of Scripture *interpose no bar* to observations and reasonings upon the mineralogical constitution of the earth, and the remains of organized creatures which its strata disclose. If those investigations should lead us to attribute to the earth, and to the other planetary and astral spheres, an antiquity which millions or ten thousand millions of years might fail to represent, *the divine records forbid not their deduction.* Let but the geologist maintain what his science so loudly proclaims, that the universe around us has been formed, at whatever epoch, or through what ever succession of epochs, to us unknown, by the power and wis dom of an Almighty First Cause; let him but reject the absurd ities of pre-existent matter, of an eternal succession of finite be ings, of formations without a former, laws without a lawgiver, ano nature without a God; let him but admit that man is but of yes terday, and that the design of revelation is to train him to the noblest purity and happiness in the immortal enjoyment of his Creator's beneficence, and he will find the doctrines of the Bible not an impediment, but his aid and his joy."—*Dr. J. Pye Smith.*

"Many of the early fathers of the Church were very explicit on this subject. Augustin, Theodoret, and others, supposed that the first verse of Genesis describes the creation of matter distinct from, and prior to, the work of six days. Justin Martyr and Gregory Nazianzen believed in an indefinite period between the creation of matter and the subsequent arrangement of all things. Still more explicit are Basil, Cæsarius, and Origen. It would be easy to quote similar opinions from more modern writers, who lived previous to the developments of geology."—*President Hitchcock.*

"The interval between the production of the matter of the chaos and the formation of light is undescribed and unknown." —*Bishop Horsley.*

"By the phrase *in the beginning*, the time is declared when something began to be. But when God produced this remarkable work, Moses does not precisely define."—*Doederlin.*

"The detailed history of creation in the first chapter of Genesis begins at the middle of the second verse."—*Dr. Chalmers.*

"Our best expositors of Scripture seem to be now pretty generally agreed that the opening verse in Genesis has no necessary connection with the verses which follow."—*Dr. Daniel King.*

"That a very long period—how long, no being but God can tell—intervened between the creation of the world and the commencement of the six days' work recorded in the following verses of the first chapter of Genesis, there can, I think, be no reasonable doubt."—*Dr. Pond.*

II. *A second apparent discrepancy between geology and revelation relates to the time of the introduction of death and the reasons therefor.*

Several passages of the sacred record teach us that the death of man was a penal infliction consequent upon the sin of our first parents.

"By one man sin entered into the world, and death by sin, and so death has passed upon all men, for that all have sinned."

What does the Bible teach in respect to the death of man?

Other passages appear to teach the same doctrine, that death and the pain which attends it were the result of sin.

The only question is, whether this penalty relates also to the brute animals; that is, whether death came into the system of organization through the apostacy of man?

There are no passages in the Bible which assert or imply that death to the inferior races was immediately connected with the fall of Adam.

It would be certainly contrary to natural justice to punish an imals generally for the fault of one particular species. Neither reason nor revelation teach or demand this; and yet it is a very common opinion. It has been suggested by an able writer on this subject that the idea was derived from Milton's Paradise Lost rather than from the Bible.

"She pluck'd, she ate!
Earth felt the wound, and Nature, from her seat,
Sighing through all her works, gave signs of woe
That all was lost."

"The sun
Had first his precept so to move, so shine,
As might affect the earth with cold and heat
Scarce tolerable, and from the north to call
Decrepit winter; from the south to bring
Solstitial summer's heat."

"Some say he bid his angels turn askance
The poles of earth twice ten degrees and more
From the sun's axle; they with labor push'd
Oblique the centric globe."

"Thus began
Outrage from lifeless things; but discord first,
Daughter of Sin, among the irrational
Death introduced, through fierce antipathy
Beast now with beast 'gan war, and fowl with fowl,
And fish with fish: to graze the herb all leaving,
Devour'd each other."

Does the Bible teach the same in respect to the brute animals? From what source have ideas on this subject been obtained?

This description seems to have been incorporated into the Christian system as of divine authority. It accords with the sentiment of the pious heart to attribute the disorders of earth and death itself to sin. But we ought to be careful, and not substitute the imaginations of the poet for the truths of revelation.

Assuming that the death of the human species was the penalty of transgression, we have the most positive proof that it occurred to the inferior animals in millions of cases before man was created. The remains of animals extend below those of men in the rocks to a depth of at least six miles, and this fact shows that death was introduced long before sin existed in this world.

1. But perhaps the most convincing argument on this subject is derived from the structure and habits of the carnivorous races. Their teeth and whole organization render it impossible for them to exist on vegetable food. They were intended to devour flesh, and this implies the necessity of the death of some to sustain others. The same organization is found in those animals whose remains are found in the rocks, and, in some cases, one animal is found in the body of another, both changed to stone, an ever-during monument that death was a law of nature long before man existed.

2. Death also results from organization itself; the parts wear out and decay, and were evidently never intended for terrestrial immortality. In fact, we can see much good resulting from death to the inferior races; for, as their powers decay, they give place to new and more vigorous forms of life, so that the aggregate of enjoyment is much greater than it could be under any other system; there is, moreover, a necessity either to restrain the multiplication of individuals, or to remove some of the species to prevent the world from being so filled that their support would be impossible.

"Put the case," says Dr. J. Pye Smith, "that there be no death. Upon this supposition, two or three modes are conceivable:

What proof that death to the inferior animals did not result from the sin of Adam? What is the most convincing argument on this subject? From what does death necessarily result?

"*a*. Life prolonged without food. But this would be irreconcilable with a system of successive production, nutrition, assimilation, and growth. Such beings would be perpetual possessors of the earth and the waters, in their own persons, without any progeny. Only imagine such a world! Shall we say one, or some number of each species? Quadruped, bird, reptile, fish, mollusc, zoophyte, insect of every kind, including all those invisible without microscopic aid: each immortal.

"*b*. Life prolonged by vegetable food alone. But this would require a differently constituted vegetable world; for there is no plant on the land or in the sea which does not nourish myriads of minute insects, which are destroyed in the eating of the plants.

"*c*. Must there be any multiplication by progeny upon any scheme? Then, either the whole number must be extremely small; or be kept down in some inconceivable way; or would, after a time, multiply to that degree that there would not be room for them. The land and the waters would be over-filled!"

3. It is not improbable that our first parents had witnessed the death of the inferior animals before their transgression. It would seem that they must have understood the nature of the threatening: "In the day thou eatest thereof thou shalt surely *die;*" or, if they had not this experience, still it is true that millions of animalcules which exist in water and vegetables must have been destroyed in the food of man and other animals previous to sin.

Some have supposed that the whole organization was changed after the fall; that, previous to that event, all animals were supported upon vegetable food; but the above fact is fatal to such an hypothesis; and, if it were not, it is too extravagant to be received; the inhabitants of the ocean, as whales for example, would have found it rather inconvenient to have obtained a sufficient amount of vegetable food, even if they could have subsisted upon it.

Death is a necessary law, both of the animal and vegetable kingdoms, and, unless the whole system were changed, it would

---

What would be the condition of the world if death had not been introduced? Did our first parents probably witness the death of the inferior animals? What hypothesis has been suggested? What objections to it?

be impossible to avoid it. It is, moreover, certain that no change of the kind has taken place since the first animals were brought upon the surface of our planet, and their bodies entombed in the lowest fossiliferous rocks. Carnivorous races have existed during all the geological periods.

Why, then, it may be asked, is it necessary to suppose that the death of mankind is an exception to this general law? what force is there in the threatening as a penalty for sin? and in what sense can it be said that "by one man came sin, and *death* by *sin?*" The true answer to these inquiries is, that *if man had not sinned, he would not have been subject to death, but would have been trans lated at the end of his probation, without experiencing the change in the same way with the inferior animals; but because he sinned, he was doomed to suffer the pains of death, even as the beasts that perish.* To him it is a much greater calamity, for he has intelligence, and can anticipate it. The sting of death is sin; he is, therefore, all his lifetime subject to bondage through fear of death. This theory relieves us from all difficulty in respect to the introduction of death previous to sin, and renders the truths of science and of revelation perfectly consistent with each other.

In order to account for the introduction of death previous to sin, another view is presented by some writers.

To the divine mind, all the events which would transpire throughout the universe were present before the work of creation commenced. This is a result of his foreknowledge; sin was, of course, included in this knowledge, and God arranged all things in view of that event. He made a world adapted to a fallen race, and introduced death as a part of the system, because sin was included in it. In this view death may be said to be the penalty for sin, and this theory may be more satisfactory to some minds than the one given above.

If, however, the views now presented do not prove that the facts of science are consistent with revelation, it would still be unreasonable tc believe that either record was false, for other

What other theory to account for the introduction of death as a consequence of sin?

modes of interpretation may yet be suggested, and future discov eries may be made, so that both shall be seen to proceed from the same source of light and truth.

Astronomy was for a long time supposed to be hostile to rev elati n, but all the difficulties which it presented have vanished as the word and the works of God have been better understood.

Ancient monuments and inscriptions have also been referred tc as conflicting with the truth of the sacred record, but these too, have all yielded a willing testimony to its truth. In fine, the word of God has been assailed from every department of human knowledge, and has come out of the contest not only victorious, but has gathered new strength by every such conflict; so that the Christian believer may never fear that science and revelation will come into such deadly hostility that a reconciliation will not be effected.

They are but parts of one great system, united by mutual affini ties. They are divergent streams, flowing from the same fount-ain, each sending forth its fertilizing influences to renovate and bless the world. Their circling currents mingle in the same great ocean; and while the philosopher lifts up his eyes to heaven with the devout exclamation, "Great and marvelous are thy works, Lord God Almighty! in wisdom hast thou made them all," the Christian may respond with deeper gratitude and in loftier strains, "Thou hast magnified thy *word above all thy works;*" and both may unite in exploring those higher mysteries which pertain to the spiritual kingdom of God.

What other sciences have been supposed to be hostile to revelation? What has been the resu t?

INDEX.

THE END.

www.ingramcontent.com/pod-product-compliance
Lightning Source LLC
LaVergne TN
LVHW010134110826
845151LV00002B/354

9781425539429